Hopf Algebras

LECTURE NOTES IN PURE AND APPLIED MATHEMATICS

1. *N. Jacobson*, Exceptional Lie Algebras
2. *L.-Å. Lindahl and F. Poulsen*, Thin Sets in Harmonic Analysis
3. *I. Satake*, Classification Theory of Semi-Simple Algebraic Groups
4. *F. Hirzebruch et al.*, *Differentiable Manifolds and Quadratic Forms*
5. *I. Chavel*, Riemannian Symmetric Spaces of Rank One
6. *R. B. Burckel*, Characterization of C(X) Among Its Subalgebras
7. *B. R. McDonald et al.*, Ring Theory
8. *Y.-T. Siu*, Techniques of Extension on Analytic Objects
9. *S. R. Caradus et al.*, Calkin Algebras and Algebras of Operators on Banach Spaces
10. *E. O. Roxin et al.*, Differential Games and Control Theory
11. *M. Orzech and C. Small*, The Brauer Group of Commutative Rings
12. *S. Thomier*, Topology and Its Applications
13. *J. M. Lopez and K. A. Ross*, Sidon Sets
14. *W. W. Comfort and S. Negrepontis*, Continuous Pseudometrics
15. *K. McKennon and J. M. Robertson*, Locally Convex Spaces
16. *M. Carmeli and S. Malin*, Representations of the Rotation and Lorentz Groups
17. *G. B. Seligman*, Rational Methods in Lie Algebras
18. *D. G. de Figueiredo*, Functional Analysis
19. *L. Cesari et al.*, Nonlinear Functional Analysis and Differential Equations
20. *J. J. Schäffer*, Geometry of Spheres in Normed Spaces
21. *K. Yano and M. Kon*, Anti-Invariant Submanifolds
22. *W. V. Vasconcelos*, The Rings of Dimension Two
23. *R. E. Chandler*, Hausdorff Compactifications
24. *S. P. Franklin and B. V. S. Thomas*, Topology
25. *S. K. Jain*, Ring Theory
26. *B. R. McDonald and R. A. Morris*, Ring Theory II
27. *R. B. Mura and A. Rhemtulla*, Orderable Groups
28. *J. R. Graef*, Stability of Dynamical Systems
29. *H.-C. Wang*, Homogeneous Branch Algebras
30. *E. O. Roxin et al.*, Differential Games and Control Theory II
31. *R. D. Porter*, Introduction to Fibre Bundles
32. *M. Altman*, Contractors and Contractor Directions Theory and Applications
33. *J. S. Golan*, Decomposition and Dimension in Module Categories
34. *G. Fairweather*, Finite Element Galerkin Methods for Differential Equations
35. *J. D. Sally*, Numbers of Generators of Ideals in Local Rings
36. *S. S. Miller*, Complex Analysis
37. *R. Gordon*, Representation Theory of Algebras
38. *M. Goto and F. D. Grosshans*, Semisimple Lie Algebras
39. *A. I. Arruda et al.*, Mathematical Logic
40. *F. Van Oystaeyen*, Ring Theory
41. *F. Van Oystaeyen and A. Verschoren*, Reflectors and Localization
42. *M. Satyanarayana*, Positively Ordered Semigroups
43. *D. L Russell*, Mathematics of Finite-Dimensional Control Systems
44. *P.-T. Liu and E. Roxin*, Differential Games and Control Theory III
45. *A. Geramita and J. Seberry*, Orthogonal Designs
46. *J. Cigler, V. Losert, and P. Michor*, Banach Modules and Functors on Categories of Banach Spaces
47. *P.-T. Liu and J. G. Sutinen*, Control Theory in Mathematical Economics
48. *C. Byrnes*, Partial Differential Equations and Geometry
49. *G. Klambauer*, Problems and Propositions in Analysis
50. *J. Knopfmacher*, Analytic Arithmetic of Algebraic Function Fields
51. *F. Van Oystaeyen*, Ring Theory
52. *B. Kadem*, Binary Time Series
53. *J. Barros-Neto and R. A. Artino*, Hypoelliptic Boundary-Value Problems
54. *R. L. Sternberg et al.*, Nonlinear Partial Differential Equations in Engineering and Applied Science
55. *B. R. McDonald*, Ring Theory and Algebra III
56. *J. S. Golan*, Structure Sheaves Over a Noncommutative Ring
57. *T. V. Narayana et al.*, Combinatorics, Representation Theory and Statistical Methods in Groups
58. *T. A. Burton*, Modeling and Differential Equations in Biology
59. *K. H. Kim and F. W. Roush*, Introduction to Mathematical Consensus Theory

60. *J. Banas and K. Goebel*, Measures of Noncompactness in Banach Spaces
61. *O. A. Nielson*, Direct Integral Theory
62. *J. E. Smith et al.*, Ordered Groups
63. *J. Cronin*, Mathematics of Cell Electrophysiology
64. *J. W. Brewer*, Power Series Over Commutative Rings
65. *P. K. Kamthan and M. Gupta*, Sequence Spaces and Series
66. *T. G. McLaughlin*, Regressive Sets and the Theory of Isols
67. *T. L. Herdman et al.*, Integral and Functional Differential Equations
68. *R. Draper*, Commutative Algebra
69. *W. G. McKay and J. Patera*, Tables of Dimensions, Indices, and Branching Rules for Representations of Simple Lie Algebras
70. *R. L. Devaney and Z. H. Nitecki*, Classical Mechanics and Dynamical Systems
71. *J. Van Geel*, Places and Valuations in Noncommutative Ring Theory
72. *C. Faith*, Injective Modules and Injective Quotient Rings
73. *A. Fiacco*, Mathematical Programming with Data Perturbations I
74. *P. Schultz et al.*, Algebraic Structures and Applications
75. *L Bican et al.*, Rings, Modules, and Preradicals
76. *D. C. Kay and M. Breen*, Convexity and Related Combinatorial Geometry
77. *P. Fletcher and W. F. Lindgren*, Quasi-Uniform Spaces
78. *C.-C. Yang*, Factorization Theory of Meromorphic Functions
79. *O. Taussky*, Ternary Quadratic Forms and Norms
80. *S. P. Singh and J. H. Burry*, Nonlinear Analysis and Applications
81. *K. B. Hannsgen et al.*, Volterra and Functional Differential Equations
82. *N. L. Johnson et al.*, Finite Geometries
83. *G. I. Zapata*, Functional Analysis, Holomorphy, and Approximation Theory
84. *S. Greco and G. Valla*, Commutative Algebra
85. *A. V. Fiacco*, Mathematical Programming with Data Perturbations II
86. *J.-B. Hiriart-Urruty et al.*, Optimization
87. *A. Figa Talamanca and M. A. Picardello*, Harmonic Analysis on Free Groups
88. *M. Harada*, Factor Categories with Applications to Direct Decomposition of Modules
89. *V. I. Istrătescu*, Strict Convexity and Complex Strict Convexity
90. *V. Lakshmikantham*, Trends in Theory and Practice of Nonlinear Differential Equations
91. *H. L. Manocha and J. B. Srivastava*, Algebra and Its Applications
92. *D. V. Chudnovsky and G. V. Chudnovsky*, Classical and Quantum Models and Arithmetic Problems
93. *J. W. Longley*, Least Squares Computations Using Orthogonalization Methods
94. *L. P. de Alcantara*, Mathematical Logic and Formal Systems
95. *C. E. Aull*, Rings of Continuous Functions
96. *R. Chuaqui*, Analysis, Geometry, and Probability
97. *L. Fuchs and L. Salce*, Modules Over Valuation Domains
98. *P. Fischer and W. R. Smith*, Chaos, Fractals, and Dynamics
99. *W. B. Powell and C. Tsinakis*, Ordered Algebraic Structures
100. *G. M. Rassias and T. M. Rassias*, Differential Geometry, Calculus of Variations, and Their Applications
101. *R.-E. Hoffmann and K. H. Hofmann*, Continuous Lattices and Their Applications
102. *J. H. Lightbourne III and S. M. Rankin III*, Physical Mathematics and Nonlinear Partial Differential Equations
103. *C. A. Baker and L. M. Batten*, Finite Geometrics
104. *J. W. Brewer et al.*, Linear Systems Over Commutative Rings
105. *C. McCrory and T. Shifrin*, Geometry and Topology
106. *D. W. Kueke et al.*, Mathematical Logic and Theoretical Computer Science
107. *B.-L. Lin and S. Simons*, Nonlinear and Convex Analysis
108. *S. J. Lee*, Operator Methods for Optimal Control Problems
109. *V. Lakshmikantham*, Nonlinear Analysis and Applications
110. *S. F. McCormick*, Multigrid Methods
111. *M. C. Tangora*, Computers in Algebra
112. *D. V. Chudnovsky and G. V. Chudnovsky*, Search Theory
113. *D. V. Chudnovsky and R. D. Jenks*, Computer Algebra
114. *M. C. Tangora*, Computers in Geometry and Topology
115. *P. Nelson et al.*, Transport Theory, Invariant Imbedding, and Integral Equations
116. *P. Clément et al.*, Semigroup Theory and Applications
117. *J. Vinuesa*, Orthogonal Polynomials and Their Applications
118. *C. M. Dafermos et al.*, Differential Equations
119. *E. O. Roxin*, Modern Optimal Control
120. *J. C. Díaz*, Mathematics for Large Scale Computing

121. *P. S. Milojević* Nonlinear Functional Analysis
122. *C. Sadosky,* Analysis and Partial Differential Equations
123. *R. M. Shortt,* General Topology and Applications
124. *R. Wong,* Asymptotic and Computational Analysis
125. *D. V. Chudnovsky and R. D. Jenks,* Computers in Mathematics
126. *W. D. Wallis et al.,* Combinatorial Designs and Applications
127. *S. Elaydi,* Differential Equations
128. *G. Chen et al.,* Distributed Parameter Control Systems
129. *W. N. Everitt,* Inequalities
130. *H. G. Kaper and M. Garbey,* Asymptotic Analysis and the Numerical Solution of Partial Differential Equations
131. *O. Arino et al.,* Mathematical Population Dynamics
132. *S. Coen,* Geometry and Complex Variables
133. *J. A. Goldstein et al.,* Differential Equations with Applications in Biology, Physics, and Engineering
134. *S. J. Andima et al.,* General Topology and Applications
135. *P Clément et al.,* Semigroup Theory and Evolution Equations
136. *K. Jarosz,* Function Spaces
137. *J. M. Bayod et al.,* *p*-adic Functional Analysis
138. *G. A. Anastassiou,* Approximation Theory
139. *R. S. Rees,* Graphs, Matrices, and Designs
140. *G. Abrams et al.,* Methods in Module Theory
141. *G. L. Mullen and P. J.-S. Shiue,* Finite Fields, Coding Theory, and Advances in Communications and Computing
142. *M. C. Joshi and A. V. Balakrishnan,* Mathematical Theory of Control
143. *G. Komatsu and Y. Sakane,* Complex Geometry
144. *I. J. Bakelman,* Geometric Analysis and Nonlinear Partial Differential Equations
145. *T. Mabuchi and S. Mukai,* Einstein Metrics and Yang–Mills Connections
146. *L. Fuchs and R. Göbel,* Abelian Groups
147. *A. D. Pollington and W. Moran,* Number Theory with an Emphasis on the Markoff Spectrum
148. *G. Dore et al.,* Differential Equations in Banach Spaces
149. *T. West,* Continuum Theory and Dynamical Systems
150. *K. D. Bierstedt et al.,* Functional Analysis
151. *K. G. Fischer et al.,* Computational Algebra
152. *K. D. Elworthy et al.,* Differential Equations, Dynamical Systems, and Control Science
153. *P.-J. Cahen, et al.,* Commutative Ring Theory
154. *S. C. Cooper and W. J. Thron,* Continued Fractions and Orthogonal Functions
155. *P. Clément and G. Lumer,* Evolution Equations, Control Theory, and Biomathematics
156. *M. Gyllenberg and L. Persson,* Analysis, Algebra, and Computers in Mathematical Research
157. *W. O. Bray et al.,* Fourier Analysis
158. *J. Bergen and S. Montgomery,* Advances in Hopf Algebras
159. *A. R. Magid,* Rings, Extensions, and Cohomology
160. *N. H. Pavel,* Optimal Control of Differential Equations
161. *M. Ikawa,* Spectral and Scattering Theory
162. *X. Liu and D. Siegel,* Comparison Methods and Stability Theory
163. *J.-P. Zolésio,* Boundary Control and Variation
164. *M. Křížek et al.,* Finite Element Methods
165. *G. Da Prato and L. Tubaro,* Control of Partial Differential Equations
166. *E. Ballico,* Projective Geometry with Applications
167. *M. Costabel et al.,* Boundary Value Problems and Integral Equations in Nonsmooth Domains
168. *G. Ferreyra, G. R. Goldstein, and F. Neubrander,* Evolution Equations
169. *S. Huggett,* Twistor Theory
170. *H. Cook et al.,* Continua
171. *D. F. Anderson and D. E. Dobbs,* Zero-Dimensional Commutative Rings
172. *K. Jarosz,* Function Spaces
173. *V. Ancona et al.,* Complex Analysis and Geometry
174. *E. Casas,* Control of Partial Differential Equations and Applications
175. *N. Kalton et al.,* Interaction Between Functional Analysis, Harmonic Analysis, and Probability
176. *Z. Deng et al.,* Differential Equations and Control Theory
177. *P. Marcellini et al.* Partial Differential Equations and Applications
178. *A. Kartsatos,* Theory and Applications of Nonlinear Operators of Accretive and Monotone Type
179. *M. Maruyama,* Moduli of Vector Bundles
180. *A. Ursini and P. Aglianò,* Logic and Algebra
181. *X. H. Cao et al.,* Rings, Groups, and Algebras
182. *D. Arnold and R. M. Rangaswamy,* Abelian Groups and Modules
183. *S. R. Chakravarthy and A. S. Alfa,* Matrix-Analytic Methods in Stochastic Models

184. *J. E. Andersen et al.*, Geometry and Physics
185. *P.-J. Cahen et al.*, Commutative Ring Theory
186. *J. A. Goldstein et al.*, Stochastic Processes and Functional Analysis
187. *A. Sorbi*, Complexity, Logic, and Recursion Theory
188. *G. Da Prato and J.-P. Zolésio*, Partial Differential Equation Methods in Control and Shape Analysis
189. *D. D. Anderson*, Factorization in Integral Domains
190. *N. L. Johnson*, Mostly Finite Geometries
191. *D. Hinton and P. W. Schaefer*, Spectral Theory and Computational Methods of Sturm–Liouville Problems
192. *W. H. Schikhof et al.*, *p*-adic Functional Analysis
193. *S. Sertöz*, Algebraic Geometry
194. *G. Caristi and E. Mitidieri*, Reaction Diffusion Systems
195. *A. V. Fiacco*, Mathematical Programming with Data Perturbations
196. *M. Křížek et al.*, Finite Element Methods: Superconvergence, Post-Processing, and A Posteriori Estimates
197. *S. Caenepeel and A. Verschoren*, Rings, Hopf Algebras, and Brauer Groups
198. *V. Drensky et al.*, Methods in Ring Theory
199. *W. B. Jones and A. Sri Ranga*, Orthogonal Functions, Moment Theory, and Continued Fractions
200. *P. E. Newstead*, Algebraic Geometry
201. *D. Dikranjan and L. Salce*, Abelian Groups, Module Theory, and Topology
202. *Z. Chen et al.*, Advances in Computational Mathematics
203. *X. Caicedo and C. H. Montenegro*, Models, Algebras, and Proofs
204. *C. Y. Yıldırım and S. A. Stepanov*, Number Theory and Its Applications
205. *D. E. Dobbs et al.*, Advances in Commutative Ring Theory
206. *F. Van Oystaeyen*, Commutative Algebra and Algebraic Geometry
207. *J. Kakol et al.*, *p*-adic Functional Analysis
208. *M. Boulagouaz and J.-P. Tignol*, Algebra and Number Theory
209. *S. Caenepeel and F. Van Oystaeyen*, Hopf Algebras and Quantum Groups
210. *F. Van Oystaeyen and M. Saorin*, Interactions Between Ring Theory and Representations of Algebras
211. *R. Costa et al.*, Nonassociative Algebra and Its Applications
212. *T.-X. He*, Wavelet Analysis and Multiresolution Methods
213. *H. Hudzik and L. Skrzypczak*, Function Spaces: The Fifth Conference
214. *J. Kajiwara et al.*, Finite or Infinite Dimensional Complex Analysis
215. *G. Lumer and L. Weis*, Evolution Equations and Their Applications in Physical and Life Sciences
216. *J. Cagnol et al.*, Shape Optimization and Optimal Design
217. *J. Herzog and G. Restuccia*, Geometric and Combinatorial Aspects of Commutative Algebra
218. *G. Chen et al.*, Control of Nonlinear Distributed Parameter Systems
219. *F. Ali Mehmeti et al.*, Partial Differential Equations on Multistructures
220. *D. D. Anderson and I. J. Papick*, Ideal Theoretic Methods in Commutative Algebra
221. *Á. Granja et al.*, Ring Theory and Algebraic Geometry
222. *A. K. Katsaras et al.*, *p*-adic Functional Analysis
223. *R. Salvi*, The Navier-Stokes Equations
224. *F. U. Coelho and H. A. Merklen*, Representations of Algebras
225. *S. Aizicovici and N. H. Pavel*, Differential Equations and Control Theory
226. *G. Lyubeznik*, Local Cohomology and Its Applications
227. *G. Da Prato and L. Tubaro*, Stochastic Partial Differential Equations and Applications
228. *W. A. Carnielli et al.*, Paraconsistency
229. *A. Benkirane and A. Touzani*, Partial Differential Equations
230. *A. Illanes et al.*, Continuum Theory
231. *M. Fontana et al.*, Commutative Ring Theory and Applications
232. *D. Mond and M. J. Saia*, Real and Complex Singularities
233. *V. Ancona and J. Vaillant*, Hyperbolic Differential Operators
234. *G. R. Goldstein et al.*, Evolution Equations
235. *A. Giambruno et al.*, Polynomial Identities and Combinatorial Methods
236. *A. Facchini et al.*, Rings, Modules, Algebras, and Abelian Groups
237. *J. Bergen et al.*, Hopf Algebras

Additional Volumes in Preparation

Hopf Algebras

Proceedings from the International Conference at DePaul University

edited by

Jeffrey Bergen
Stefan Catoiu
William Chin

DePaul University
Chicago, Illinois, U.S.A.

CRC Press
Taylor & Francis Group
Boca Raton London New York

CRC Press is an imprint of the
Taylor & Francis Group, an **informa** business

CRC Press
Taylor & Francis Group
6000 Broken Sound Parkway NW, Suite 300
Boca Raton, FL 33487-2742

First issued in paperback 2019

ISBN-13: 978-0-8247-5566-9 (hbk)
ISBN-13: 978-0-367-39457-8 (pbk)

Library of Congress Cataloging-in-Publication Data
A catalog record for this book is available from the Library of Congress.

Visit the Taylor & Francis Web site at
http://www.taylorandfrancis.com

and the CRC Press Web site at
http://www.crcpress.com

Preface

These proceedings are an outgrowth of an International Conference on Hopf Algebras held at DePaul University, Chicago, Illinois. Participants came from seven different countries. We would like to thank Susan Montgomery of the University of Southern California for her help in organizing this meeting. This conference was one of four algebra conferences held at DePaul University during the 2001-2002 academic year. We would also like to thank our DePaul colleagues Allan Berele, Barbara Cortzen, Jerry Goldman, and Leonid Krop for their help in organizing these four conferences. In addition, we thank the College of Liberal Arts & Sciences at DePaul University for its financial support of these conferences.

All of the papers in these proceedings were refereed. We thank our referees: Margaret Beattie, Christopher Bendel, Allan Berele, Stefan Catoiu, William Chin, Walter Ferrer Santos, Jerry Goldman, Matias Grana, Piotr Grzeszczuk, Yevgenia Kashina, Leonid Krop, Richard Larson, Jean-Louis Loday, Ian Musson, Declan Quinn, David Radford, and Dragos Stefan.

Finally, we would like to thank Maria Allegra of Marcel Dekker, Inc., for her help in preparing these proceedings.

Jeffrey Bergen
Stefan Catoiu
William Chin

Contents

Contributors

Marcelo Aguiar Texas A&M University, College Station, Texas, U.S.A.

Nicolás Andruskiewitsch Universidad Nacional de Córdoba, Córdoba, Argentina

M. Beattie Mount Allison University, Sackville, New Brunswick, Canada

Georgia Benkart University of Wisconsin, Madison, Wisconsin, U.S.A.

Jeffrey Bergen DePaul University, Chicago, Illinois, U.S.A.

S. Caenepeel Vrije Universiteit Brussel, Brussels, Belgium

William Chin DePaul University, Chicago, Illinois, U.S.A.

Sorin Dăscălescu Kuwait University, Safat, Kuwait

Yukio Doi Okayama University, Okayama, Japan

Robert L. Grossman University of Illinois at Chicago, Chicago, Illinois, U.S.A.

T. Guedenon Vrije Universiteit Brussel, Brussels, Belgium

Mikhail Kochetov University of Saskatchewan, Saskatoon, Saskatchewan, Canada

R. G. Larson University of Illinois at Chicago, Chicago, Illinois, U.S.A.

Akira Masuoka University of Tsukuba, Ibaraki, Japan

Siu-Hung Ng Towson University, Baltimore, Maryland, U.S.A.

Julia Pevtsova Institute for Advanced Study, Princeton, New Jersey, U.S.A.

Serban Raianu California State University, Carson, California, U.S.A.

Walter Ricardo Ferrer Santos Centro de Matemática, Montevideo, Uruguay

Axel Schüler Universität Leipzig, Leipzig, Germany

Mitsuhiro Takeuchi University of Tsukuba, Tsukuba, Ibaraki, Japan

Sarah Witherspoon Amherst College, Amherst, Massachusetts, U.S.A.

INFINITESIMAL BIALGEBRAS, PRE-LIE AND DENDRIFORM ALGEBRAS

MARCELO AGUIAR

Department of Mathematics,
Texas A&M University,
College Station, TX 77843, USA.
maguiar@math.tamu.edu

ABSTRACT. We introduce the categories of infinitesimal Hopf modules and bimodules over an infinitesimal bialgebra. We show that they correspond to modules and bimodules over the infinitesimal version of the double. We show that there is a natural, but non-obvious way to construct a pre-Lie algebra from an arbitrary infinitesimal bialgebra and a dendriform algebra from a quasitriangular infinitesimal bialgebra. As consequences, we obtain a pre-Lie structure on the space of paths on an arbitrary quiver, and a striking dendriform structure on the space of endomorphisms of an arbitrary infinitesimal bialgebra, which combines the convolution and composition products. We extend the previous constructions to the categories of Hopf, pre-Lie and dendriform bimodules. We construct a brace algebra structure from an arbitrary infinitesimal bialgebra; this refines the pre-Lie algebra construction. In two appendices, we show that infinitesimal bialgebras are comonoid objects in a certain monoidal category and discuss a related construction for counital infinitesimal bialgebras.

1. INTRODUCTION

The main results of this paper establish connections between infinitesimal bialgebras, pre-Lie algebras and dendriform algebras, which were a priori unexpected.

An infinitesimal bialgebra (abbreviated ϵ-bialgebra) is a triple (A, μ, Δ) where (A, μ) is an associative algebra, (A, Δ) is a coassociative coalgebra, and Δ is a derivation (see Section 2). We write $\Delta(a) = a_1 \otimes a_2$, omitting the sum symbol.

Infinitesimal bialgebras were introduced by Joni and Rota [17, Section XII]. The basic theory of these objects was developed in [1, 3], where analogies with the theories of ordinary Hopf algebras and Lie bialgebras were found; among which we remark the existence of a "double" construction analogous to that of Drinfeld for ordinary Hopf algebras or Lie bialgebras. On the other hand, infinitesimal bialgebras have found important applications in combinatorics [4, 11].

A pre-Lie algebra is a vector space P equipped with an operation $x \circ y$ satisfying a certain axiom (3.1), which guarantees that $x \circ y - y \circ x$ defines a Lie algebra structure on P. These objects were introduced by Gerstenhaber [13], whose terminology we follow, and independently by Vinberg [29]. See [8, 7] for more references, examples, and some of the general theory of pre-Lie algebras.

We show that any ϵ-bialgebra can be turned into a pre-Lie algebra by defining

$$a \circ b = b_1 a b_2 .$$

This is Theorem 3.2. As an application, we construct a canonical pre-Lie structure on the space of paths on an arbitrary quiver. We also note that the Witt Lie algebra arises in this way from the ϵ-bialgebra of divided differences (Examples 3.4). Other properties of this construction are provided in Section 3.

A dendriform algebra is a space D equipped with two operations $x \succ y$ and $x \prec y$ satisfying certain axioms (4.1), which guarantee that $x \succ y + x \prec y$ defines an associative algebra structure on D. Dendriform algebras were introduced by Loday [20, Chapter 5]. See [6, 26, 21, 22] for additional recent work on this subject.

There is a special class of ϵ-bialgebras for which the derivation Δ is principal, called quasitriangular ϵ-bialgebras. These are defined from solutions $r = \sum u_i \otimes v_i$ of the associative Yang-Baxter equation, introduced in [1] and reviewed in Section 2 of this paper. In Theorem 4.6, we show that any quasitriangular ϵ-bialgebra can be made into a dendriform algebra by defining

$$x \succ y = \sum_i u_i x v_i y \quad \text{and} \quad x \prec y = \sum_i x u_i y v_i .$$

This is derived from a more general construction of dendriform algebras from associative algebras equipped with a Baxter operator, given in Proposition 4.5. (Baxter operators should not be confused with Yang-Baxter operators, see Remark 4.4.)

As a main application of this construction, we work out the dendriform algebra structure associated to the Drinfeld double of an ϵ-bialgebra A. This construction, introduced in [1] and reviewed here in Section 2, produces a quasitriangular ϵ-bialgebra structure on the space $(A \otimes A^*) \oplus A \oplus A^*$. We provide explicit formulas for the resulting dendriform structure in Theorem 4.9. This is one of the main results of this paper. It turns out that the subspace $A \otimes A^*$ is closed under the dendriform operations. The resulting dendriform algebra structure on the space $\mathsf{End}(A)$ of linear endomorphisms of A is (Corollary 4.14)

$$T \succ S = (id * T * id)S + (id * T)(S * id) \quad \text{and} \quad T \prec S = T(id * S * id) + (T * id)(id * S) .$$

In this formula, T and S are arbitrary endomorphisms of A, $T * S = \mu(T \otimes S)\Delta$ is the convolution, and the concatenation of endomorphisms denotes composition. When A is a quasitriangular ϵ-bialgebra, our results give dendriform structures on A and $\mathsf{End}(A)$. In Proposition 4.13, we show that they are related by a canonical morphism of dendriform algebras $\mathsf{End}(A) \to A$.

Other properties of the construction of dendriform algebras are given in Section 4. In particular, it is shown that the constructions of pre-Lie algebras from ϵ-bialgebras and of dendriform algebras from quasitriangular ϵ-bialgebras are compatible, in the sense that the diagram

$$\begin{array}{ccc} \text{Quasitriangular } \epsilon\text{-bialgebras} & \longrightarrow & \epsilon\text{-bialgebras} \\ \downarrow & & \downarrow \\ \text{Dendriform algebras} & \longrightarrow & \text{Pre-Lie algebras} \end{array}$$

commutes.

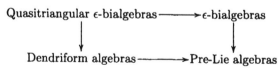

This paper also introduces the appropriate notion of modules over infinitesimal bialgebras. These are called infinitesimal Hopf modules (ϵ-Hopf bimodules for short). They are defined in Section 2. In the same section, it is shown that ϵ-Hopf bimodules are precisely modules over the double, when the ϵ-bialgebra is finite dimensional (Theorem 2.5), and that any module can be turned into an ϵ-Hopf bimodule, when the ϵ-bialgebra is quasitriangular (Proposition 2.7).

The constructions of dendriform and pre-Lie algebras are extended to the corresponding categories of bimodules in Section 5. A commutative diagram of the form

$$\begin{array}{ccc} \text{Associative bimodules} & \longrightarrow & \epsilon\text{-Hopf bimodules} \\ \downarrow & & \downarrow \\ \text{Dendriform bimodules} & \longrightarrow & \text{Pre-Lie bimodules} \end{array}$$

is obtained.

A brace algebra is a space B equipped with a family of higher degree operations satisfying certain axioms(6.1). Brace algebras originated in work of Kadeishvili [18], Getzler [16] and Gerstenhaber and Voronov [14, 15]. In this paper we deal with the ungraded, unsigned version of these objects, as in the recent works of Chapoton [6] and Ronco [26]. Brace algebras sit between dendriform and pre-Lie; as explained in [6, 26], the functor from dendriform to pre-Lie algebras factors through the category of brace algebras. Following a suggestion of Ronco, we show in Section 6 that the construction of pre-Lie algebras from ϵ-bialgebras can be refined accordingly. We associate a brace algebra to any ϵ-bialgebra (Theorem 6.2) and obtain a commutative diagram

$$\begin{array}{ccccc} \text{Quasitriangular } \epsilon\text{-bialgebras} & \longrightarrow & \epsilon\text{-bialgebras} & & \\ \downarrow & & \downarrow & \searrow & \\ \text{Dendriform algebras} & \longrightarrow & \text{Brace algebras} & \longrightarrow & \text{Pre-Lie algebras} \end{array}$$

The brace algebra associated to the ϵ-bialgebra of divided differences is explicitly described in Example 6.3. The higher braces are given by

$$\langle \mathbf{x}^{p_1}, \ldots, \mathbf{x}^{p_n}; \mathbf{x}^r \rangle = \binom{r}{n} \mathbf{x}^{r+p_1+\cdots+p_n-n},$$

where $\binom{r}{n}$ is the binomial coefficient.

In Appendix A we construct a certain monoidal category of algebras for which the comonoid objects are precisely ϵ-bialgebras, and we discuss how ϵ-bialgebras differ from bimonoid objects in certain related braided monoidal categories.

In Appendix B we study certain special features of counital ϵ-bialgebras. We construct another monoidal category of algebras and show that comonoid objects in this category are precisely counital ϵ-bialgebras (Proposition B.5). The relation to the constructions of Appendix A is explained. We also describe counital ϵ-Hopf modules in terms of this monoidal structure (Proposition B.9).

Notation and basic terminology. All spaces and algebras are over a fixed field k, often omitted from the notation. Sum symbols are omitted from Sweedler's notation: we write $\Delta(a) = a_1 \otimes a_2$ when Δ is a coassociative comultiplication, and similarly for comodule structures. The composition of maps $f : U \to V$ with $g : V \to W$ is denoted by $gf : U \to W$.

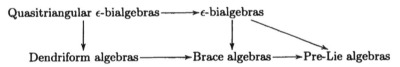

2. Infinitesimal modules over infinitesimal bialgebras

An infinitesimal bialgebra (abbreviated ϵ-bialgebra) is a triple (A, μ, Δ) where (A, μ) is an algebra, (A, Δ) is a coalgebra, and for each $a, b \in A$,

$$\Delta(ab) = ab_1 \otimes b_2 + a_1 \otimes a_2 b. \tag{2.1}$$

We do not require the algebra to be unital or the coalgebra to be counital.

A derivation of an algebra A with values in a A-bimodule M is a linear map $D : A \to M$ such that

$$D(ab) = a \cdot D(b) + D(a) \cdot b \quad \forall\, a, b \in A\,.$$

We view $A \otimes A$ as an A-bimodule via

$$a \cdot (b \otimes c) = ab \otimes c \text{ and } (b \otimes c) \cdot a = b \otimes ca\,.$$

A coderivation from a C-bicomodule M to a coalgebra C is a map $D : M \to C$ such that

$$\Delta D = (id_C \otimes D)t + (D \otimes id_C)s\ ,$$

where $t : M \to C \otimes M$ and $s : M \to M \otimes C$ are the bicomodule structure maps [Doi]. We view $C \otimes C$ as a C-bicomodule via

$$t = \Delta \otimes id_C \text{ and } s = id_C \otimes \Delta\,.$$

The compatibility condition (2.1) may be written as

$$\Delta\mu = (\mu \otimes id_A)(id_A \otimes \Delta) + (id_A \otimes \mu)(\Delta \otimes id_A)$$

This says that $\Delta : A \to A \otimes A$ is a derivation of the algebra (A, μ) with values in the A-bimodule $A \otimes A$, or equivalently, that $\mu : A \otimes A \to A$ is a coderivation from the A-bicomodule $A \otimes A$ with values in the coalgebra (A, Δ).

Definition 2.1. Let (A, μ, Δ) be an ϵ-bialgebra. A left infinitesimal Hopf module (abbreviated ϵ-Hopf module) over A is a space M endowed with a left A-module structure $\lambda : A \otimes M \to M$ and a left A-comodule structure $\Lambda : M \to A \otimes M$, such that

$$\Lambda\lambda = (\mu \otimes id_M)(id_A \otimes \Lambda) + (id_A \otimes \lambda)(\Delta \otimes id_M)\,.$$

We will often write

$$\lambda(a \otimes m) = am \text{ and } \Lambda(m) = m_{-1} \otimes m_0$$

The compatibility condition above may be written as $\Lambda(am) = a\Lambda(m) + \Delta(a)m$, or more explicitly,

$$(am)_{-1} \otimes (am)_0 = am_{-1} \otimes m_0 + a_1 \otimes a_2 m\,, \quad \text{for each } a \in A \text{ and } m \in M. \tag{2.2}$$

The notion of ϵ-Hopf modules bears a certain analogy to the notion of Hopf modules over ordinary Hopf algebras. The basic examples of Hopf modules from [25, 1.9.2-3] admit the following versions in the context of ϵ-bialgebras.

Examples 2.2. Let (A, μ, Δ) be an ϵ-bialgebra.

(1) A itself is an ϵ-Hopf module via μ and Δ.
(2) More generally, for any space V, $A \otimes V$ is an ϵ-Hopf module via

$$\mu \otimes id : A \otimes A \otimes V \to A \otimes V \text{ and } \Delta \otimes id : A \otimes V \to A \otimes A \otimes V\,.$$

(3) A more interesting example follows. Assume that the coalgebra (A, Δ) admits a counit $\eta : A \to k$. Let N be a left A-module. Then there is an ϵ-Hopf module structure on the space $A \otimes N$ defined by

$$a \cdot (a' \otimes n) = aa' \otimes n + \eta(a')\, a_1 \otimes a_2 n \quad \text{and} \quad \Lambda(a \otimes n) = a_1 \otimes a_2 \otimes n\,.$$

This can be checked by direct calculations. A more conceptual proof will be given later (Corollary B.10). Note that if N is a trivial A-module ($an \equiv 0$) then this structure reduces to that of example 2.

When H is a finite dimensional (ordinary) Hopf algebra, left Hopf modules over H are precisely left modules over the *Heisenberg double* of H [25, Examples 4.1.10 and 8.5.2].

There is an analogous result for infinitesimal bialgebras which, as it turns out, involves the Drinfeld double of ϵ-bialgebras.

We first recall the construction of the Drinfeld double $D(A)$ of a finite dimensional ϵ-bialgebra (A, μ, Δ) from [1, Section 7]. Consider the following version of the dual of A

$$A' := (A^*, \Delta^{*^{op}}, -\mu^{*^{cop}})\,.$$

Explicitly, the structure on A' is:

$$(f \cdot g)(a) = g(a_1)f(a_2)\ \forall\, a \in A,\ f, g \in A' \text{ and} \tag{2.3}$$

$$\Delta(f) = f_1 \otimes f_2 \iff f(ab) = -f_2(a)f_1(b)\ \forall\, f \in A',\ a, b \in A\,. \tag{2.4}$$

Below we always refer to this structure when dealing with multiplications or comultiplications of elements of A'. Consider also the actions of A' on A and A on A' defined by

$$f \rightharpoonup a = f(a_1)a_2 \text{ and } f \leftharpoonup a = -f_2(a)f_1 \tag{2.5}$$

or equivalently

$$g(f \rightharpoonup a) = (gf)(a) \text{ and } (f \leftharpoonup a)(b) = f(ab)\,. \tag{2.6}$$

Proposition 2.3. *Let A be a finite dimensional ϵ-bialgebra, consider the vector space*

$$D(A) := (A \otimes A') \oplus A \oplus A'$$

and denote the element $a \otimes f \in A \otimes A' \subseteq D(A)$ by $a \bowtie f$. Then $D(A)$ admits a unique ϵ-bialgebra structure such that:

(a) *A and A' are subalgebras, $a \cdot f = a \bowtie f$, $f \cdot a = f \rightharpoonup a + f \leftharpoonup a$, and*
(b) *A and A' are subcoalgebras.*

Proof. See [3, Theorem 7.3]. □

We will make use of the following universal property of the double.

Proposition 2.4. *Let A be a finite dimensional ϵ-bialgebra, B an algebra and $\rho : A \to B$ and $\rho' : A' \to B$ morphisms of algebras such that $\forall\, a \in A,\ f \in A'$,*

$$\rho'(f)\rho(a) = \rho(f \rightharpoonup a) + \rho'(f \leftharpoonup a)\,. \tag{2.7}$$

Then there exists a unique morphism of algebras $\hat{\rho} : D(A) \to B$ such that $\hat{\rho}_{|_A} = \rho$ and $\hat{\rho}_{|'_A} = \rho'$.

Proof. This follows from Propositions 6.5 and 7.1 in [1]. □

We can now show that ϵ-Hopf modules are precisely modules over the double.

Theorem 2.5. *Let A be a finite dimensional ϵ-bialgebra and M a space. If M is a left ϵ-Hopf module over A via $\lambda(a\otimes m) = am$ and $\Lambda(m) = m_{-1}\otimes m_0$, then M is a left module over $D(A)$ via*

$$a \cdot m = am, \quad f \cdot m = f(m_{-1})m_0 \quad \text{and} \quad (a \bowtie f) \cdot m = f(m_{-1})am_0 .$$

Conversely, if M is a left module over $D(A)$, then M is a left ϵ-Hopf module over A and the structures are related as above.

Proof. Suppose first that M is a left ϵ-Hopf module over A.

Since (M, Λ) is a left A-comodule, it is also a left A'-module via $f \cdot m := f(m_{-1})m_0$. Let $\rho : A \to \mathsf{End}(M)$ and $\rho' : A' \to \mathsf{End}(M)$ be the morphisms of algebras corresponding to the left module structures:

$$\rho(a)(m) = am, \quad \rho'(f)(m) = f(m_{-1})m_0 .$$

We will apply Proposition 2.4 to deduce the existence of a morphism of algebras $\hat{\rho} : D(A) \to \mathsf{End}(M)$ extending ρ and ρ'. We need to check (2.7). We have

$$\begin{aligned}\rho'(f)\rho(a)(m) &= f\big((am)_{-1}\big)(am)_0 \overset{(2.2)}{=} f(a_1)a_2m + f(am_{-1})m_0 \\ &\overset{(2.5,\,2.6)}{=} (f \to a)m + (f \leftarrow a)(m_{-1})m_0 \\ &= \rho(f \to a)(m) + \rho'(f \leftarrow a)(m)\end{aligned}$$

as needed. Thus, $\hat{\rho}$ exists and M becomes a left $D(A)$-module via $\alpha \cdot m = \hat{\rho}(\alpha)(m)$. Since $\hat{\rho}$ extends ρ and ρ', we have

$$a \cdot m = \rho(a)(m) = am, \quad f \cdot m = \rho'(f)(m) = f(m_{-1})m_0$$

and, from the description of the multiplication in $D(A)$ in Proposition 2.3,

$$(a \bowtie f) \cdot m = \hat{\rho}(af)(m) = \rho(a)\rho'(f)(m) = f(m_{-1})am_0 .$$

This completes the proof of the first assertion.

Conversely, if M is a left $D(A)$-module, then restricting via the morphisms of algebras $A \hookrightarrow D(A)$ and $A' \hookrightarrow D(A)$, M becomes a left A-module and left A'-module. As above, the latter structure is equivalent to a left A-comodule structure on M. From the associativity axiom

$$f \cdot (a \cdot m) = (fa) \cdot m = (f \to a) \cdot m + (f \leftarrow a) \cdot m$$

we deduce

$$f\big((am)_{-1}\big)(am)_0 = f(a_1)a_2m + f(am_{-1})m_0 .$$

Since this holds for every $f \in A'$, we obtain the ϵ-Hopf module Axiom (2.2). Also,

$$(a \bowtie f) \cdot m = (af) \cdot m = a \cdot (f \cdot m) = f(m_{-1})am_0 ,$$

so the structures of left module over $D(A)$ and left ϵ-Hopf module over A are related as stated. □

We close the section by showing that when A is a quasitriangular ϵ-bialgebra, any A-module carries a natural structure of ϵ-Hopf module over A.

We first recall the definition of quasitriangular ϵ-bialgebras. Let A be an associative algebra. An element $r = \sum_i u_i \otimes v_i \in A\otimes A$ is a solution of the associative Yang-Baxter equation [1, Section 5] if

$$r_{13}r_{12} - r_{12}r_{23} + r_{23}r_{13} = 0 \tag{2.8}$$

or, more explicitly,

$$\sum_{i,j} u_i u_j \otimes v_j \otimes v_i - \sum_{i,j} u_i \otimes v_i u_j \otimes v_j + \sum_{i,j} u_j \otimes u_i \otimes v_i v_j = 0\,.$$

This condition implies that the *principal* derivation $\Delta : A \to A \otimes A$ defined by

$$\Delta(a) = r \cdot a - a \cdot r = \sum_i u_i \otimes v_i a - \sum_i a u_i \otimes v_i\,, \tag{2.9}$$

is coassociative [1, Proposition 5.1]. Thus, endowed with this comultiplication, A becomes an ϵ-bialgebra. We refer to the pair (A, r) as a quasitriangular ϵ-bialgebra [1, Definition 5.3].

Remark 2.6. In our previous work [1, 3], we have used the comultiplication

$$-\Delta(a) = a \cdot r - r \cdot a = \sum_i a u_i \otimes v_i - \sum_i u_i \otimes v_i a$$

instead of Δ. Both Δ and $-\Delta$ endow A with a structure of ϵ-bialgebra, and there is no essential difference in working with one or the other. The choice we adopt in (2.9), however, is more convenient for the purposes of this work, particularly in relating quasitriangular ϵ-bialgebras and their bimodules to dendriform algebras and their bimodules (Sections 4 and 5).

It is then necessary to make the corresponding sign adjustments to the results on quasitriangular ϵ-bialgebras from [1, 3] before applying them in the present context. For instance, Proposition 5.5 in [1] translates as

$$\sum_{i,j} u_i \otimes u_j \otimes v_j v_i = r_{23} r_{13} = (\Delta \otimes id)(r) = \Delta(u_i) \otimes v_i\,. \tag{2.10}$$

Proposition 2.7. *Let (A, r) be a quasitriangular ϵ-bialgebra and M a left A-module. Then M becomes a left ϵ-Hopf module over A via $\Lambda : M \to A \otimes M$,*

$$\Lambda(m) = \sum_i u_i \otimes v_i m\,.$$

Proof. We first check that Λ is coassociative, i.e., $(id \otimes \Lambda)\Lambda = (\Delta \otimes id)\Lambda$. We have

$$(id \otimes \Lambda)\Lambda(m) = \sum_i u_i \otimes \Lambda(v_i m) = \sum_{i,j} u_i \otimes u_j \otimes v_j v_i m$$

and

$$(\Delta \otimes id)\Lambda(m) = \sum \Delta(u_i) \otimes v_i m\,.$$

According to (2.10), these two expressions agree.

It only remains to check Axiom (2.2). Since $\Delta(a) = \sum_i u_i \otimes v_i a - \sum_i a u_i \otimes v_i$, we have

$$\Delta(a)m + a\Lambda(m) = \sum_i u_i \otimes v_i a m - \sum_i a u_i \otimes v_i m + \sum_i a u_i \otimes v_i m = \sum_i u_i \otimes v_i a m = \Lambda(am)\,,$$

as needed. □

Remark 2.8. If A is a finite dimensional quasitriangular ϵ-bialgebra, then there is a canonical morphism of ϵ-bialgebras $\pi : D(A) \to A$, which is the identity on A [1, Proposition 7.5]. Therefore, any left A-module M can be first made into a left $D(A)$-module by restriction via π, and then, by Theorem 2.5, into a left

ϵ-Hopf module over A. It is easily seen that this structure coincides with the one of Proposition 2.7. Note that the construction of the latter proposition is more general, since it does not require finite dimensionality of A.

3. Pre-Lie algebras

Definition 3.1. A (left) pre-Lie algebra is a vector space P together with a map $\circ : P \otimes P \to P$ such that

$$x \circ (y \circ z) - (x \circ y) \circ z = y \circ (x \circ z) - (y \circ x) \circ z\,. \tag{3.1}$$

There is a similar notion of right pre-Lie algebras. In this paper, we will only deal with left pre-Lie algebras and we will refer to them simply as pre-Lie algebras.

Defining a new operation $P \otimes P \to P$ by $[x, y] = x \circ y - y \circ x$ one obtains a Lie algebra structure on P [13, Theorem 1].

Next we show that every ϵ-bialgebra A gives rise to a structure of pre-Lie algebra, and hence also of Lie algebra, on the underlying space of A.

Theorem 3.2. *Let (A, μ, Δ) be an ϵ-bialgebra. Define a new operation on A by*

$$a \circ b = b_1 a b_2\,. \tag{3.2}$$

Then $(A, \circ)$ is a pre-Lie algebra.

Proof. By repeated use of (2.1) we find

$$\Delta(abc) = ab \cdot \Delta(c) + \Delta(ab) \cdot c = abc_1 \otimes c_2 + ab_1 \otimes b_2 c + a_1 \otimes a_2 bc\,.$$

Together with coassociativity this gives

$$\Delta(c_1 b c_2) = c_1 b c_2 \otimes c_3 + c_1 b_1 \otimes b_2 c_2 + c_1 \otimes c_2 b c_3\,.$$

Combining this with (3.2) we obtain

$$a \circ (b \circ c) = a \circ (c_1 b c_2) = c_1 b c_2 a c_3 + c_1 b_1 a b_2 c_2 + c_1 a c_2 b c_3\,.$$

On the other hand,

$$(a \circ b) \circ c = (b_1 a b_2) \circ c = c_1 b_1 a b_2 c_2\,.$$

Therefore,

$$a \circ (b \circ c) - (a \circ b) \circ c = c_1 b c_2 a c_3 + c_1 a c_2 b c_3\,.$$

Since this expression is invariant under $a \leftrightarrow b$, Axiom (3.1) holds and $(A, \circ)$ is a pre-Lie algebra. □

For a vector space V, let $\mathfrak{gl}(V)$ denote the space of all linear maps $V \to V$, viewed as a Lie algebra under the commutator bracket $[T, S] = TS - ST$.

If P is a pre-Lie algebra and $x \in P$, let $L_x : P \to P$ be $L_x(y) = x \circ y$. The map $L : P \to \mathfrak{gl}(P)$, $x \mapsto L_x$ is a morphism of Lie algebras. This statement is just a reformulation of Axiom (3.1).

In the case when the pre-Lie algebra comes from an ϵ-bialgebra (A, μ, Δ), more can be said about this canonical map. Let $\mathsf{Der}(A, \mu)$ denote the space of all derivations $D : A \to A$ of the associative algebra A. Recall that this is a Lie subalgebra of $\mathfrak{gl}(A)$.

Proposition 3.3. *Let (A, μ, Δ) be an ϵ-bialgebra and consider the associated pre-Lie and Lie algebra structures on A. The canonical morphism of Lie algebras $L : (A, [\ ,\ ]) \to \mathfrak{gl}(A)$ actually maps to the Lie subalgebra $\mathsf{Der}(A, \mu)$ of $\mathfrak{gl}(A)$.*

Proof. We must show that each $L_c \in \mathfrak{gl}(A)$ is a derivation of the associative algebra A. We have $\Delta(ab) \overset{(2.1)}{=} ab_1 \otimes b_2 + a_1 \otimes a_2 b$ and hence

$$L_c(ab) = c \circ (ab) \overset{(3.2)}{=} ab_1 cb_2 + a_1 ca_2 b \overset{(3.2)}{=} a(c \circ b) + (c \circ a)b = aL_c(b) + L_c(a)b \ ,$$

as needed.

□

Examples 3.4.

(1) Consider the ϵ-bialgebra of *divided differences*. This is the algebra $k[\mathbf{x}, \mathbf{x}^{-1}]$ of Laurent polynomials, with $\Delta(f(\mathbf{x})) = \frac{f(\mathbf{x})-f(\mathbf{y})}{\mathbf{x}-\mathbf{y}}$. This was the example that motivated Joni and Rota to abstract the notion of ϵ-bialgebras [17, Section XII]. More explicitly,

$$\Delta(\mathbf{x}^n) = \sum_{i=0}^{n-1} \mathbf{x}^i \otimes \mathbf{x}^{n-1-i} \quad \text{and} \quad \Delta(\frac{1}{\mathbf{x}^n}) = -\sum_{i=1}^{n} \frac{1}{\mathbf{x}^i} \otimes \frac{1}{\mathbf{x}^{n+1-i}} \ , \quad \text{for } n \geq 0.$$

The corresponding pre-Lie algebra structure is

$$\mathbf{x}^m \circ \mathbf{x}^n = n\mathbf{x}^{m+n-1} \ , \quad \text{for any } n \in \mathbf{Z},$$

and the Lie algebra structure on $k[\mathbf{x}, \mathbf{x}^{-1}]$ is

$$[\mathbf{x}^m, \mathbf{x}^n] = (n-m)\mathbf{x}^{m+n-1} \quad \text{for } n, m \in \mathbf{Z}.$$

This is the so called Witt Lie algebra. The canonical map $k[\mathbf{x}, \mathbf{x}^{-1}] \to \mathsf{Der}(k[\mathbf{x}, \mathbf{x}^{-1}], \mu)$ of Proposition 3.3 sends $\mathbf{x}^m$ to $\mathbf{x}^m \frac{d}{d\mathbf{x}}$, so it is an isomorphism of Lie algebras.

(2) The algebra of matrices $M_2(k)$ is an ϵ-bialgebra under

$$\Delta \begin{bmatrix} a & b \\ c & d \end{bmatrix} = \begin{bmatrix} 0 & a \\ 0 & c \end{bmatrix} \otimes \begin{bmatrix} 0 & 1 \\ 0 & 0 \end{bmatrix} - \begin{bmatrix} 0 & 1 \\ 0 & 0 \end{bmatrix} \otimes \begin{bmatrix} c & d \\ 0 & 0 \end{bmatrix}$$

[1, Example 2.3.7]. One finds easily that the corresponding Lie algebra splits as a direct sum of Lie algebras

$$\mathfrak{g} = \mathfrak{h} \oplus \mathfrak{o}$$

where $\mathfrak{h} = k\{x, y, z\}$ is the 3-dimensional *Heisenberg* algebra

$$\{x, y\} = z, \quad \{x, z\} = \{y, z\} = 0 \ ,$$

and $\mathfrak{o} = k\{i\}$ is the 1-dimensional Lie algebra. To realize this isomorphism explicitly, one may take

$$x = \begin{bmatrix} 0 & 0 \\ 1 & 0 \end{bmatrix}, \ y = \begin{bmatrix} 1 & 0 \\ 0 & 0 \end{bmatrix}, \ z = \begin{bmatrix} 0 & 1 \\ 0 & 0 \end{bmatrix} \text{ and } i = \begin{bmatrix} 1 & 0 \\ 0 & 1 \end{bmatrix} .$$

(3) The *path algebra* of a quiver carries a canonical ϵ-bialgebra structure [1, Example 2.3.2]. Let Q be an arbitrary quiver (i.e., an oriented graph). Let Q_n be the set of paths α in Q of length n:

$$\alpha : e_0 \xrightarrow{a_1} e_1 \xrightarrow{a_2} e_2 \xrightarrow{a_3} \dots e_{n-1} \xrightarrow{a_n} e_n \ .$$

In particular, Q_0 is the set of vertices and Q_1 is the set of arrows. Recall that the path algebra of Q is the space $kQ = \oplus_{n=0}^{\infty} kQ_n$ where multiplication is concatenation of paths whenever possible; otherwise is zero. The comultiplication is defined on a path $\alpha = a_1 a_2 \dots a_n$ as above by

$$\Delta(\alpha) = e_0 \otimes a_2 a_3 \dots a_n + a_1 \otimes a_3 \dots a_n + \dots + a_1 \dots a_{n-1} \otimes e_n \ .$$

In particular, $\Delta(e) = 0$ for every vertex e and $\Delta(a) = e_0 \otimes e_1$ for every arrow $e_0 \xrightarrow{a} e_1$.

In order to describe the corresponding pre-Lie algebra structure on kQ, consider pairs (α, b) where α is a path from e_0 to e_n (as above) and b is an arrow from e_0 to e_n. Let us call such a pair a *shortcut*. The pre-Lie algebra structure on kQ is

$$\alpha \circ \beta = \sum_{b_i \in \beta} b_1 \dots b_{i-1} \alpha b_{i+1} \dots b_m \,,$$

where the sum is over all arrows b_i in the path $\beta = b_1 \dots b_m$ such that (α, b_i) is a shortcut.

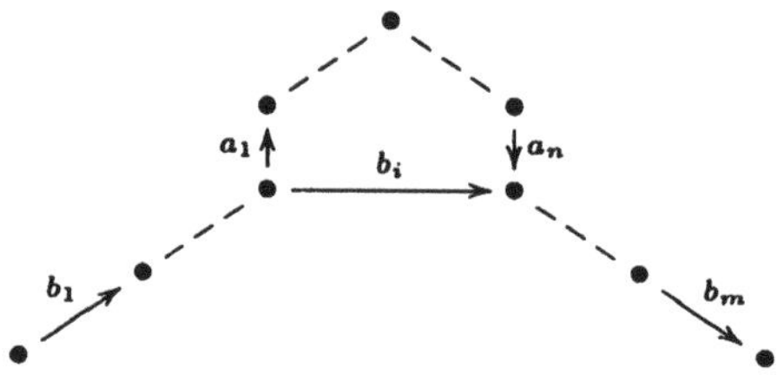

A *biderivation* of an ϵ-bialgebra (A, μ, Δ) is a map $B : A \to A$ that is both a derivation of (A, μ) and a coderivation of (A, Δ), i.e.,

(3.3) $$B(ab) = aB(b) + B(a)b \quad \text{and} \quad \Delta(B(a)) = a_1 \otimes B(a_2) + B(a_1) \otimes a_2 \,.$$

A derivation of a pre-Lie algebra is a map $D : P \to P$ such that

$$D(x \circ y) = x \circ D(y) + D(x) \circ y \,.$$

Such a map D is always a derivation of the associated Lie algebra.

Proposition 3.5. *Let B be a biderivation of an ϵ-bialgebra A. Then B is a derivation of the associated pre-Lie algebra (and hence also of the associated Lie algebra).*

Proof. We have

$$B(a) \circ b = b_1 B(a) b_2 \quad \text{and} \quad a \circ B(b) = b_1 a B(b_2) + B(b_1) a b_2 \,.$$

Hence,

$$B(a) \circ b + a \circ B(b) = b_1 B(a) b_2 + b_1 a B(b_2) + B(b_1) a b_2 = B(b_1 a b_2) = B(a \circ b) \,.$$

□

The construction of a pre-Lie algebra from and ϵ-bialgebra can be extended to the categories of modules. This will be discussed in the appropriate generality in Section 5. A first result in this direction is discussed next.

Let $(P, \circ)$ be a pre-Lie algebra. A left P-module is a space M together with a map $P \otimes M \to M$, $x \otimes m \mapsto x \circ m$, such that

(3.4) $$x \circ (y \circ m) - (x \circ y) \circ m = y \circ (x \circ m) - (y \circ x) \circ m \,.$$

Proposition 3.6. *Let A be an ϵ-bialgebra and M a left ϵ-Hopf module over A via*

$$\lambda(a \otimes m) = am \quad \text{and} \quad \Lambda(m) = m_{-1} \otimes m_0 .$$

Then M is a left pre-Lie module over the pre-Lie algebra $(A, \circ)$ of Theorem 3.2 via

$$a \circ m = m_{-1} a m_0 .$$

Proof. We first compute

$$\Lambda(a \circ m) = \Lambda(m_{-1} am) \overset{(2.2)}{=} \Delta(m_{-1} a) m_0 + m_{-1} a \Lambda(m_0)$$
$$\overset{(2.1)}{=} m_{-1} \otimes m_{-1} a m_0 + m_{-1} a_1 \otimes a_2 m_0 + m_{-2} a m_{-1} \otimes m_0 ,$$

where we have used the coassociativity axiom for the comodule structure Λ. It follows that

$$b \circ (a \circ m) = m_{-2} b m_{-1} a m_0 + m_{-1} a_1 b a_2 m_0 + m_{-2} a m_{-1} b m_0 .$$

On the other hand,

$$(b \circ a) \circ m = m_{-1} (b \circ a) m_0 \overset{(3.2)}{=} m_{-1} a_1 b a_2 m_0 .$$

Therefore,

$$b \circ (a \circ m) - (b \circ a) \circ m = m_{-2} b m_{-1} a m_0 + m_{-2} a m_{-1} b m_0 .$$

Since this expression is invariant under $a \leftrightarrow b$, Axiom (3.4) holds. □

Remark 3.7. Since the notion of ϵ-bialgebras is self-dual [1, Section 2], one should expect dual constructions to those of Theorem 3.2 and Propositions 3.5 and 3.6. This is indeed the case. Namely, if A is an arbitrary ϵ-bialgebra, then the map $\gamma : A \to A \otimes A$ defined by

$$\gamma(a) = a_2 \otimes a_1 a_3$$

endows A with a structure of *left pre-Lie coalgebra*. Also, if $B : A \to A$ is a biderivation of A then it is also a coderivation of (A, γ). Moreover, if M is a left ϵ-Hopf module over A, then M is left pre-Lie comodule over (A, γ) via $\psi : M \to A \otimes M$ defined by

$$\psi(m) = m_{-1} \otimes m_{-2} m_0 .$$

If A is an ϵ-bialgebra, then A carries structures of pre-Lie algebra and pre-Lie coalgebra, as just explained. Hence, it also carries structures of Lie algebra and Lie coalgebra, by

$$[a, b] = b_1 a b_2 - a_1 b a_2 \quad \text{and} \quad \delta(a) = a_2 \otimes a_1 a_3 - a_1 a_3 \otimes a_2 .$$

In general, these structures are not compatible, in the sense that they do *not* define a structure of Lie bialgebra on A.

4. Dendriform algebras

Definition 4.1. A dendriform algebra is a vector space D together with maps $\succ : D\otimes D \to D$ and $\prec : D \times D \to D$ such that

$$
\begin{aligned}
(x \prec y) \prec z &= x \prec (y \prec z) + x \prec (y \succ z)\\
x \succ (y \prec z) &= (x \succ y) \prec z \qquad (4.1)\\
x \succ (y \succ z) &= (x \prec y) \succ z + (x \succ y) \succ z\,.
\end{aligned}
$$

Dendriform algebras were introduced by Loday [20, Chapter 5]. There is also a notion of *dendriform trialgebras*, which involves three operations [23]. When it is necessary to distinguish between these two notions, one uses the name *dendriform dialgebras* to refer to what in this paper (and in [20]) are called dendriform algebras. Since only dendriform algebras (in the sense of Definition 4.1) will be considered in this paper, this usage will not be adopted.

Let $(D, \succ, \prec)$ be a dendriform algebra. Defining $x \cdot y = x \succ y + x \prec y$ one obtains an associative algebra structure on D. In addition, defining

$$x \circ y = x \succ y - y \prec x \qquad (4.2)$$

one obtains a (left) pre-Lie algebra structure on D. Moreover, the Lie algebras canonically associated to $(D, \cdot)$ and $(D, \circ)$ coincide; namely,

$$x \cdot y - y \cdot x = x \succ y + x \prec y - y \succ x - y \prec x = x \circ y - y \circ x\,.$$

If $f : D \to D'$ is a morphism of dendriform algebras, then it is also a morphism with respect to any of the other three structures on D. The situation may be summarized by means of the following commutative diagram of categories

$$
\begin{array}{ccc}
\text{Dendriform algebras} & \longrightarrow & \text{Pre-Lie algebras}\\
\downarrow & & \downarrow\\
\text{Associative algebras} & \longrightarrow & \text{Lie algebras}
\end{array}
$$

Recall the notion of quasitriangular ϵ-bialgebras from Section 2. In this section we show that there is a commutative diagram as follows:

$$
\begin{array}{ccc}
\text{Quasitriangular } \epsilon\text{-bialgebras} & \longrightarrow & \epsilon\text{-bialgebras}\\
\downarrow & & \downarrow\\
\text{Dendriform algebras} & \longrightarrow & \text{Pre-Lie algebras}
\end{array}
\qquad (4.3)
$$

In this diagram, the right vertical arrow is the functor constructed in Section 3, the bottom horizontal arrow is the construction just discussed (4.2) and the top horizontal arrow is simply the inclusion. It remains to discuss the construction of a dendriform algebra from a quasitriangular ϵ-bialgebra, and to verify the commutativity of the diagram.

This construction is best understood from the point of view of *Baxter operators*.

Definition 4.2. Let A be an associative algebra. A Baxter operator is a map $\beta : A \to A$ that satisfies the condition

$$\beta(x)\beta(y) = \beta\Big(x\beta(y) + \beta(x)y\Big)\,. \qquad (4.4)$$

Baxter operators arose in probability theory [5] and were a subject of interest to Gian-Carlo Rota [27, 28].

We start by recalling a basic result from [2], which provides us with the examples of Baxter operators that are most relevant for our present purposes.

Proposition 4.3. *Let $r = \sum_i u_i \otimes v_i$ be a solution of the associative Yang-Baxter equation* (2.8) *in an associative algebra A. Then the map $\beta : A \to A$ defined by*

$$\beta(x) = \sum_i u_i x v_i$$

is a Baxter operator.

Proof. Replacing the tensor symbols in the associative Yang-Baxter equation (2.8) by x and y one obtains precisely (4.4). □

Remark 4.4. The associative Yang-Baxter equation is analogous to the *classical Yang-Baxter equation*, which is named after C. N. Yang and R. J. Baxter. Baxter operators, on the other hand, are named after Glen Baxter.

The following result provides the second step in the construction of dendriform algebras from quasitriangular ϵ-bialgebras.

Proposition 4.5. *Let A be an associative algebra and $\beta : A \to A$ a Baxter operator. Define new operations on A by*

$$x \succ y = \beta(x)y \quad \text{and} \quad x \prec y = x\beta(y) \ .$$

Then $(A, \succ, \prec)$ is a dendriform algebra.

Proof. We verify the last axiom in (4.1); the others are similar. We have

$$\begin{aligned} x \succ (y \succ z) &= \beta(x)(y \succ z) = \beta(x)\beta(y)z \\ &\overset{(4.4)}{=} \beta\Big(x\beta(y) + \beta(x)y\Big)z = \beta(x \prec y + x \succ y)z \\ &= \beta(x \prec y)z + \beta(x \succ y)z = (x \prec y) \succ z + (x \succ y) \succ z\,. \end{aligned}$$

□

Extensions of the above result appear in [2, Propositions 5.1 and 5.2].

Finally, we have the desired construction of dendriform algebras from quasitriangular ϵ-bialgebras.

Theorem 4.6. *Let (A, r) be a quasitriangular ϵ-bialgebra, $r = \sum_i u_i \otimes v_i$. Define new operations on A by*

$$x \succ y = \sum_i u_i x v_i y \quad \text{and} \quad x \prec y = \sum_i x u_i y v_i\,. \tag{4.5}$$

Then, $(A, \succ, \prec)$ is a dendriform algebra.

Proof. Combine Propositions 4.3 and 4.5. □

A morphism between quasitriangular ϵ-bialgebras (A, r) and (A', r') is a morphism of algebras $f : A \to A'$ such that $(f \otimes f)(r) = r'$. Clearly, such a map f preserves the dendriform structures on A and A'. Thus, we have constructed a functor from quasitriangular ϵ-bialgebras to dendriform algebras.

We briefly discuss the functoriality of the construction with respect to derivations.

A derivation of a quasitriangular ϵ-bialgebra (A, r) is a map $D : A \to A$ that is a derivation of the associative algebra A such that

$$(D \otimes id + id \otimes D)(r) = 0 .$$

This implies that D is a biderivation of the ϵ-bialgebra associated to (A, r), in the sense of (3.3). In fact, it is easy to see that D is a biderivation if and only $(D \otimes id + id \otimes D)(r)$ is an invariant element in the A-bimodule $A \otimes A$. These two conditions are analogous to the ones encountered in the definition of quasitriangular and coboundary ϵ-bialgebras [3, Section 1]. The stronger condition guarantees the following:

Proposition 4.7. *Let D be a derivation of a quasitriangular ϵ-bialgebra (A, r). Then D is also a derivation of the associated dendriform algebra, i.e.,*

$$D(a \succ b) = a \succ D(b) + D(a) \succ b \quad \text{and} \quad D(a \prec b) = a \prec D(b) + D(a) \prec b .$$

Proof. Similar to the other proofs in this section. □

It remains to verify the commutativity of diagram (4.3). Starting from (A, r) and going clockwise, we pass through the ϵ-bialgebra with comultiplication $\Delta(b) = \sum_i u_i \otimes v_i b - \sum_i b u_i \otimes v_i$, according to (2.9) (see also Remark 2.6). The associated pre-Lie algebra structure is, by (3.2), $a \circ b = \sum_i u_i a v_i b - \sum_i b u_i a v_i$. According to (4.5), this expression is equal to $a \succ b - b \prec a$, which by (4.2) is the pre-Lie algebra structure obtained by going counterclockwise around the diagram.

Examples 4.8.

(1) Let A be an associative unital algebra and $b \in A$ an element such that $b^2 = 0$. Then, $r := 1 \otimes b$ is a solution of the associative Yang-Baxter equation (2.8) [1, Example 5.4.1]. The corresponding dendriform structure on A is simply

$$x \succ y = xyb \quad \text{and} \quad x \prec y = xby .$$

This structure is well defined even if A does not have a unit.

(2) The element $r := \begin{bmatrix} 1 & 0 \\ 0 & 0 \end{bmatrix} \otimes \begin{bmatrix} 0 & 1 \\ 0 & 0 \end{bmatrix} - \begin{bmatrix} 0 & 1 \\ 0 & 0 \end{bmatrix} \otimes \begin{bmatrix} 1 & 0 \\ 0 & 0 \end{bmatrix}$ is a solution of (2.8) in the algebra of matrices $M_2(k)$ [1, Example 5.4.5.e] and [3, Examples 2.3.1 and 2.8]. The corresponding dendriform structure on $M_2(k)$ is

$$\begin{bmatrix} a & b \\ c & d \end{bmatrix} \succ \begin{bmatrix} x & y \\ z & w \end{bmatrix} = \begin{bmatrix} az - cx & aw - cy \\ 0 & 0 \end{bmatrix} \quad \text{and} \quad \begin{bmatrix} a & b \\ c & d \end{bmatrix} \prec \begin{bmatrix} x & y \\ z & w \end{bmatrix} = \begin{bmatrix} -az & ax \\ -cz & cx \end{bmatrix} .$$

(3) The most important example is provided by Drinfeld's double, which is a quasitriangular ϵ-bialgebra canonically associated to an arbitrary finite dimensional ϵ-bialgebra. This will occupy the rest of the section.

Let A be a finite dimensional ϵ-bialgebra. Recall the definition of the double $D(A)$ from Section 2. Let $\{e_i\}$ be a linear basis of A and $\{f_i\}$ the dual basis of A^*. Let $r \in D(A) \otimes D(A)$ be the element

$$r = \sum_i e_i \otimes f_i \in A \otimes A^* \subseteq D(A) \otimes D(A) .$$

According to [1, Theorem 7.3], $(D(A), r)$ is a quasitriangular ϵ-bialgebra (see Remark 2.6). By Theorem 4.6, there is a dendriform algebra structure on the space $D(A) = (A \otimes A^*) \oplus A \oplus A^*$.

In order to make this structure explicit, we introduce some notation. We identify $A\otimes A^*$ with $\mathsf{End}(A)$ via

$$(a\bowtie f)(b) = f(b)a\,.$$

For each $a \in A$ and $f \in A^*$, define linear endomorphisms of A by

$$\begin{aligned} L_a(x) &= ax & L_f(x) &= f(x_2)x_1 \\ R_a(x) &= xa & R_f(x) &= f(x_1)x_2 \\ P_a(x) &= x_1 a x_2 & P_f(x) &= f(x_2)x_1x_3\,. \end{aligned}$$

The composition of linear maps $\phi : U \to V$ and $\psi : V \to W$ is denoted by $\psi\phi$, or $\psi(\phi)$, if the expression for ϕ is complicated. This should not be confused with the evaluation of an endomorphism T on an element $a \in A$, denoted by $T(a)$. The convolution of linear endomorphisms T and S of A is

$$T * S = \mu(T\otimes S)\Delta\,.$$

Theorem 4.9. *Let A be an arbitrary ϵ-bialgebra. There is a dendriform structure on the space $\mathsf{End}(A)\oplus A\oplus A^*$, given explicitly as follows. For a, $b \in A$, f, $g \in A^*$ and T, $S \in \mathsf{End}(A)$,*

$$\begin{aligned} a \succ b &= P_a(b) + R_aL_b & a \prec b &= L_aR_b \\ f \succ a &= P_f(a) + L_fL_a & a \prec f &= L_aL_f \\ f \prec a &= fP_a + R_fR_a & a \succ f &= R_aR_f \\ f \prec g &= fP_g + R_fL_g & f \succ g &= L_fR_g \end{aligned}$$

$$\begin{aligned} a \succ T &= P_aT + R_a(T * id) & a \prec T &= L_a(id * T) \\ f \succ T &= P_fT + L_f(T * id) & f \prec T &= f(id * T * id) + R_f(id * T) \\ T \prec a &= TP_a + (T * id)R_a & T \succ a &= (id * T * id)(a) + (id * T)L_a \\ T \prec f &= TP_f + (T * id)L_f & T \succ f &= (id * T)R_f \end{aligned}$$

$$\begin{aligned} T \succ S &= (id * T * id)S + (id * T)(S * id) \\ T \prec S &= T(id * S * id) + (T * id)(id * S) \end{aligned}$$

Proof. Assume that A is finite dimensional, so $D(A)$ and $r = \sum e_i\otimes f_i$ are well defined, and hence there is a dendriform structure on the space $D(A)$. For the details of the infinite dimensional case see Remark 4.10.

We provide the derivations of the first and last formulas, the others are similar. We make use of the ϵ-bialgebra structure of $D(A)$ as described in Proposition 2.3.

For the first formula we have

$$\begin{aligned} a \succ b &= \sum_i e_iaf_ib = \sum_i e_ia(f_i \to b) + \sum_i e_ia(f_i \leftarrow b) \\ &\overset{(2.5)}{=} \sum_i e_iaf_i(b_1)b_2 + \sum_i e_ia\bowtie(f_i \leftarrow b) \\ &= b_1ab_2 + \sum_i e_ia\bowtie(f_i \leftarrow b)\,. \end{aligned}$$

Now, for any $x \in A$,

$$\sum_i (f_i \leftarrow b)(x)e_i a \overset{(2.6)}{=} \sum_i f_i(bx)e_i a = bxa = R_a L_b(x)\,.$$

Thus, $a \succ b = P_a(b) + R_a L_b$, as claimed.

For the last formula, let $T = a \bowtie f$ and $S = b \bowtie g$. We have

$$\begin{aligned} T \prec S &= (a \bowtie f) \prec (b \bowtie g) = \sum_i (a \bowtie f) e_i (b \bowtie g) f_i \\ &= \sum_i a(f \rightarrow e_i b) \bowtie g f_i + \sum_i a \bowtie (f \leftarrow e_i b) g f_i\,. \end{aligned}$$

Hence, for any $x \in A$,

$$\begin{aligned} (T \prec S)(x) &\overset{(2.6)}{=} \sum_i f_i(x_1)g(x_2)a(f \rightarrow e_i b) + \sum_i f_i(x_1)g(x_2)(f \leftarrow e_i b)(x_3)a \\ &= g(x_2)a(f \rightarrow x_1 b) + g(x_2)(f \leftarrow x_1 b)(x_3)a \\ &\overset{(2.6)}{=} a(f \rightarrow x_1 S(x_2)) + g(x_2)f(x_1 b x_3)a \\ &\overset{(2.5)}{=} f\big((x_1 S(x_2))_1\big)a(x_1 S(x_2))_2 + f(x_1 S(x_2)x_3)a \\ &= (T * id)(id * S)(x) + T(id * S * id)(x)\,. \end{aligned}$$

Thus, $T \prec S = (T * id)(id * S) + T(id * S * id)$, as claimed. □

Remark 4.10. The ϵ-bialgebra structure on $D(A)$ and the element $r \in D(A) \otimes D(A)$ are well defined only if A is finite dimensional. However, all formulas in Theorem 4.9 make sense and the theorem holds even if A is infinite dimensional. This may be seen as follows. There is always an algebra structure on the space $\mathsf{End}(A) \oplus A \oplus A^*$, extending that of $D(A)$. Moreover, there is always a Baxter operator on this algebra, well defined by

$$\beta(a) = R_a, \quad \beta(f) = L_f \quad \text{and} \quad \beta(T) = id * T\,.$$

It is easy to see that this coincides with the operator corresponding to r, when A is finite dimensional. In the general case, it may be checked directly that β satisfies (4.4). The result then follows from Proposition 4.5.

Remark 4.11. In order to fully appreciate the symmetry in the previous formulas, the following relations should be kept in mind:

$$\begin{aligned} T * R_a &= R_a(T * id) & T * L_f &= (T * id)L_f \\ L_a * T &= L_a(id * T) & R_f * T &= (id * T)R_f \\ L_a R_b &= R_b L_a & L_f R_g &= R_g L_f \end{aligned}$$

Remark 4.12. Consider the pre-Lie algebra structures on A and A' corresponding to their ϵ-bialgebra structures by means of Theorem 3.2. Since A and A' are ϵ-subbialgebras of $D(A)$ (Proposition 2.3), the functoriality of the construction implies that A and A' are pre-Lie subalgebras of $D(A)$, with respect to the pre-Lie structure associated to the dendriform structure as in (4.2). Let us verify this fact explicitly. The pre-Lie structure on A is

$$a \circ b = a \succ b - b \prec a = P_a(b) + R_a L_b - L_b R_a = P_a(b) = b_1 a b_2\,,$$

as expected. The pre-Lie structure on A' is

$$f \circ g = f \succ g - g \prec f = L_f R_g - gP_f - R_g L_f = -gP_f\,.$$

Thus,

$$(f \circ g)(a) = -g(f(a_2)a_1a_3) \overset{(2.4)}{=} f(a_2)g_2(a_1)g_1(a_3) \overset{(2.3)}{=} (g_1 f g_2)(a)\,,$$

also as expected (the ϵ-bialgebra structure on A' was described in Section 2).

When A is a quasitriangular ϵ-bialgebra, there are dendriform algebra structures both on A and $\mathsf{End}(A)\oplus A\oplus A^*$, by Theorems 4.6 and 4.9. These two structures are related by a canonical morphism of dendriform algebras.

Proposition 4.13. *Let (A, r) be a quasitriangular ϵ-bialgebra, $r = \sum_i u_i \otimes v_i$. Then, the map*

$$\pi : \mathit{End}(A)\oplus A\oplus A^* \to A, \quad \pi(a) = a, \ \pi(f) = \sum_i f(u_i)v_i \ \text{ and } \pi(T) = \sum_i T(u_i)v_i$$

is a morphism of dendriform algebras.

Proof. Assume that A is finite dimensional. According to [1, Proposition 7.5], the above formulas define a morphism of ϵ-bialgebras $\pi : D(A) \to A$ (see Remark 2.6). By the functoriality of the construction of dendriform algebras, π is also a morphism of dendriform algebras.

The general case may be obtained by showing that π commutes with the Baxter operators on $\mathsf{End}(A)\oplus A\oplus A^*$ and A. This follows from (2.8) and (2.10). □

The formulas in Theorem 4.9 show that $\mathsf{End}(A)$ is closed under the dendriform operations. Together with Proposition 4.13, this gives the following:

Corollary 4.14. *Let A be an arbitrary ϵ-bialgebra. Then there is a dendriform algebra structure on the space $\mathit{End}(A)$ of linear endomorphisms of A, defined by*

$$T \succ S = (id*T*id)S + (id*T)(S*id) \quad \text{and} \quad T \prec S = T(id*S*id) + (T*id)(id*S)\,.$$

Moreover, if A is quasitriangular, with $r = \sum_i u_i \otimes v_i$, then there is a morphism of dendriform algebras $\mathit{End}(A) \to A$ given by

$$T \mapsto \sum_i T(u_i)v_i$$

Proof. □

Remark 4.15. There are in fact other, more primitive, dendriform structures on $\mathsf{End}(A)$ whenever A is an ϵ-bialgebra. These will be studied in future work.

5. Infinitesimal Hopf bimodules, pre-Lie bimodules, dendriform bimodules

In previous sections, we have shown how to construct a pre-Lie algebra from an ϵ-bialgebra and a dendriform algebra from a quasitriangular ϵ-bialgebra. These constructions are compatible, in the sense of (4.3). In this section we extend these constructions to the corresponding categories of bimodules.

The first step is to define the appropiate notion of bimodules over ϵ-bialgebras. Recall the notion of left infinitesimal Hopf modules from Definition 2.1. Right infinitesimal Hopf modules are defined similarly. We combine these two notions in the following:

Definition 5.1. Let (A,μ,Δ) be an ϵ-bialgebra. An infinitesimal Hopf bimodule (abbreviated ϵ-Hopf bimodule) over A is a space M endowed with maps

$$\lambda : A\otimes M \to M,\ \ \Lambda : M \to A\otimes M,\ \ \xi : M\otimes A \to M \text{ and } \Xi : M \to M\otimes A$$

such that

(a) (M,λ,Λ) is a left ϵ-Hopf module over (A,μ,Δ),
(b) (M,ξ,Ξ) is a right ϵ-Hopf module over (A,μ,Δ),
(c) (M,λ,ξ) is a bimodule over (A,μ),
(d) (M,Λ,Ξ) is a bicomodule over (A,Δ), and
(e) the following diagrams commute:

$$\begin{array}{ccc} A\otimes M & \xrightarrow{id\otimes\Xi} & A\otimes M\otimes A \\ {\scriptstyle\lambda}\downarrow & & \downarrow{\scriptstyle\lambda\otimes id} \\ M & \xrightarrow[\Xi]{} & M\otimes A \end{array} \qquad \begin{array}{ccc} M\otimes A & \xrightarrow{\Lambda\otimes id} & A\otimes M\otimes A \\ {\scriptstyle\xi}\downarrow & & \downarrow{\scriptstyle id\otimes\xi} \\ M & \xrightarrow[\Lambda]{} & A\otimes M \end{array}$$

Example 5.2. For any ϵ-bialgebra (A,μ,Δ), the space $M = A\otimes A$ may be endowed with the following ϵ-Hopf bimodule structure:

$$\lambda = \mu\otimes id,\ \ \Lambda = \Delta\otimes id,\ \ \xi = id\otimes\mu \text{ and } \Xi = id\otimes\Delta\,.$$

Note that A itself, with the canonical bimodule and bicomodule structures, is not an ϵ-Hopf bimodule.

We will often use the following notation, for an ϵ-Hopf bimodule $(M,\lambda,\Lambda,\xi,\Xi)$:

$$\text{(5.1)}\quad \lambda(a\otimes m) = am,\ \ \xi(m\otimes a) = ma,\ \ \Lambda(m) = m_{-1}\otimes m_0 \text{ and } \Xi(m) = m_0\otimes m_1\,.$$

As is well known, this notation efficiently encodes the bicomodule axioms. For instance, $m_{-2}\otimes m_{-1}\otimes m_0$ stands for $(\Delta\otimes id)\Lambda(m) = (id\otimes\Lambda)\Lambda(m)$, and $m_{-1}\otimes m_0\otimes m_1$ for $(id\otimes\Xi)\Lambda(m) = (\Lambda\otimes id)\Xi(m)$.

Just as left ϵ-Hopf modules over A are left modules over $D(A)$ (Theorem 2.5), ϵ-Hopf bimodules over A are bimodules over $D(A)$.

Proposition 5.3. *Let A be a finite dimensional ϵ-bialgebra and M a space. If $(M,\lambda,\Lambda,\xi,\Xi)$ is an ϵ-Hopf bimodule over A as in* (5.1), *then M is a bimodule over $D(A)$ via*

$$\text{(5.2)}\qquad a\cdot m = am,\ \ f\cdot m = f(m_{-1})m_0,\ \ (a\bowtie f)\cdot m = f(m_{-1})am_0$$

and

$$\text{(5.3)}\qquad m\cdot a = ma,\ \ m\cdot f = f(m_1)m_0,\ \ m\cdot(a\bowtie f) = f(m_1)m_0a\,.$$

Conversely, if M is a bimodule over $D(A)$ then M is an ϵ-Hopf bimodule over A and the structures are related as above.

Proof. Suppose M is an ϵ-Hopf bimodule over A. Then (M,λ,Λ) is a left ϵ-Hopf module and (M,ξ,Ξ) is a right ϵ-Hopf module over A. By Theorem 2.5, M is a left and right module over $D(A)$ by means of (5.2) and (5.3). It only remains to check that these structures commute.

By assumption (c) (resp. (d)) in Definition 5.1, the left action of A (resp. A') commutes with the right action of A (resp. A'). Similarly, by assumption (e), the left action of A (resp. A') commutes with the right action of A' (resp. A). Since, by Proposition 2.3, $D(A)$ is generated as an algebra by $A \oplus A'$, the previous facts guarantee that the left and right actions of $D(A)$ on M commute.

The converse is similar. □

Next, we will relate ϵ-Hopf bimodules over an ϵ-bialgebra A to bimodules over the associated pre-Lie algebra, and similarly for the case of a quasitriangular ϵ-bialgebra and the associated dendriform algebra. For this purpose, we recall the definition of bimodules over these types of algebras. There is a general notion from the theory of operads that dictates the bimodule axioms in each case [12]. In the cases of present interest, it turns out that the bimodule axioms are obtained from the axioms for the corresponding type of algebras by the following simple procedure. Each axiom for the given type of algebras yields three bimodule axioms, obtained by choosing one of the variables x, y or z and replacing it by a variable m from the bimodule (this may yield repeated axioms). This leads to the following definitions.

Definition 5.4. Let $(P, \circ)$ be a (left) pre-Lie algebra. A P-bimodule is a space M endowed with maps $P \otimes M \to M$, $x \otimes m \mapsto x \circ m$ and $M \otimes P \to M$, $m \otimes x \mapsto m \circ x$, such that

$$x \circ (y \circ m) - (x \circ y) \circ m = y \circ (x \circ m) - (y \circ x) \circ m, \tag{5.4}$$

$$x \circ (m \circ z) - (x \circ m) \circ z = m \circ (x \circ z) - (m \circ x) \circ z. \tag{5.5}$$

In Section 3 we encountered left P-modules (3.4). Note that any such can be turned into a P-bimodule by choosing the trivial right action $m \circ x \equiv 0$.

Definition 5.5. Let $(D, \succ, \prec)$ be a dendriform algebra. A D-bimodule is a vector space M together with four maps

$$\begin{array}{cccc} D \otimes M \to M & D \otimes M \to M & M \otimes D \to M & M \otimes D \to M \\ x \otimes m \mapsto x \succ m & x \otimes m \mapsto x \prec m & m \otimes x \mapsto m \succ x & m \otimes x \mapsto m \prec x \end{array}$$

such that

$$(x \prec y) \prec m = x \prec (y \prec m) + x \prec (y \succ m),$$
$$x \succ (y \prec m) = (x \succ y) \prec m,$$
$$x \succ (y \succ m) = (x \prec y) \succ m + (x \succ y) \succ m,$$

$$(x \prec m) \prec z = x \prec (m \prec z) + x \prec (m \succ z),$$
$$x \succ (m \prec z) = (x \succ m) \prec z,$$
$$x \succ (m \succ z) = (x \prec m) \succ z + (x \succ m) \succ z,$$

$$(m \prec y) \prec z = m \prec (y \prec z) + m \prec (y \succ z),$$
$$m \succ (y \prec z) = (m \succ y) \prec z,$$
$$m \succ (y \succ z) = (m \prec y) \succ z + (m \succ y) \succ z.$$

It is easy to verify that if M is dendriform bimodule over D then it is also a pre-Lie bimodule over the associated pre-Lie algebra (4.2) by means of

$$x \circ m = x \succ m - m \prec x \quad \text{and} \quad m \circ x = m \succ x - x \prec m. \tag{5.6}$$

Next we show that the construction of pre-Lie algebras from ϵ-bialgebras can be extended to bimodules.

Proposition 5.6. *Let A be an ϵ-bialgebra and $(M, \lambda, \Lambda, \xi, \Xi)$ an ϵ-Hopf bimodule over A. Then M is a bimodule over the pre-Lie algebra $(A, \circ)$ of Theorem 3.2 via*

$$a \circ m = m_{-1}am_0 + m_0am_1 \quad \text{and} \quad m \circ a = a_1ma_2 . \tag{5.7}$$

Proof. Consider the first axiom in Definition 5.4. We have $b \circ m = m_{-1}bm_0 + m_0bm_1$. Using the axioms in Definition 5.1 we calculate

$$\Lambda(b \circ m) = m_{-2} \otimes m_{-1}bm_0 + m_{-1}b_1 \otimes b_2m_0 + m_{-2}bm_{-1} \otimes m_0 + m_{-1} \otimes m_0bm_1 ,$$
$$\Xi(b \circ m) = m_{-1}bm_0 \otimes m_1 + m_0b_1 \otimes b_2m_1 + m_0bm_1 \otimes m_2 + m_0 \otimes m_1bm_2 .$$

Hence,

$$a \circ (b \circ m) = m_{-2}am_{-1}bm_0 + m_{-1}b_1ab_2m_0 + m_{-2}bm_{-1}am_0 + m_{-1}am_0bm_1$$
$$+ m_{-1}bm_0am_1 + m_0b_1ab_2m_1 + m_0bm_1am_2 + m_0am_1bm_2 .$$

On the other hand, $a \circ b = b_1ab_2$, so

$$(a \circ b) \circ m = m_{-1}b_1ab_2m_0 + m_0b_1ab_2m_1 .$$

Therefore,

$$a \circ (b \circ m) - (a \circ b) \circ m = m_{-2}am_{-1}bm_0 + m_{-2}bm_{-1}am_0 + m_{-1}am_0bm_1$$
$$+ m_{-1}bm_0am_1 + m_0bm_1am_2 + m_0am_1bm_2 .$$

Since this expression is symmetric under $a \leftrightarrow b$, Axiom (5.4) holds.

Consider now the second axiom. We have $m \circ b = b_1mb_2$. Using the axioms in Definition 5.1 we calculate $\Lambda(m \circ b) = b_1m_{-1} \otimes m_0b_2 + b_1 \otimes b_2mb_3$ and $\Xi(m \circ b) = b_1m_0 \otimes m_1b_2 + b_1mb_2 \otimes b_3$. It follows that

$$a \circ (m \circ b) = b_1m_{-1}am_0b_2 + b_1ab_2mb_3 + b_1m_0am_1b_2 + b_1mb_2ab_3 .$$

Since

$$(a \circ m) \circ b = b_1m_{-1}am_0b_2 + b_1m_0am_1b_2 .$$

we deduce that

$$a \circ (m \circ b) - (a \circ m) \circ b = b_1ab_2mb_3 + b_1mb_2ab_3 . \tag{$*$}$$

On the other hand, since $\Delta(a \circ b) = \Delta(b_1ab_2) = b_1 \otimes b_2ab_3 + b_1a_1 \otimes a_2b_2 + b_1ab_2 \otimes b_3$, we have that

$$m \circ (a \circ b) = b_1mb_2ab_3 + b_1a_1ma_2b_2 + b_1ab_2mb_3 ,$$

and since

$$(m \circ a) \circ b = b_1a_1ma_2b_2 ,$$

we obtain that,

$$m \circ (a \circ b) - (m \circ a) \circ b = b_1mb_2ab_3 + b_1ab_2mb_3 . \tag{$**$}$$

Comparing $(*)$ with $(**)$ with we see that Axiom (5.5) holds as well. □

Remark 5.7. Proposition 3.6 may be seen as the particular case of Proposition 5.6 when the right module and comodule structures on M are trivial (i.e., zero).

We set now to extend the construction of dendriform algebras from quasitriangular ϵ-bialgebras to the categories of bimodules. As in Section 4, it is convenient to consider the more general context of Baxter operators.

Definition 5.8. Let A be an associative algebra and $\beta_A : A \to A$ a Baxter operator (4.4). A Baxter operator on a A-bimodule M (relative to β_A) is a map $\beta_M : M \to M$ such that

$$\beta_A(a)\beta_M(m) = \beta_M\Big(a\beta_M(m) + \beta_A(a)m\Big)\,, \tag{5.8}$$

$$\beta_M(m)\beta_A(a) = \beta_M\Big(m\beta_A(a) + \beta_M(m)a\Big)\,. \tag{5.9}$$

Proposition 5.9. *Let A be an associative algebra, $\beta_A : A \to A$ a Baxter operator on A, M an A-bimodule and β_M a Baxter operator on M. Define new actions of A on M by*

$$a \succ m = \beta_A(a)m\,,\quad m \succ a = \beta_M(m)a\,,\quad a \prec m = a\beta_M(m) \quad \text{and} \quad m \prec a = m\beta_A(a)\,.$$

Equipped with actions, M is a bimodule over the dendriform algebra of Proposition 4.5.

Proof. Similar to the proof of Proposition 4.5. □

Proposition 5.10. *Let $r = \sum_i u_i \otimes v_i$ be a solution of the associative Yang-Baxter equation* (2.8) *in an associative algebra A. Let M be an A-bimodule. Then the map $\beta_M : M \to M$ defined by*

$$\beta_M(m) = \sum_i u_i m v_i$$

is a Baxter operator on M, relative to the Baxter operator on A of Proposition 4.3

Proof. Replacing the first tensor symbol by a and the second by m in the associative Yang-Baxter equation (2.8) one obtains (5.8). Replacing them in the other order yields (5.9). □

Finally, we have the desired construction of dendriform bimodules from bimodules over quasitriangular ϵ-bialgebras.

Corollary 5.11. *Let (A, r) be a quasitriangular ϵ-bialgebra, $r = \sum_i u_i \otimes v_i$. Let M be an arbitrary A-bimodule. Define new actions of A on M by*

$$a \succ m = \sum_i u_i a v_i m\,, \qquad m \succ a = \sum_i u_i m v_i a\,, \tag{5.10}$$

$$a \prec m = \sum_i a u_i m v_i\,, \qquad m \prec a = \sum_i m u_i a v_i\,.$$

Equipped with these actions, M is a bimodule over the dendriform algebra of Theorem 4.6.

Proof. Combine Propositions 5.10 and 5.9. □

We have thus constructed, from an ϵ-Hopf bimodule over an ϵ-bialgebra, a bimodule over the associated pre-Lie algebra (Proposition 5.6), and from a bimodule over a quasitriangular ϵ-bialgebra, a bimodule over the associated dendriform algebra (Corollary 5.11). Also, a bimodule over a dendriform algebra always yields a bimodule over the associated pre-Lie algebra (5.6). In order to close the circle, it remains to construct an ϵ-Hopf bimodule from a bimodule over a quasitriangular ϵ-bialgebra.

Proposition 5.12. *Let (A, r) be a quasitriangular ϵ-bialgebra, M an A-bimodule. Then M becomes a ϵ-Hopf bimodule over A via $\Lambda : M \to A \otimes M$ and $\Xi : M \to M \otimes A$ defined by*

$$\Lambda(m) = \sum_i u_i \otimes v_i m \quad \text{and} \quad \Xi(m) = -\sum_i m u_i \otimes v_i \,. \tag{5.11}$$

Proof. We already know, from Proposition 2.7, that Λ turns M into a left ϵ-Hopf module over A. Thus, axiom (a) in Definition 5.1 holds. Axiom (b) can be checked similarly: Ξ is coassociative because of the fact that $(id \otimes \Delta)(r) = -r_{13} r_{12}$, which holds according to [1, Proposition 5.5] (one must take into account Remark 2.6). Note that the minus sign in the definition of Ξ is essential, to ensure both this and the right ϵ-Hopf module axiom.

Axiom (c) holds by hypothesis. Axiom (d) holds because both $(id \otimes \Xi)\Lambda(m)$ and $(\Lambda \otimes id)\Xi(m)$ are equal to

$$-\sum_{i,j} u_i \otimes v_i m u_j \otimes v_i \,.$$

Axiom (e) holds because both $\Xi\lambda(a \otimes m)$ and $(\lambda \otimes id)(id \otimes \Xi)(a \otimes m)$ are equal to

$$-\sum_i a m u_i \otimes v_i \,,$$

and similarly for $\Lambda\xi$ and $(id \otimes \xi)(\Lambda \otimes id)$. This completes the proof. □

Let (A, r) be a quasitriangular ϵ-bialgebra. Consider the associated algebras as in diagram (4.3). In this section, we have constructed the corresponding diagram at the level of bimodules:

$$\begin{array}{ccc} \text{(Associative) bimodules} & \longrightarrow & \epsilon\text{-Hopf bimodules} \\ \downarrow & & \downarrow \\ \text{Dendriform bimodules} & \longrightarrow & \text{Pre-Lie bimodules} \end{array} \tag{5.12}$$

Each arrow is a functor, as morphisms are clearly preserved. The diagram is indeed commutative. Going around clockwise, we pass through the ϵ-Hopf bimodule with coactions $\Lambda(m) = \sum_i u_i \otimes v_i m$ and $\Xi(m) = -\sum_i m u_i \otimes v_i$, according to (5.11). The associated pre-Lie bimodule actions are, by (5.7),

$$a \circ m = \sum_i u_i a v_i m - \sum_i m u_i a v_i \quad \text{and} \quad m \circ a = \sum_i u_i m v_i a - \sum_i a u_i m v_i \,.$$

According to (5.10), these expressions are respectively equal to $a \succ m - m \prec a$ and $m \succ a - a \prec m$, which by (5.6) is the pre-Lie bimodule structure obtained by going counterclockwise around the diagram.

6. Brace algebras

In this section we explain how one may associate a brace algebra to an arbitrary ϵ-bialgebra, in a way that refines the pre-Lie algebra construction of Section 3 and that is compatible with the dendriform algebra construction of Section 4.

We provide the left version of the definition of brace algebras given in [6].

Definition 6.1. A (left) brace algebra is a space B equipped with multilinear operations $B^n \times B \to B$, $(x_1, \dots, x_n, z) \mapsto \langle x_1, \dots, x_n; z\rangle$, one for each $n \geq 0$, such that

$$\langle z\rangle = z$$

and for any n, $m \geq 1$,

$$(6.1)\quad \langle x_1, \dots, x_n; \langle y_1, \dots, y_m; z\rangle\rangle = \sum \langle X_0, \langle X_1; y_1\rangle, X_2, \langle X_3; y_2\rangle, X_4, \dots, X_{2m-2}, \langle X_{2m-1}; y_m\rangle, X_{2m}; z\rangle\,,$$

where the sum takes place over all partitions of the ordered set $\{x_1, \dots, x_n\}$ into (possibly empty) consecutive intervals $X_0 \sqcup X_1 \sqcup \cdots \sqcup X_{2m}$.

The case $n = m = 1$ of Axiom 6.1 says

$$(6.2)\qquad \langle x; \langle y; z\rangle\rangle = \langle x, y; z\rangle + \langle\langle x; y\rangle; z\rangle + \langle y, x; z\rangle\,.$$

The three terms on the right hand side correspond respectively to the partitions $(\{x\}, \emptyset, \emptyset)$, $(\emptyset, \{x\}, \emptyset)$ and $(\emptyset, \emptyset, \{x\})$.

The operation $x \circ y := \langle x; y\rangle$ endows B with a pre-Lie algebra structure. In fact, (6.2) shows that $x\circ(y\circ z) - (x\circ y)\circ z$ is symmetric under $x \leftrightarrow y$, so Axiom (3.1) holds. This defines a functor from brace algebras to pre-Lie algebras. The construction of pre-Lie algebras from ϵ-bialgebras in Section 3 can be refined accordingly, as we explain next.

In order to describe this refined construction, we must depart from our notational convention for coproducts and revert to Sweedler's original notation. Thus, in this section, the coproducts of an element b will be denoted by

$$\Delta(b) = \sum_{(b)} b_{(1)} \otimes b_{(2)}$$

and the n-th iteration of the coproduct by

$$\Delta^{(n)}(b) = \sum_{(b)} b_{(1)} \otimes \cdots \otimes b_{(n+1)}\,.$$

Theorem 6.2. *Let A be an ϵ-bialgebra. Define operations $A^n \times A \to A$ by*

$$\langle a_1, \dots, a_n; b\rangle = \sum_{(b)} b_{(1)} a_1 b_{(2)} a_2 \dots b_{(n)} a_n b_{(n+1)}\,.$$

These operations turn A into a brace algebra.

Proof. The complete details of the proof will be provided elsewhere. The idea is simple: each term on the right hand side of (6.1) corresponds to a term in the expansion of

$$\Delta^{(n)}\Big(\sum_{(z)} z_{(1)} y_1 z_{(2)} y_2 \dots z_{(m)} y_m z_{(m+1)}\Big)$$

obtained by successive applications of (2.1). For instance, when $n = 2$ and $m = 1$, one has

$$\Delta^{(2)}\Big(\sum_{(z)} z_{(1)}yz_{(2)}\Big)$$
$$= \sum_{(z)} z_{(1)}\otimes z_{(2)}\otimes z_{(3)}yz_{(4)} + z_{(1)}\otimes z_{(2)}yz_{(3)}\otimes z_{(4)} + z_{(1)}yz_{(2)}\otimes z_{(3)}\otimes z_{(4)}$$
$$+ \sum_{(z),(y)} z_{(1)}y_{(1)}\otimes y_{(2)}z_{(2)}\otimes z_{(3)} + z_{(1)}y_{(1)}\otimes y_{(2)}\otimes y_{(3)}z_{(2)} + z_{(1)}\otimes z_{(2)}y_{(1)}\otimes y_{(2)}z_{(3)}$$

Therefore,

$$\langle x_1, x_2; \langle y; z\rangle\rangle$$
$$= \sum_{(z)} z_{(1)}x_1z_{(2)}x_2z_{(3)}yz_{(4)} + z_{(1)}x_1z_{(2)}yz_{(3)}x_2z_{(4)} + z_{(1)}yz_{(2)}x_1z_{(3)}x_2z_{(4)}$$
$$+ \sum_{(z),(y)} z_{(1)}y_{(1)}x_1y_{(2)}z_{(2)}x_2z_{(3)} + z_{(1)}y_{(1)}x_1y_{(2)}x_2y_{(3)}z_{(2)} + z_{(1)}x_1z_{(2)}y_{(1)}x_2y_{(2)}z_{(3)}$$
$$= \langle x_1, x_2, y; z\rangle + \langle x_1, y, x_2; z\rangle + \langle y, x_1, x_2; z\rangle$$
$$+ \langle\langle x_1; y\rangle, x_2; z\rangle + \langle\langle x_1, x_2; y\rangle; z\rangle + \langle x_1, \langle x_2; y\rangle; z\rangle$$

which is Axiom (6.1). □

By construction, the first brace operation on A is simply

$$\langle a; b\rangle = \sum_{(b)} b_{(1)}ab_{(2)},$$

which agrees with the pre-Lie operation (3.2). In this sense, the constructions of Theorems 3.2 and 6.2 are compatible.

Example 6.3. Consider the ϵ-bialgebra $k[\mathbf{x}, \mathbf{x}^{-1}]$ of divided differences (Examples 3.4). It is easy to see that for any $n \geq 0$ and $r \in \mathbf{Z}$,

$$\mu^{(n)}\Delta^{(n)}(\mathbf{x}^r) = \binom{r}{n}\mathbf{x}^{r-n},$$

where it is understood, as usual, that $\binom{r}{n} = 0$ if $n > r \geq 0$ and $\binom{r}{n} = (-1)^n\binom{-r+n-1}{n}$ if $r < 0$. It follows that the brace algebra structure on $k[\mathbf{x}, \mathbf{x}^{-1}]$ is

$$\langle \mathbf{x}^{p_1}, \ldots, \mathbf{x}^{p_n}; \mathbf{x}^r\rangle = \binom{r}{n}\mathbf{x}^{r+p_1+\cdots+p_n-n}.$$

The brace axioms (6.1) boil down to a set of interesting identities involving binomial coefficients.

Frédéric Chapoton made us aware of the fact that if one applies the general construction of [15, Proposition 1] (dropping all signs) to the associative operad, one obtains precisely the brace subalgebra $k[\mathbf{x}]$ of our brace algebra $k[\mathbf{x}, \mathbf{x}^{-1}]$.

This example may be generalized in another direction. Namely, if A is a commutative algebra and $D : A \to A$ a derivation, then one obtains a brace algebra structure on A by defining

$$\langle x_1, \ldots, x_n; z\rangle = x_1 \cdots x_n \frac{D^n(z)}{n!}$$

(assuming $\mathsf{char}(k) = 0$). The example above corresponds to $A = k[\mathbf{x}, \mathbf{x}^{-1}]$, $D = \frac{d}{d\mathbf{x}}$.

Brace algebras sit between dendriform and pre-Lie algebras: Ronco has shown that one can associate a brace algebra to a dendriform algebra, by means of certain operations [26, Theorem 3.4]. Our constructions of dendriform and brace algebras from Theorems 4.6 and 6.2 are compatible with this functor.

In summary, one obtains a commutative diagram

$$\begin{array}{ccccc} \text{Quasitriangular } \epsilon\text{-bialgebras} & \longrightarrow & \epsilon\text{-bialgebras} & & \\ \downarrow & & \downarrow & \searrow & \\ \text{Dendriform algebras} & \longrightarrow & \text{Brace algebras} & \longrightarrow & \text{Pre-Lie algebras} \end{array}$$

The details will be provided elsewhere.

Appendix A. Infinitesimal bialgebras as comonoid objects

Ordinary bialgebras are bimonoid objects in the braided monoidal category of vector spaces, where the monoidal structure is the usual tensor product $V \otimes W$ and the braiding is the trivial symmetry $x \otimes y \mapsto y \otimes x$. In this appendix, we construct a certain monoidal category of algebras for which the comonoid objects are precisely ϵ-bialgebras. Related notions of bimonoid objects are discussed as well.

For the basics on monoidal categories the reader is referred to [24, Chapters VII and XI] and [19, Chapter XI]. The monoidal categories we consider possess a unit object, and whenever we refer to monoid objects these are assumed to be unital, even if not explicitly stated. Similarly, comonoid objects are assumed to be counital.

We start by recalling the well known *circle* tensor product of vector spaces.

Definition A.1. The circle product of two vector spaces V and W is

$$V \circ W = V \oplus W \oplus (V \otimes W) .$$

We denote the elements of this space by triples $(v, w, x \otimes y)$. The circle product of maps $f : V \to X$ and $g : W \to Y$ is

$$(f \circ g)(v, w, x \otimes y) = \big(f(v), g(w), (f \otimes g)(x \otimes y)\big) .$$

Both spaces $(U \circ V) \circ W$ and $U \circ (V \circ W)$ can be canonically identified with

$$U \oplus V \oplus W \oplus (U \otimes V) \oplus (U \otimes W) \oplus (V \otimes W) \oplus (U \otimes V \otimes W) .$$

This gives rise to a natural isomorphism $(U \circ V) \circ W \cong U \circ (V \circ W)$ which satisfies the pentagon for associativity. This endows the category of vector spaces with a monoidal structure, for which the unit object is the zero space. We denote this monoidal category by $(\mathsf{Vec}, \circ, 0)$.

Let $(\mathsf{Vec}, \otimes, k)$ denote the usual monoidal category of vector spaces, where the monoid objects are unital associative algebras and the comonoid objects are counital coassociative coalgebras. There is an obvious monoidal functor $\alpha : (\mathsf{Vec}, \circ, 0) \to (\mathsf{Vec}, \otimes, k)$ defined by

$$V \mapsto V \oplus k .$$

It is the so called *augmentation* functor.

Monoids and comonoids in $(\mathsf{Vec}, \circ, 0)$ are easy to describe: they are, respectively, non unital algebras and non counital coalgebras (Proposition A.2, below). Monoids and comonoids are preserved by monoidal functors. In the present situation this simply says that a non unital algebra can be canonically augmented into a unital algebra, and similarly for coalgebras.

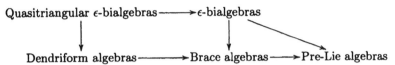

Proposition A.2. *A unital monoid object in* $(\mathsf{Vec}, \circ, 0)$ *is precisely an associative algebra, not necessarily unital. A counital comonoid object is precisely a coassociative coalgebra, not necessarily counital.*

Proof. Let (A, μ) be an associative algebra, $\mu(a \otimes a') = aa'$. Define a map $\tilde{\mu} : A \circ A \to A$ by

$$(a, a', x \otimes x') \mapsto a + a' + xx' \,. \tag{A.1}$$

Let $u : 0 \to A$ be the unique map. Then, the diagrams

$$\begin{array}{ccc} A \circ A \circ A & \xrightarrow{\tilde{\mu} \circ id} & A \circ A \\ {\scriptstyle id \circ \tilde{\mu}} \downarrow & & \downarrow {\scriptstyle \tilde{\mu}} \\ A \circ A & \xrightarrow[\tilde{\mu}]{} & A \end{array} \quad \text{and} \quad \begin{array}{ccccc} 0 \circ A & \xrightarrow{u \otimes id} & A \circ A & \xleftarrow{id \circ u} & A \circ 0 \\ & {\scriptstyle \cong} \searrow & \downarrow {\scriptstyle \tilde{\mu}} & \swarrow {\scriptstyle \cong} & \\ & & A & & \end{array}$$

commute. Thus, $(A, \tilde{\mu}, u)$ is a unital monoid in $(\mathsf{Vec}, \circ, 0)$.

Conversely, if $(A, \tilde{\mu}, u)$ is a unital monoid in $(\mathsf{Vec}, \circ, 0)$, then $\tilde{\mu}$ must be of the form (A.1) for an associative multiplication μ on A, by the commutativity of the diagrams above.

The assertion for comonoids is similar. The comultiplication $\Delta : A \to A \otimes A$ is related to the comonoid structure $\tilde{\Delta} : A \to A \circ A$ by

$$\tilde{\Delta}(a) = (a, a, \Delta(a)) \tag{A.2}$$

and $\epsilon : A \to 0$ is the unique map. □

Remark A.3. It is natural to wonder if there is a braiding on the monoidal category $(\mathsf{Vec}, \circ, 0)$ for which the bimonoid objects are precisely ϵ-bialgebras. We know of two braidings on $(\mathsf{Vec}, \circ, 0)$. The corresponding notions of bimonoid objects are briefly discussed next. Neither yields ϵ-bialgebras.

(1) For any spaces V and W, consider the map $\sigma_{V,W} : V \circ W \to W \circ V$ defined by

$$(v, w, x \otimes y) \mapsto (w, v, y \otimes x) \,.$$

This family of maps clearly satisfies the axioms for a braiding on the monoidal category $(\mathsf{Vec}, \circ, 0)$. Under the monoidal functor α, the braiding σ corresponds to the usual braiding on $(\mathsf{Vec}, \otimes, k)$ (the trivial symmetry). For this reason, a bimonoid object in $(\mathsf{Vec}, \circ, 0, \sigma)$ can be canonically augmented into an ordinary bialgebra.

It follows from Proposition A.2) that a bimonoid object in $(\mathsf{Vec}, \circ, 0, \sigma)$ is a space A, equipped with an associative algebra structure $A \otimes A \to A$, $a \otimes a' \mapsto aa'$, and a coassociative coalgebra structure $\Delta : A \to A \otimes A$, $a \mapsto a_1 \otimes a_2$, related by the axiom

$$\Delta(aa') = a \otimes a' + a' \otimes a + aa'_1 \otimes a'_2 + a'_1 \otimes aa'_2 + a_1 a' \otimes a_2 + a_1 \otimes a_2 a' + a_1 a'_1 \otimes a_2 a'_2 \,.$$

This is *not* the axiom which defines ϵ-bialgebras (2.1).

The axiom above is a translation of the fact that the map $A \to A \circ A$ must be a morphism of monoids. We omit this calculation, but provide an explicit description of the monoid structure on $A \circ A$. More generally, we describe the tensor product of two monoids A and B in $(\mathsf{Vec}, \circ, 0, \sigma)$.

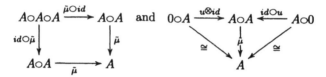

According to Proposition A.2, the monoid structure on $A \circ B$ is uniquely determined by an associative multiplication on the space $A \circ B$. We describe this multiplication, in terms of those of A and B. It is

$$(a,b,x \otimes y) \cdot (a',b',x' \otimes y') \\ = (aa', bb', a \otimes b' + a' \otimes b + ax' \otimes y' + x' \otimes by' + xa' \otimes y + x \otimes yb' + xy \otimes x'y').$$

This is the result of composing

$$(A \circ B) \otimes (A \circ B) \hookrightarrow (A \circ B) \circ (A \circ B) \xrightarrow{id \circ \sigma_{B,A} \circ id} (A \circ A) \circ (B \circ B) \xrightarrow{\bar{\mu}_A \circ \bar{\mu}_B} A \circ B.$$

(2) There exists a second braiding on the monoidal category $(\mathsf{Vec}, \circ, 0)$, for which bimonoid objects are somewhat closer to ϵ-bialgebras. It is the family of maps $\beta_{V,W} : V \circ W \to W \circ V$ defined by

$$(v, w, x \otimes y) \mapsto (w, v, 0).$$

Apart from the fact that β is *not* an isomorphism, the braiding axioms are satisfied by β. This allows us to construct a monoid structure on the circle product of two monoids in $(\mathsf{Vec}, \circ, 0)$, and therefore to speak of bimonoid objects in $(\mathsf{Vec}, \circ, 0, \beta)$, as usual.

Since the monoidal functor α does not preserve this braiding, the augmentation of a bimonoid in $(\mathsf{Vec}, \circ, 0, \beta)$ is not an ordinary bialgebra. Neither is it true that these bimonoid objects are ϵ-bialgebras. In fact, a bimonoid object in $(\mathsf{Vec}, \circ, 0, \beta)$ is a space A, equipped with an associative algebra structure and a coassociative coalgebra structure, as above, related by the axiom

$$\Delta(aa') = a \otimes a' + aa'_1 \otimes a'_2 + a_1 \otimes a_2 a'.$$

Compare with Axiom (2.1) for infinitesimal bialgebras.

This can be deduced from the following description of the tensor product in $(\mathsf{Vec}, \circ, 0, \beta)$ of two monoid objects A and B. This structure is determined by the following (associative) multiplication on $A \circ B$:

$$(a, b, x \otimes y) \cdot (a', b', x' \otimes y') = (aa', bb', a \otimes b' + ax' \otimes y' + x \otimes yb').$$

This is the result of composing

$$(A \circ B) \otimes (A \circ B) \hookrightarrow (A \circ B) \circ (A \circ B) \xrightarrow{id \circ \beta_{B,A} \circ id} (A \circ A) \circ (B \circ B) \xrightarrow{\bar{\mu}_A \circ \bar{\mu}_B} A \circ B.$$

Let Alg denote the category of monoids in $(\mathsf{Vec}, \circ, 0)$, that is, associative algebras which are not necessarily unital. We define a new monoidal structure on this category, independent of any braiding on $(\mathsf{Vec}, \circ, 0)$. We will show that ϵ-bialgebras are precisely comonoid objects in the resulting monoidal category.

Proposition A.4. *Let A and B be associative algebras, not necessarily unital. Then $A \circ B$ is an associative algebra via*

$$(a, b, x \otimes y) \cdot (a', b', x' \otimes y') = (aa', bb', ax' \otimes y' + x \otimes yb'). \tag{A.3}$$

Proof. Consider the algebra $R = A \oplus B$ and the R-bimodule $M = A \otimes B$, with

$$(a, b) \cdot x \otimes y = ax \otimes y \quad \text{and} \quad x \otimes y \cdot (a, b) = x \otimes yb.$$

The algebra $A \circ B$ is precisely the trivial extension $R \oplus M$, where the multiplication is

$$(r, m) \cdot (r', m') = (rr', rm' + mr').$$

If $f : A \to A'$ and $g : B \to B'$ are morphisms of algebras, then so is $f \circ g : A \circ B \to A' \circ B'$. In this way, $(\mathsf{Alg}, \circ, 0)$ becomes a monoidal category.

Proposition A.5. *A counital comonoid object in the monoidal category $(Alg, \circ, 0)$ is precisely an ϵ-bialgebra.*

Proof. Let (A, μ, Δ) be an ϵ-bialgebra. By Proposition A.2, A may be seen as a monoid and comonoid in $(\mathsf{Alg}, \circ, 0)$. It only remains to verify that the comonoid structure $\tilde{\Delta} : A \to A \circ A$, $\tilde{\Delta}(a) = (a, a, \Delta(a))$, is a morphism of algebras. This is clear from (A.3) and (2.1).

The converse is similar. □

The category of modules over an ordinary bialgebra H is monoidal: the tensor product of two H-modules acquires an H-module structure by restricting the natural $H \otimes H$-module structure via the comultiplication $\Delta : H \to H \otimes H$. There is no analogous construction for arbitrary ϵ-bialgebras. However, it is possible to construct tensor products of certain modules over ϵ-bialgebras, as discussed next.

Proposition A.6. *Let A and B be associative algebras. Let M be a right A-module and N a left B-module. Then $A \circ N$ is a left $A \circ B$-module via*

$$(a, b, x \otimes y) \cdot (a', n, x' \otimes v) := (aa', bn, ax' \otimes v + x \otimes yn),$$

and $M \circ B$ is a right $A \circ B$-module via

$$(m, b', u \otimes y') \cdot (a, b, x \otimes y) := (ma, b'b, ub \otimes y' + mx \otimes y).$$

Proof. Similar to the proof of Proposition A.4. □

Let A be an ϵ-bialgebra and N a left A-module. It is possible to define a left A-module structure on $A \circ N$, by restricting the structure of Proposition A.6 along the morphism of algebras $\tilde{\Delta} : A \to A \circ A$. The resulting action of A on $A \circ N$ is

$$a \cdot (a', n, x \otimes v) = (aa', an + ax \otimes v, a_1 \otimes a_2 n). \tag{A.4}$$

Appendix B. Counital infinitesimal bialgebras

Definition B.1. An ϵ-bialgebra (A, μ, Δ) is said to be counital if the underlying coalgebra is counital, that is, if there exists a map $\eta : A \to k$ such that $(id \otimes \eta)\Delta = id = (\eta \otimes id)\Delta$.

The map η is necessarily unique and is called the counit of A. We use η instead of the customary ε to avoid confusion with the abbreviation for infinitesimal bialgebras.

Recall that if an ϵ-bialgebra A is both unital *and* counital then $A = 0$ [1, Remark 2.2]. Nevertheless, many ϵ-bialgebras arising in practice are either unital or counital. In this appendix we study counital ϵ-bialgebras; all constructions and results admit a dual version that applies to unital ϵ-bialgebras.

We first show that counital ϵ-bialgebras can be seen as comonoid objects in a certain monoidal category of algebras. This construction is parallel to that for arbitrary ϵ-bialgebras discussed in Appendix A. The two constructions are related by means of a pair of monoidal functors, but neither is more general than the other.

Lemma B.2. *Let A be a counital ϵ-bialgebra with counit η. Then*

$$\eta(aa') = 0 \text{ for all } a,\ a' \in A.$$

Proof. We show that any coderivation $D : M \to C$ from a counital bicomodule (M, s, t) to a counital coalgebra (C, Δ, η) maps to the kernel of η. The result follows by applying this remark to the coderivation $\mu : A \otimes A \to A$.

We have

$$\begin{aligned}\eta D &= (\eta \otimes \eta)\Delta D = (\eta \otimes \eta)\big((id_C \otimes D)t + (D \otimes id_C)s\big) \\ &= (id_k \otimes \eta D)(\eta \otimes id_M)t + (\eta D \otimes id_k)(id_M \otimes \eta)s \\ &= \eta D + \eta D = 2 \cdot \eta D,\end{aligned}$$

whence $\eta D = 0$. □

This motivates the following definition.

Definition B.3. Let (A, μ) be an algebra over k, not necessarily unital. We say that it is *augmented* if there is given a map $\eta : A \to k$ such that

$$\eta(aa') = 0 \text{ for all } a,\ a' \in A.$$

A morphism between augmented algebras (A, η_A) and (B, η_B) is a morphism of algebras $f : A \to B$ such that $\eta_B f = \eta_A$.

Proposition B.4. *Let (A, η_A) and (B, η_B) be augmented algebras. Then $A \otimes B$ is an associative algebra with multiplication*

$$(a \otimes b) \cdot (a' \otimes b') := \eta_B(b) aa' \otimes b' + \eta_A(a') a \otimes bb'. \tag{B.1}$$

Moreover, $A \otimes B$ is augmented by

$$\eta_{A \otimes B}(a \otimes b) := \eta_A(a)\eta_B(b).$$

Proof. The first assertion is Lemma 3.5.b in [1] and the second is straightforward. □

We denote the resulting augmented algebra by $A \otimes_\epsilon B$. This operation defines a monoidal structure on the category of augmented algebras over k. The unit object is the base field k equipped with the zero multiplication and the identity augmentation. We denote this monoidal category by $(\mathsf{AAlg}, \otimes_\epsilon, k)$.

Proposition B.5. *A counital comonoid object in the monoidal category $(\mathsf{AAlg}, \otimes_\epsilon, k)$ is precisely a counital ϵ-bialgebra.*

Proof. Start from a counital ϵ-bialgebra (A, μ, Δ, η). By Lemma B.2, (A, η) is an augmented algebra. Moreover, by Lemma 3.6.b in [1], $\Delta : A \to A \otimes_\epsilon A$ is a morphism of algebras, and it preserves the augmentations by counitality. Clearly, $\eta : A \to k$ is also a morphism of augmented algebras. Thus, (A, μ, Δ, η) is a counital comonoid in $(\mathsf{Alg}, \otimes_\epsilon, k)$.

Conversely, let A be a counital comonoid in $(\mathsf{AAlg}, \otimes_\epsilon, k)$. First of all, the counit $A \to k$ must preserve the augmentations of A and k, so it must coincide with the augmentation of A. The comultiplication must be a morphism of algebras $A \to A \otimes_\epsilon A$. This implies Axiom (2.1), by definition of the algebra structure on $A \otimes_\epsilon A$ and counitality. Thus A is a counital ϵ-bialgebra. □

Remark B.6. An augmented algebra may be seen as a monoid in a certain monoidal category of "augmented vector spaces". However, the monoidal structure on the category $(\mathsf{AAlg}, \otimes_\epsilon, k)$ does not come from a braiding on the larger category of augmented vector spaces. For this reason, we cannot view counital ϵ-bialgebras as bimonoid objects. The situation parallels that encountered in Appendix A for arbitrary ϵ-bialgebras. In fact, there is pair of monoidal functors relating the two situations, as we discuss next.

Remark B.7. Given a non unital algebra A, let $A^+ := A \oplus k$, with algebra structure

$$(a,x)\cdot(b,y) = (ab,0)\,. \tag{B.2}$$

Note that A^+ is not the usual augmentation of A; in fact, A^+ is non unital. Define $\eta(a,x) = x$. Then $\eta((a,x)\cdot(b,y)) = 0$, so (A^+,η) is an augmented algebra in the sense of Definition B.3. Moreover, it is easy to see that there is a natural isomorphism of augmented algebras

$$(A \circ B)^+ \cong A^+ \otimes_\epsilon B^+\,.$$

The application $A \mapsto A^+$ is thus a monoidal functor

$$(\mathsf{Alg}, \circ, 0) \to (\mathsf{AAlg}, \otimes_\epsilon, k)\,.$$

The fact that comonoid objects are preserved by this monoidal functor simply says that any ϵ-bialgebra A can be made into a counital ϵ-bialgebra A^+, by extending the comultiplication via $\Delta(1) = 1 \otimes 1$ and the multiplication as in (B.2).

In the other direction, consider the forgetful functor

$$(\mathsf{AAlg}, \otimes_\epsilon, k) \to (\mathsf{Alg}, \circ, 0), \quad (A,\eta) \mapsto A\,.$$

It is easy to see that the map

$$A \otimes_\epsilon B \to A \circ B, \quad a \otimes b \mapsto \eta_B(b)a + \eta_A(a)b + a \otimes b \tag{B.3}$$

is a (natural) morphism of algebras. It follows that the forgetful functor is *lax monoidal*. The fact that comonoid objects are preserved by this type of functors simply says in this case that any counital ϵ-bialgebra is in particular an ϵ-bialgebra.

Neither functor between these two categories of algebras is a monoidal equivalence. For this reason, neither situation in Appendices A and B is more general than the other.

The following is the analog of Proposition A.6 for augmented algebras.

Proposition B.8. *Let A and B be augmented algebras. Let M be a right A-module and N a left B-module. Then $A \otimes N$ is a left $A \otimes_\epsilon B$-module via*

$$(a \otimes b)\cdot(a' \otimes n) := \eta_B(b)aa' \otimes n + \eta_A(a')a \otimes bn$$

and $M \otimes B$ is a right $A \otimes_\epsilon B$-module via

$$(m \otimes b')\cdot(a \otimes b) := \eta_A(a)m \otimes b'b + \eta_B(b')ma \otimes b\,.$$

Proof. Similar to the proof of Proposition B.4. □

One may similarly show that the map

$$A \otimes N \to A \circ N, \quad a \otimes n \mapsto \eta_A(a)n + a \otimes n$$

is a morphism of left $A \otimes_\epsilon B$-modules, where $A \circ N$ is viewed as a left $A \otimes_\epsilon B$-module by restriction via the morphism of algebras (B.3).

Finally, we discuss the analog of the construction (A.4) for counital ϵ-bialgebras, and apply these general considerations to the construction of an ϵ-Hopf module.

Let A be a counital ϵ-bialgebra and N a left A-module. It is possible to define a left A-module structure on $A \otimes N$, by restricting the structure of Proposition B.8 along the morphism of augmented algebras $\Delta : A \to A \otimes_\epsilon A$. By counitality, the action of A on $A \otimes N$ reduces to

$$a \cdot (a' \otimes n) = aa' \otimes n + \eta(a') a_1 \otimes a_2 n \,. \tag{B.4}$$

We denote this module structure on the space $A \otimes N$ by $A \otimes_\epsilon N$.

Our next result describes ϵ-Hopf modules over counital ϵ-bialgebras in a way that is analogous to the definition of Hopf modules over an ordinary Hopf algebra.

Recall that a left Hopf module over a Hopf algebra H is a space M that is both a left module and comodule over H and for which the comodule structure map $M \to H \otimes M$ is a morphism of left H-modules [25, Definition 1.9.1]. It is understood that $H \otimes M$ is a left H-module by restriction via the comultiplication of H.

Proposition B.9. *Let A be a counital ϵ-bialgebra. Let $\lambda : A \otimes N \to N$ be a left A-module structure on N and $\Lambda : N \to A \otimes N$ a counital comodule structure on N. Then (N, λ, Λ) is an ϵ-Hopf module over A if and only if $\Lambda : N \to A \otimes_\epsilon N$ is a morphism of left A-modules.*

Proof. Write $\lambda(a \otimes n) = an$ and $\Lambda(n) = n_{-1} \otimes n_0$. According to (B.4),

$$\begin{aligned} a \cdot \Lambda(n) &= an_{-1} \otimes n_0 + \eta_A(n_{-1})\, a_1 \otimes a_2 n_0 \\ &= an_{-1} \otimes n_0 + a_1 \otimes a_2 n \,, \end{aligned}$$

by counitality for N. Thus, Λ is a morphism of A-modules if and only if

$$\Lambda(an) = an_{-1} \otimes n_0 + a_1 \otimes a_2 n \,,$$

which is Axiom (2.2) in the definition of ϵ-Hopf module. □

Next, we make use of Proposition B.9 to obtain the general construction of ϵ-Hopf modules of Example 2.2. 3.

First, note that the tensor product construction of Proposition B.8 is associative, in the sense that if A, B and C are augmented algebras and N is a left C-module, then

$$(A \otimes_\epsilon B) \otimes_\epsilon N \cong A \otimes_\epsilon (B \otimes_\epsilon N)$$

as left $A \otimes_\epsilon B \otimes_\epsilon C$-modules. In fact, one has

$$\begin{aligned} &\big((a \otimes b) \otimes c\big) \cdot \big((a' \otimes b') \otimes n\big) = \\ &\qquad \eta_C(c)\eta_B(b) aa' \otimes b' \otimes n + \eta_C(c)\eta_A(a') a \otimes bb' \otimes n + \eta_A(a')\eta_B(b') a \otimes b \otimes cn \\ &\qquad\qquad = \big(a \otimes (b \otimes c)\big) \cdot \big(a' \otimes (b' \otimes n)\big) \,. \end{aligned}$$

On the other hand, if $f : A \to B$ is a morphism of augmented algebras and N is a left C-module, then $f \otimes id_N : A \otimes_\epsilon N \to B \otimes_\epsilon N$ is a morphism of left $A \otimes_\epsilon C$-modules, where $B \otimes_\epsilon N$ is an $A \otimes_\epsilon C$-module by restriction via the morphism of algebras $f \otimes id_C : A \otimes_\epsilon C \to B \otimes_\epsilon C$.

Let us apply these considerations to a left module N over a counital ϵ-bialgebra A, $B = A \otimes_\epsilon A$, $C = A$ and $f = \Delta$. We obtain that

$$\Delta \otimes id_N : A \otimes_\epsilon N \to A \otimes_\epsilon A \otimes_\epsilon N$$

is a morphism of left $A\otimes_\epsilon A$-modules. Hence, it is also a morphism of left A-modules, by restriction via Δ. An application of Proposition B.9 then yields the following

Corollary B.10. *Let A be a counital ϵ-bialgebra and N a left N-module. Let $M = A\otimes_\epsilon N$, an A-module as in* (B.4). *Define $\Lambda : M \to A\otimes M$ by*

$$\Lambda(a\otimes n) = a_1\otimes a_2\otimes n\,.$$

With these module and comodule structures, M is a left ϵ-Hopf module over A.

In this paper, quasitriangular ϵ-bialgebras play an important role (Section 4). Our last result shows that the classes of counital ϵ-bialgebras and quasitriangular ϵ-bialgebras are disjoint.

Proposition B.11. *If a quasitriangular ϵ-bialgebra A is counital then $A = 0$.*

Proof. Let $r = \sum u_i\otimes v_i$ be the canonical element and η the counit. According to (2.9), we have

$$\Delta(a) = \sum_i u_i\otimes v_i a - a u_i\otimes v_i\,.$$

Applying $\eta\otimes id$ and using Lemma B.2 we deduce

$$a = \sum_i \eta(u_i)v_i a \text{ for every } a \in A\,.$$

Similarly, applying $id\otimes\eta$ we deduce

$$a = -\sum_i a u_i \eta(v_i) \text{ for every } a \in A\,.$$

Thus, A has a left unit and a right unit. These must therefore coincide and A must be unital. But an ϵ-bialgebra A that is both unital and counital must be 0 [1, Remark 2.2]. □

References

[1] Marcelo Aguiar, *Infinitesimal Hopf algebras*, New trends in Hopf algebra theory (La Falda, 1999), 1–29, Contemp. Math., 267, Amer. Math. Soc., Providence, RI, 2000. MR **2001k:**16066

[2] Marcelo Aguiar, *Pre-Poisson algebras*, Lett. Math. Phys. 54 (2000), no. 4, 263–277. MR **2002k:**17041

[3] Marcelo Aguiar, *On the associative analog of Lie bialgebras*, J. Algebra 244 (2001), no. 2, 492–532. MR 1 859 038

[4] Marcelo Aguiar, *Infinitesimal Hopf algebras and the cd-index of polytopes*, Geometric combinatorics (San Francisco, CA/Davis, CA, 2000). Discrete Comput. Geom. 27 (2002), no. 1, 3–28. MR 1 871 686

[5] Glen Baxter, *An analytic problem whose solution follows from a simple algebraic identity.* Pacific J. Math. 10 1960 731–742. MR 22 #9990

[6] Frédéric Chapoton, *Un théorème de Cartier-Milnor-Moore-Quillen pour les bigèbres dendriformes et les algèbres braces*, J. Pure Appl. Algebra 168 (2002), no. 1, 1–18. MR **2002k:**18013

[7] Frédéric Chapoton and Muriel Livernet, *Pre-Lie algebras and the rooted trees operad*, Internat. Math. Res. Notices 2001, no. 8, 395–408. MR **2002e:**17003

[8] Askar Dzhumadil'daev, *Cohomologies and deformations of right-symmetric algebras*, Algebra, 11. J. Math. Sci. (New York) 93 (1999), no. 6, 836–876. MR **2000e:**17002

[9] Yukio Doi, *Homological coalgebra*, J. Math. Soc. Japan 33 (1981), no. 1, 31–50. MR **82g:**16014

[10] V. G. Drinfeld, *Quantum Groups*, Proceedings of the International Congress of Mathematicians, Vol. 1, 2 (Berkeley, Calif., 1986), 798–820, Amer. Math. Soc., Providence, RI, 1987. MR **89f:**17017

[11] Richard Ehrenborg and Margaret Readdy, *Coproducts and the cd-index*, J. Algebraic Combin. 8 (1998), no. 3, 273–299. MR **2000b:**52009

[12] Thomas F. Fox and Martin Markl, *Distributive laws, bialgebras, and cohomology* Operads: Proceedings of Renaissance Conferences (Hartford, CT/Luminy, 1995), 167–205, Contemp. Math., 202, Amer. Math. Soc., Providence, RI, 1997. MR **98a**:18006

[13] Murray Gerstenhaber, *The cohomology structure of an associative ring*, Ann. of Math. (2) 78 1963 267–288. MR 28 #5102

[14] M. Gerstenkhaber and A. A.Voronov, *Higher-order operations on the Hochschild complex*, Funktsional. Anal. i Prilozhen. 29 (1995), no. 1, 1–6, 96; translation in Funct. Anal. Appl. 29 (1995), no. 1, 1–5. MR **96g**:18006

[15] M. Gerstenkhaber and A. A.Voronov, *Homotopy G-algebras and moduli space operad*, Internat. Math. Res. Notices 1995, no. 3, 141–153. MR 1 321 701

[16] Ezra Getzler, *Cartan homotopy formulas and the Gauss-Manin connection in cyclic homology*, Quantum deformations of algebras and their representations (Ramat-Gan, 1991/1992; Rehovot, 1991/1992), 65–78, Israel Math. Conf. Proc., 7, Bar-Ilan Univ., Ramat Gan, 1993. MR **95c**:19002

[17] S. A. Joni and G. C. Rota, *Coalgebras and Bialgebras in Combinatorics*, Stud. Appl. Math. 61 (1979), no. 2, 93-139. Reprinted in *Gian-Carlo Rota on Combinatorics: Introductory papers and commentaries* (Joseph P.S. Kung, Ed.), Birkhäuser, Boston (1995). MR **81c**:05002

[18] T. V. Kadeishvili, *The structure of the $A(\infty)$-algebra, and the Hochschild and Harrison cohomologies*, (Russian) Trudy Tbiliss. Mat. Inst. Razmadze Akad. Nauk Gruzin. SSR 91 (1988), 19–27. MR **91a**:18016

[19] Christian Kassel, *Quantum groups*, Graduate Texts in Mathematics, 155. Springer-Verlag, New York, 1995. xii+531 pp. MR **96e**:17041

[20] Jean-Louis Loday, *Dialgebras*, Dialgebras and Related Operads, Lecture Notes in Mathematics, no. 1763, Springer-Verlag, 2001, pp. 7–66. MR **2002i**:17004

[21] Jean-Louis Loday and María O. Ronco, *Hopf algebra of the planar binary trees*, Adv. Math. **139** (1998), no. 2, 293–309. MR **99m**:16063

[22] Jean-Louis Loday and María O. Ronco, *Order structure on the algebra of permutations and of planar binary trees*, J. Alg. Combinatorics **15** (2002), 253–270. MR 1 900 627

[23] Jean-Louis Loday and María O. Ronco, *Trialgebras and families of polytopes*, (2002).

[24] Saunders Mac Lane, *Categories for the working mathematician*, Second edition. Graduate Texts in Mathematics, 5. Springer-Verlag, New York, 1998. xii+314 pp. MR **2001j**:18001

[25] Susan Montgomery, *Hopf algebras and their actions on rings*, Published for the Conference Board of the Mathematical Sciences, Washington, DC, 1993. MR **94i**:16019

[26] María O. Ronco, *Primitive elements in a free dendriform algebra*, New trends in Hopf algebra theory (La Falda, 1999), 245–263, Contemp. Math., 267, Amer. Math. Soc., Providence, RI, 2000. MR **2001k**:16076

[27] Gian-Carlo Rota, *Baxter Operators, an Introduction*, in *Gian-Carlo Rota on Combinatorics: Introductory papers and commentaries* (Joseph P.S. Kung, Ed.), Birkhäuser, Boston, 1995. 504-512. MR 1 392 973

[28] Gian-Carlo Rota, *Ten Mathematics Problems I will never solve*, DMV Mitteilungen, Heft 2 (1998) 45-52. MR 1 631 412

[29] È. B. Vinberg, *The theory of homogeneous convex cones*, Trudy Moskov. Mat. Obšč. 12 1963 303–358. MR 28 #1637

SOME REMARKS ON NICHOLS ALGEBRAS

NICOLÁS ANDRUSKIEWITSCH

Facultad de Matemática, Astronomía y Física, CIEM - CONICET
Universidad Nacional de Córdoba
(5000) Ciudad Universitaria, Córdoba, Argentina
E-mail: andrus@mate.uncor.edu

ABSTRACT. Two algebras can be attached to a braided vector space (V, c) in an intrinsic way; the FRT-bialgebra and the Nichols algebra $\mathfrak{B}(V, c)$. The FRT-bialgebra plays the rôle of the algebra of quantum matrices, whereas the rôle of the Nichols algebra is less understood. Some authors call $\mathfrak{B}(V)$ a quantum symmetric algebra. The purpose of this paper is to discuss some properties of certain Nichols algebras, in an attempt to establish classes of Nichols algebras which are worth further study.

1. **Definitions and examples of Nichols algebras.** Let (V, c) be a braided vector space; that is, V is a finite-dimensional complex vector space and $c : V \otimes V \to V \otimes V$ is an invertible solution of the braid equation: $(c \otimes \mathrm{id})(\mathrm{id} \otimes c)(c \otimes \mathrm{id}) = (\mathrm{id} \otimes c)(c \otimes \mathrm{id})(\mathrm{id} \otimes c)$. There is a remarkable braided graded Hopf algebra $\mathfrak{B}(V, c) = \oplus_{n \geq 0} \mathfrak{B}^n(V, c)$, which is connected, generated in degree one, with $\mathfrak{B}^1(V, c) \simeq (V, c)$ as braided vector spaces, such that all its primitive elements have degree one; and which is unique with respect to these properties. Algebras of this kind appeared naturally in our approach to classification of pointed Hopf algebras [AS1, AS3] but we quickly realized they were already known to several authors under various presentations. We first briefly recall different definitions of Nichols algebras and survey examples of classes of Nichols algebras that are known. A detailed exposition on Nichols algebras can be found in [AS3].

We shall simply write $\mathfrak{B}(V) = \mathfrak{B}(V, c)$ omitting the reference to c unless it is needed. We shall always assume that c is rigid, *i. e.* the associated map $c^\flat : V^* \otimes V \to V \otimes V^*$ is also bijective (this is the case in all the examples below). Here $c^\flat = (\mathrm{ev}_V \otimes \mathrm{id}_{V \otimes V^*})(\mathrm{id}_{V^*} \otimes c \otimes \mathrm{id}_{V^*})(\mathrm{id}_{V^* \otimes V} \otimes \mathrm{ev}_V^*)$.

This work was partially supported by CONICET, Agencia Córdoba Ciencia, ANPCyT and Secyt (UNC).

This remarkable braided Hopf algebra was first described by W. Nichols in his thesis [N], as the invariant part of his "bialgebras of type one". In his honor, $\mathfrak{B}(V)$ is called the Nichols algebra of the braided vector space (V, c). There are several ways to present $\mathfrak{B}(V)$.

Consider $T(V)\otimes T(V)$ as an algebra with the product 'twisted' by c. Then $T(V)$ is a braided Hopf algebra, with the comultiplication uniquely defined by $\Delta(v) = v\otimes 1 + 1\otimes v$, $v \in V$. Let $I(V)$ be the largest Hopf ideal generated by homogeneous elements of degree greater than 1; then $\mathfrak{B}(V) := T(V)/I(V)$ satisfies all the properties listed above [AS3, Prop. 2.2].

The vector space $T(V)$ has another structure of coalgebra, the free coalgebra over V; let us denote it by $t(V)$. M. Rosso observed that it admits a 'quantum shuffle product', so that $t(V)$ is also a braided Hopf algebra, called the quantum shuffle algebra. The canonical map $\Omega : T(V) \to t(V)$ turns out to be a map of braided Hopf algebras; the image of Ω, that is the subalgebra of $t(V)$ generated by V, is the Nichols algebra of V. The nilpotent part $U_q^+(\mathfrak{g})$ of a quantized enveloping algebra was characterized in this way by Rosso as the Nichols algebra of a suitable braided vector space $\mathfrak{h}$ [Ro1, Ro2]. Results in the same spirit were also obtained by J. A. Green [Gr].

Now, the components of the graded map Ω, that is $\Omega^{(m)} : T^m(V) \to T^m(V)$ are the so-called "quantum symmetrizers" defined through the action of the braid group on $T^m(V)$. Therefore, the Nichols algebra of (V, c) coincides with the quantum exterior algebra of $(V, -c)$, defined by S. L. Woronowicz [Wo]. Indeed, the quantum symmetrizers of $-c$ are the quantum antisymmetrizers of c.

G. Lusztig characterized $U_q^+(\mathfrak{g})$ as the quotient of a $T(\mathfrak{h})$ by the radical of an invariant bilinear form [L]. This is indeed a general fact; the ideal $I(V)$ is always the radical of an invariant bilinear form [AG1].

M. Rosso found that the Nichols algebra $\mathfrak{B}(V)$ of a braided vector space of diagonal type has always a "PBW-basis" in terms of the so-called Lyndon words [Ro3]. Related work was done by V. K. Kharchenko, who also studied abstractly Nichols algebras from various points of view [Kh1, Kh2, Kh3].

The following two questions arise from classification problems of Hopf algebras [AS1, AS3]. Answers to both questions are needed to classify Hopf algebras of certain types.

- Under which conditions on (V, c) is $\mathfrak{B}(V)$ finite-dimensional, respectively of finite Gelfand-Kirillov dimension?
- For those pairs with a positive answer to the preceding question, give an explicit presentation of $\mathfrak{B}(V)$; that is, find a minimal set of generators of the ideal $I(V)$.

The study of $\mathfrak{B}(V)$ is very difficult; neither the subalgebra of the quantum shuffle algebra generated by V, nor the Lyndon words, nor the ideal $I(V)$ have an explicit description.

∘ In particular, we do not know if the ideal $I(V)$ is finitely generated.

There is little hope of performing explicit computations with a computer program without a positive answer to this question. However, let $\widehat{\mathfrak{B}}_r(V) = T(V)/J_r$, where J_r is the two-sided ideal generated by the kernels of $\Omega^{(m)}$, $m \leq r$; these are braided Hopf algebras and we have epimorphisms $\widehat{\mathfrak{B}}_r(V) \to \mathfrak{B}(V)$ for all $r \geq 2$. Hence, if one of the algebras $\widehat{\mathfrak{B}}_r(V)$ is finite-dimensional, or has finite Gelfand-Kirillov dimension, so does $\mathfrak{B}(V)$. In the first case, under favorable hypothesis we may conclude that $\widehat{\mathfrak{B}}_r(V) \simeq \mathfrak{B}(V)$, see [AG2, Th. 6.4].

There are several classes of braided vector spaces which seem to be of special interest.

- We say that (V, c) is *of diagonal type* if there exists a basis $x_1, \dots, x_\theta$ of V, and non-zero scalars q_{ij} such that $c(x_i \otimes x_j) = q_{ij} x_j \otimes x_i$, $1 \leq i, j \leq \theta$.

Nichols algebras of these braided vector spaces appear naturally in the classification of pointed Hopf algebras with abelian coradical, and also in the theory of quantum groups.

Namely, let $(a_{ij})_{1 \leq i,j \leq \theta}$ be a generalized Cartan matrix; let $\mathfrak{h}$ be a vector space with a a basis $x_1, \dots, x_\theta$, let q be a non-zero scalar and let c be given by

$$c(x_i \otimes x_j) = q^{a_{ij}} x_j \otimes x_i, \qquad 1 \leq i, j \leq \theta. \tag{1.1}$$

Then $\mathfrak{B}(\mathfrak{h}) = U_q^+(\mathfrak{g})$ if q is not a root of 1 [L, Ro1, Ro2], and $\mathfrak{B}(\mathfrak{h}) = \mathfrak{u}_q^+(\mathfrak{g})$ if $q \neq 1$ is a root of 1 (under some hypothesis on the order of q) [Ro1, Ro2, Mü1].

- We say that (V, c) is *of rack type* if there exists a basis X of V, a function $\triangleright : X \times X \to X$ and non-zero scalars q_{ij} such that

  $$c(i \otimes j) = q_{ij}\, i \triangleright j \otimes i, \tag{1.2}$$

  $i, j \in X$. Then $(X, \triangleright)$ is a rack and q_{ij} is a rack 2-cocycle with coefficients in $\mathbb{k}^\times$, see for example [Gñ1, AG2]. These braided vector spaces appear naturally in the classification of pointed Hopf algebras.

- We say that (V, c) is *of Jordanian type* if there exists a basis $x_1, \dots, x_\theta$ of V, and a non-zero scalar q such that $c(x_i \otimes x_1) = q x_1 \otimes x_i$, $c(x_i \otimes x_j) = (q x_j + x_{j-1}) \otimes x_i$, $1 \leq i \leq \theta$, $2 \leq j \leq \theta$.

These braided vector spaces appear in the classification of pointed Hopf algebras with coradical $\mathbb{Z}$.

- We say that (V, c) is *of Hecke type* if $(c - q)(c + 1) = 0$, for some non-zero scalar q.
- We say that (V, c) is *of quantum group type* if V is a module over some quantized enveloping algebra $U_q(\mathfrak{g})$ and c arises from the action of the corresponding universal R-matrix.

Here is what is known about the problems stated above.

- $\mathfrak{B}(V) = T(V)$ generically. That is, consider the locally closed space of all $c \in \operatorname{End}(V \otimes V)$ which are invertible solutions of the braid equation; then the subset of those c such that $\mathfrak{B}(V, c) = T(V)$ contains a non-empty open subset.
- Assume that (V, c) is of diagonal type, where the q_{ii}'s are positive and different from one, $1 \le i \le \theta$. Then $\mathfrak{B}(V)$ has finite Gelfand-Kirillov dimension if and only if $q_{ij}q_{ji} = q_{ii}^{a_{ij}}$ for some Cartan matrix of finite type [Ro2].
- Assume that (V, c) is of diagonal type, that the q_{ii}'s are are roots of 1 but not 1, and that $q_{ij}q_{ji} = q_{ii}^{a_{ij}}$, $1 \le i, j \le \theta$, where $a_{ij} \in \mathbb{Z}$, $a_{ii} = 2$, $\operatorname{ord} q_{ii} < a_{ij} \le 0$ if $1 \le i \neq j \le \theta$. Then $(a_{ij})_{1 \le i,j \le \theta}$ is a generalized Cartan matrix and, under suitable conditions, $\mathfrak{B}(V)$ has finite dimension if and only if $(a_{ij})_{1 \le i,j \le \theta}$ is of finite type [AS2].

In these two cases, the calculation of $\mathfrak{B}(V)$ is reduced to the calculation of $\mathfrak{B}(\mathfrak{h})$ as in (1.1); but the last requires deep facts on representation theory, and the action of the quantum Weyl group defined by Lusztig.

- There are a few examples of (V, c) of diagonal type with finite-dimensional $\mathfrak{B}(V)$, due to Nichols [N] and Graña [Gñ1], besides those in the last item (they are listed in [AS3, Section 3.3]).
- If c is of Hecke type and q is not a root of 1, or if $q = 1$, then the Nichols algebra is quadratic: $\mathfrak{B}(V) = \widehat{\mathfrak{B}}_2(V)$. Furthermore, the quadratic dual is also a Nichols algebra: $\mathfrak{B}(V)^! = \mathfrak{B}(V^*, -q^{-1}c^t)$; see Proposition 2.3 below.
- If c is of diagonal type, information about $\det \Omega^{(m)}$ is given in [FG].
- There are a few examples of (V, c) of rack type with finite-dimensional $\mathfrak{B}(V)$ [MS, Gñ1, AG2, Gñ2]. See the table in the Appendix for a flavor of the kind of algebras obtained.

∘ Almost nothing is known about Nichols algebras of quantum group type, except when they are of Hecke type. To begin with, it would be interesting to know what happens when $V = L(n)$ is a highest weight module over $U_q(\mathfrak{sl}(2))$, $n \ge 3$ (if $n = 1$ it is of Hecke type, if $n = 2$ it seems to be known but I do not have a reference).

- Nichols algebras of Jordanian type were not considered in the literature, to my knowledge. It is likely that the quantum Jordanian plane is a Nichols algebra of Jordanian type.

In conclusion, a Nichols algebra may be finite-dimensional or not, have finite Gelfand-Kirillov dimension or not, and there is no general technique, up to now, to explicitly decide for a given braided vector space, what is the case for its Nichols algebra[1].

This indicates that there is no general approach to study Nichols algebras; it is necessary to split the category of braided vector spaces in classes to be considered separately.

Remark 1.1. Another important question in the classification of Hopf algebras is the following. Let $B = \oplus_{n\geq 0} B^n$ be a braided graded Hopf algebra, connected, generated in degree one, and denote by V the braided vector subspace B^1 of B. Is it possible to conclude that B is the Nichols algebra of V, *i. e.* that all its primitive elements have degree one, under some abstract conditions? Partial positive answers to this question are given in [AS4, Th. 7.6] (finite-dimensional case) and [AS5, Lemma 5.1] (finite Gelfand-Kirillov dimension case).

2. **Some properties of some Nichols algebras.** It is clear that no fine ring theoretical properties can be established for Nichols algebras in general. But there might be a suitable class of braided vector spaces whose Nichols algebras deserve attention from this point of view.

2.1. *Graded algebras.* We shall only consider finitely generated graded algebras $R = \oplus_{i\geq 0} R_i$, with $R_0 = \mathbb{C}$.

Definition 2.1. A graded algebra R is *AS-regular* if it has finite global dimension d, finite Gelfand-Kirillov dimension and is AS-Gorenstein. Thus, $\mathrm{Ext}^i_R(\mathbb{C}, R) = 0$, if $i \neq d$, and $= \mathbb{C}$ if $i = d$.

This class of graded algebras has been intensively investigated in the last years; AS is in honor of Artin and Schelter. The study of the category of graded modules of such an algebra has a strong geometrical flavor; this is usually called a noncommutative projective space. In particular, the space of all "point modules" is a genuine projective space which provides important information on the full category. See [St, SV]. The homological conditions are designed to ensure good regularity properties; we refer again to [St, SV] and references therein.

---

[1]Some techniques are available for specific classes of Nichols algebras, *e. g.* for Nichols algebras of diagonal type, as already said.

- When is a Nichols algebra AS-regular?
- Let $\mathfrak{B}(V)$ be a Nichols algebra which is a domain with finite Gelfand-Kirillov dimension. Is it AS-regular?

Some insight about these questions is explained in the next subsections.

2.2. *Koszul algebras.* Let us first recall some well-known facts about Koszul algebras.

A *quadratic algebra* is a graded algebra $A = \oplus_{n\geq 0} A_n$ generated in degree one with relations in degree 2; that is $A \simeq T(V)/\langle R\rangle$, where $V = A_1$ and $R \subset V \otimes V$ is the kernel of the multiplication. We shall denote $A = (V, R)$. The quadratic dual of a quadratic algebra $A = (V, R)$ is $A^! = (V^*, R^\perp)$. By [Lö], $A^! \simeq E(A) :=$ the subalgebra of $\operatorname{Ext}^*_A(\mathbb{C}, \mathbb{C})$ generated by $\operatorname{Ext}^1_A(\mathbb{C}, \mathbb{C})$.

A *graded Koszul* algebra is a quadratic algebra A such that

$$A^! \simeq \operatorname{Ext}^\bullet(\mathbb{C}, \mathbb{C})$$

as graded algebras.

Lemma 2.2. *Let A be any graded connected algebra.*

(a). [Sm, 1.4 and 5.9]. *If* $\operatorname{gldim} A < \infty$ *then* $\dim A^! < \infty$. *The converse holds if A is Koszul.*

(b). [Sm, 5.10]. *If A is Koszul and has finite global dimension, then A is AS-Gorenstein if and only if $A^!$ is Frobenius.* □

We can now decide when a Nichols algebra of Hecke type is AS-regular.

Proposition 2.3. *Let (V, c) be a braided vector space and assume that c satisfies a Hecke-type condition with label q, $q = 1$ or not a root of 1. Then $\mathfrak{B}(V)$ is AS-regular if and only if it has finite Gelfand-Kirillov dimension and the dimension of $\mathfrak{B}(V^*)$ is finite.*

Proof. By [AA, Prop. 3.3.1], see also [AS3, Prop. 3.4], the Nichols algebra $\mathfrak{B}(V)$ is quadratic, and its quadratic dual is $\mathfrak{B}(V)^! = \mathfrak{B}(V^*)$, the Nichols algebra with respect to $-q^{-1}c^t$. By [Gu, Wa], $\mathfrak{B}(V)$ is Koszul. By Lemma 2.2 (a), if $\mathfrak{B}(V)$ is AS-regular then $\dim \mathfrak{B}(V^*) < \infty$. Conversely, assume that $\dim \mathfrak{B}(V^*) < \infty$. By Lemma 2.2 part (a), $\operatorname{gldim} A < \infty$; and by part (b), $\mathfrak{B}(V)$ is AS-Gorenstein. Indeed, $\mathfrak{B}(V^*)$ is a braided Hopf algebra; hence it is Frobenius whenever finite-dimensional. □

Remark 2.4. I do not know if the hypothesis on the Gelfand-Kirillov dimension can be removed. There are examples of Koszul algebras where $A^!$ is finite-dimensional and Frobenius but A has infinite GK-dimension. I am indebted to James Zhang for pointing out this to me. We do know the

Hilbert series of $\mathfrak{B}(V)$: $H(\mathfrak{B}(V))(t) = \dfrac{1}{H(\mathfrak{B}(V^*), -q^{-1}c^t)(-t)}$, by [BGS, Th. 2.11.1], and $H(\mathfrak{B}(V^*))(t)$ is a polynomial.

Remark 2.5. D. Gurevich studied intensively Nichols algebras of Hecke type, provided ways to construct explicit examples, and classified those such that $H(\mathfrak{B}(V^*), -q^{-1}c^t)$ is a polynomial of degree two [Gu].

Remark 2.6. A quadratic Nichols algebra is not necessarily Koszul; see [R].

2.3. *Nichols algebras related to quantum groups.*

Proposition 2.7. *Let* $(a_{ij})_{1\le i,j\le\theta}$ *be a Cartan matrix of finite type, let* q *be a non-zero scalar and let* $(\mathfrak{h}, c)$ *be the braided vector space as in* (1.1). *Then*

(a). [GL, 4.7] $\mathfrak{B}(\mathfrak{h}) = U_q^+(\mathfrak{g})$ *is AS-regular when* q *is not a root of 1.*

(b). [GK, 2.5] $\mathfrak{B}(\mathfrak{h}) = \mathfrak{u}_q^+(\mathfrak{g})$ *is* not *AS-regular if* $1 \neq q$ *is a root of 1.* □

It would be interesting to have another proof of (a) in the spirit of [DCK, GK]. Namely, to consider the algebra filtration given by the PBW-basis; the associated graded algebra is a quantum linear space [DCK], hence a Nichols algebra of Hecke type; therefore it has finite global dimension. Then lift this information by a spectral sequence argument. Note also that Proposition 2.7 extends to the multiparametric case without difficulties. In view of Propositions 2.3 and 2.7, it is tempting to suggest the following questions.

- If $A = \mathfrak{B}(V)$, when is $E(A)$ also a Nichols algebra?
- Is the graded algebra associated to the filtration given by the PBW-basis on the Lyndon words of $A = \mathfrak{B}(V)$, also a Nichols algebra?

Incidentally, it seems that the determination of the space of point modules for $U_q^+(\mathfrak{g})$ does not appear explicitly in the literature.

2.4. *Invariants.* A natural question in noncommutative geometry is the study of spaces of invariants under group (or Hopf algebra) actions. We believe that a suitable setting to discuss it is when the noncommutative space corresponds to a Nichols algebra $\mathfrak{B}(V)$ with suitable properties. Specifically, we propose to study the subalgebra of invariants of $\mathfrak{B}(V)$ under a coaction of a Hopf algebra H.

The first step is to find a good number of examples of Hopf algebras H such that $\mathfrak{B}(V)$ is an H-comodule algebra. It is well-known that $\mathfrak{B}(V)$ is a comodule braided Hopf algebra over the FRT-Hopf algebra $H(c)$ associated to the braided vector space (V, c), see for example [T]; thus, $\mathfrak{B}(V)$ is a comodule algebra over any Hopf algebra quotient of $H(c)$. See [Mü2] for the classification of finite-dimensional quotients of $\mathbb{C}_q[G]$, G a simple algebraic group.

We stress that these comodule algebra structures are "linear", that is, they preserve also the comultiplication of $\mathfrak{B}(V)$; many other structures may arise. Assume for example that $(V,c) = (V,\tau)$, where τ is the usual transposition. Then the Nichols algebra $\mathfrak{B}(V)$ is the symmetric algebra $S(V)$. The automorphism group of a polynomial algebra is much larger than the group of linear automorphisms, and the determination of the former is a classical open problem.

In the quantum case the situation is much more rigid. Indeed, assume that q is not a root of 1. Then $\mathrm{Aut}_{\mathrm{alg}}\, U_q^+(\mathfrak{g})$ coincides with $\mathrm{Aut}_{\ \mathrm{Hopf\ alg}}\, U_q^+(\mathfrak{g}) \simeq (T \rtimes \mathrm{Aut}\,\Delta)$, where T is a maximal torus and Δ is the Dynkin diagram, if $\mathfrak{g}$ is of type A_2 [AlD] or of type B_2 [AD], and conjecturally for all the types. Again, one is tempted to ask for the class of braided vector spaces (V,c) such that $\mathrm{Aut}_{\mathrm{alg}}\, \mathfrak{B}(V) = \mathrm{Aut}_{\ \mathrm{Hopf\ alg}}\, \mathfrak{B}(V)$.

Appendix. For illustration, we collect some information about finite dimensional Nichols algebras of rack type. Below we consider braided vector spaces of rack type, with $q_{ij} \equiv -1$, see (1.2). The rank of a Nichols algebra $\mathfrak{B}(V)$ is the dimension of V.

An affine rack is a rack (A,g) where A is a finite abelian group and $g \in \mathrm{Aut}\, A$; then $a \triangleright b = g(b) + (\mathrm{id} - g)(a)$. The first four racks listed below are affine. A subset of a group stable under conjugation is a rack; so is the

Rack	**rk**	**Relations**	$\dim \mathfrak{B}(V)$	**top**
$(\mathbb{Z}/3, \triangleright^g), g=2$ (Transpositions in $\mathbb{S}_3$)	3	5 relations in deg 2	$12 = 3.2^2$	$4 = 2^2$
$(\mathbb{Z}/5, \triangleright^g), g=2$	5	10 relations in deg 2 1 relation in deg 4	$1280 = 5.4^4$	$16 = 4^2$
$(\mathbb{Z}/7, \triangleright^g), g=3$	7	21 relations in deg 2 1 relation in deg 6	$326592 = 7.6^6$	$36 = 6^2$
$(\mathbb{Z}/2 \times \mathbb{Z}/2, \triangleright^g)$, $g = \begin{pmatrix} 0 & 1 \\ 1 & 1 \end{pmatrix}$	4	8 relations in deg 2 1 relation in deg 6	72	$9 = 3^2$
Transpositions in $\mathbb{S}_4$	6	16 relations in deg 2	576	12
Faces of the cube	6	16 relations in deg 2	576	12
Transpositions in $\mathbb{S}_5$	10	45 relations in deg 2	8294400	40

set of transpositions in $\mathbb{S}_n$. The action for the rack of faces of the cube can be described either geometrically or as an extension.

There is no problem to find the space of relations in degree 2; it is the kernel of $c+\mathrm{id}$. Relations in higher degree (not coming from those in degree 2) are more difficult to find, as said. For affine racks, a first step is given in [AG2, 6.13]. Typical relations in degree 4 and 6 are respectively

$$x_0x_1x_0x_1 + x_1x_0x_1x_0 = 0,$$
$$x_0x_1x_2x_0x_1x_2 + x_2x_0x_1x_2x_0x_1 + x_1x_2x_0x_1x_2x_0 = 0.$$

These relations depend upon the order of $-g$. Most of the computations were done with help of a computer program. See [Gñ2] for details. No explanation of the numbers appearing in the table is available until now, but there are some evident patterns.

Except for the racks of transpositions in $\mathbb{S}_4$ and faces of the cube, all the other racks are simple (they do not project properly onto a non-trivial rack). Those two racks are extensions with the same base and fiber but they are not isomorphic. The similarities between the corresponding Nichols algebras are explained by a kind of Fourier transform, see [AG2, Ch. 5].

More examples of finite-dimensional Nichols algebras of rack type are given in [AG2, Prop. 6.8]; they are not of diagonal type but they arise from Nichols algebras of diagonal type by the same kind of Fourier transform.

Acknowledgements. The author is grateful to J. Alev, F. Dumas and S. Natale for many conversations about various aspects of Nichols algebras; several of the questions in the text arise from discussions with them; and also to J. T. Stafford and J. Zhang for answers to some consultations. The author also thanks F. Dumas for his warm hospitality during a visit to the University of Clermont-Ferrand in March 2002 (when this work was began); to S. Catoiu for the kind invitation to the International Conference in Chicago; and to the IHES, where this paper was written.

References

[AA] A. Abella and N. Andruskiewitsch, *Compact quantum groups and the FRT-construction*, Bol. Acad. Ciencias (Córdoba) **63** (1999), 15-44.

[AlD] J. Alev and F. Dumas, *Rigidité des plongements des quotients primitifs minimaux de $U_q(\mathrm{sl}(2))$ dans l'algèbre quantique de Weyl-Hayashi*, Nagoya Math. J. **143** (1996), 119–146.

[AD] N. Andruskiewitsch and F. Dumas, *Sur les automorphismes de $U_q^+(\mathfrak{g})$*, Beiträge Algebra Geom., to appear, preprint (2002).

[AG1] N. Andruskiewitsch and M. Graña, *Braided Hopf algebras over non-abelian groups*, Bol. Acad. Ciencias (Córdoba) **63** (1999), 45-78. Also in `http://arxiv.org/9802074`.

[AG2] ———, *From racks to pointed Hopf algebras*, Adv. Math., to appear. Also in `http://arxiv.org/0202084`.

[AS1] N. Andruskiewitsch and H.-J. Schneider, *Lifting of Quantum Linear Spaces and Pointed Hopf Algebras of order p^3*, J. Algebra **209** (1998), 658–691.

[AS2] ———, *Finite quantum groups and Cartan matrices*, Adv. Math. **154** (2000), 1–45.

[AS3] ———, *Pointed Hopf Algebras*, in "New directions in Hopf algebras", 1–68, Math. Sci. Res. Inst. Publ. **43**, Cambridge Univ. Press, Cambridge, 2002.

[AS4] ———, *Finite quantum groups over abelian groups of prime exponent*, Ann. Sci. Ec. Norm. Super. **35** (2002), 1–26.

[AS5] ———, *A characterization of quantum groups*, `math.QA/0201095`, 21 pages.

[BGS] A. Beilinson, V. Ginzburg and W. Sörgel, *Koszul duality patterns in representation theory*, J. Amer. Math. Soc. **9**, (1996), 473–527.

[DCK] C. De Concini and V. G. Kac, *Representations of quantum groups at roots of 1*, in "Operator Algebras, Unitary Representations, Enveloping Algebras, and Invariant Theory", ed. A. Connes *et al* (2000); Birkhäuser, 471–506.

[FG] D. Flores de Chela and J. Green, *Quantum symmetric algebras*, Algebr. Represent. Theory **4** (2001), 55-76.

[GK] Ginzburg, V.; Kumar, S., *Cohomology of quantum groups at roots of unity*, Duke Math. J. **69** (1993), 179–198.

[GL] K.Goodearl and T. Lenagan, *Catenarity in quantum algebras*, J. Pure Appl. Algebra **111** (1996), 123–142.

[Gñ1] M. Graña, *On Nichols algebras of low dimension*, in New trends in Hopf algebra theory (La Falda, 1999), Contemp. Math. **267** (2000), 111–134.

[Gñ2] ———, *Zoo of Nichols algebras of nonabelian group type*, available at `http://mate.dm.uba.ar/ matiasg/zoo.html`.

[Gr] J. Green, *Quantum groups, Hall algebras and quantized shuffles*, in Finite reductive groups (Luminy, 1994), Progr. Math. **141**, Birkhäuser, (1997), 273–290.

[Gu] D. Gurevich, *Algebraic aspects of the quantum Yang-Baxter equation*, Leningrad J. Math. **2**, (1991), 801–828.

[Kh1] V. Kharchenko, *An Existence Condition for Multilinear Quantum Operations*, J. Algebra **217** (1999), 188-228.

[Kh2] ———, *Skew primitive elements in Hopf algebras and related identities*, J. Algebra **238** (2001), 534-559.

[Kh3] ———, *A combinatorial approach to the quantification of Lie algebras*, Pacific J. Math. **203** (2002), 191–233.

[Lö] C. Löfwall, *On the subalgebra generated by the one-dimensional elements in the Yoneda Ext-algebra* in Algebra, algebraic topology and their interactions (Stockholm, 1983), 291–338, Lecture Notes in Math. **1183**, Springer, 1986.

[L] G. Lusztig, *Introduction to quantum groups*, Birkhäuser, 1993.

[MS] A. Milinski and H-J. Schneider, *Pointed Indecomposable Hopf Algebras over Coxeter Groups*, in New trends in Hopf algebra theory (La Falda, 1999), Contemp. Math. **267** (2000), 215–236.

[Mü1] E. Müller, *Some topics on Frobenius-Lusztig kernels, I*, J. Algebra **206** (1998), 624–658.

[Mü2] ———, *Finite subgroups of the quantum general linear group*, Proc. London Math. Soc. (3) **81** (2000), 190–210.

[N] W.D. Nichols, *Bialgebras of type one*, Commun. Alg. **6** (1978), 1521–1552.

[R] J.-E. Roos, *Some non-Koszul algebras*, Progr. Math. 172, Birkhauser, (1999), 385–389.

[Ro1] M. Rosso, *Groupes quantiques et algebres de battage quantiques*, C.R.A.S. (Paris) **320** (1995), 145–148.

[Ro2] ———, *Quantum groups and quantum shuffles*, Inventiones Math. **133** (1998), 399–416.

[Ro3] ———, *Lyndon words and Universal R-matrices*, talk at MSRI, October 26, 1999, available at `http://www.msri.org`.

[Sm] Smith, S. Paul, *Some finite-dimensional algebras related to elliptic curves*, in Representation theory of algebras and related topics (Mexico City, 1994), CMS Conf. Proc. **19** (1996), 315–348, Amer. Math. Soc., Providence, RI.

[St] J. T. Stafford, *Noncommutative projective geometry*, in Proc. of the International Congress of Mathematicians, Beijing 2002, vol. II (2002), 93–103.

[SV] J. T. Stafford and M. Van den Bergh, *Noncommutative curves and noncommutative surfaces*, Bull. Amer. Math. Soc. **38** (2001), 171–216.

[T] M. Takeuchi, *Survey of braided Hopf algebras*, in New trends in Hopf algebra theory (La Falda, 1999), Contemp. Math. **267** (2000), 301–324.

[VdB] Van den Bergh, M., *Existence theorems for dualizing complexes over non-commutative graded and filtered rings*, J. Algebra **195** (1997), 662–679.

[Wa] M. Wambst, *Complex de Koszul quantiques*, Ann. Inst. Fourier (Grenoble) **43** (1993), 1089–1156.

[Wo] S. L. Woronowicz, *Differential calculus on compact matrix pseudogroups (quantum groups)*, Comm. Math. Phys. **122** (1989), 125–170.

The coradical of the dual of a lifting of a quantum plane

N. Andruskiewitsch *
Facultad de Matemática, Astronomía y Física,
Universidad Nacional de Córdoba
CIEM - CONICET
(5000) Ciudad Universitaria, Córdoba, Argentina
email: andrus@mate.uncor.edu
and
M. Beattie †
Department of Mathematics and Computer Science
Mount Allison University
Sackville, NB, Canada E4L 1E6
email: mbeattie@mta.ca

Abstract

In this short note we compute the coradical of the dual A^* of a lifting A of a quantum linear space in the case that A^* has only trivial grouplike elements.

1 Introduction

The classification of finite dimensional Hopf algebras is known to be a difficult problem, for which only a few general techniques are available. One technique for studying Hopf algebras whose coradical is a sub Hopf algebra is the "lifting method" of Andruskiewitsch and Schneider. The simplest application of the lifting method is the description of all pointed Hopf algebras with finite abelian group of grouplikes Γ and with space V of $(1,g)$ primitives, $V \in_{K[\Gamma]}^{K[\Gamma]} \mathcal{YD}$, a quantum linear space.

*This work was partially supported by CONICET, Agencia Córdoba Ciencia, ANPCyT and Secyt (UNC).
†Research supported by NSERC.

A lifting of a quantum linear space is a pointed Hopf algebra A whose associated graded Hopf algebra is isomorphic to the Radford biproduct $\mathcal{B}(V)\#K[\Gamma]$ where $\mathcal{B}(V)$ is the Nichols algebra of the quantum linear space V, and is a Hopf algebra in ${}^{K[\Gamma]}_{K[\Gamma]}\mathcal{YD}$. Liftings of quantum linear spaces were constructed and completely described in [2], and also independently, using repeated Ore extensions, in [5].

Liftings of quantum linear spaces, although fairly easy to describe, have enough interesting structure to prove useful in various situations. They provided counterexamples to Kaplansky's Tenth Conjecture [2], [5], [9]. Suppose K is an algebraically closed field of characteristic 0. It is shown in [2] or [6] that every pointed nonsemisimple Hopf algebra over K of dimension p^3, p prime, is a lifting of a quantum linear space. In [10] it is shown that every pointed Hopf algebra of index p^2, p a prime, is the lifting of a quantum line or a quantum linear space; this result generalizes [1, Theorem 1.7] and [8, Theorem 1.2]. In [7], it is shown that over K, every pointed nonsemisimple Hopf algebra of dimension 16 is a lifting of a quantum linear space.

However, although the structure of liftings of quantum linear spaces is well understood, the coradicals of the duals (and thus the irreducible representations) of these Hopf algebras have yet to be completely described. In [12], and more recently in [3], for V a quantum linear space of dimension 1 or 2, the coradicals of the duals of some liftings of $\mathcal{B}(V)\#K[\Gamma]$ are described. In particular, if V has dimension 1, then the coradical of the dual A^* of a lifting A of $\mathcal{B}(V)\#K[\Gamma]$ is completely described. If V has dimension 2, then the coradical of the dual A^* is described in some cases where A^* has nontrivial grouplikes.

In this short note, we study some cases where V has dimension 2 and A^* has only the trivial grouplike element and we explicitly construct the coradical of the dual of a Hopf algebra in the family of pointed Hopf algebras of dimension $81 = 3^4$ which provided one of the early counterexamples to Kaplansky's Tenth Conjecture.

2 Preliminaries

Throughout, K will denote an algebraically closed field of characteristic 0, Γ a finite abelian group and ${}^{\Gamma}_{\Gamma}\mathcal{YD} = {}^{K[\Gamma]}_{K[\Gamma]}\mathcal{YD}$ the category of Yetter-Drinfel'd modules over the group algebra $K[\Gamma]$. For $V \in {}^{\Gamma}_{\Gamma}\mathcal{YD}$, $g \in \Gamma, \chi \in \hat{\Gamma}$, we write V_g^{χ} for the set of $v \in V$ with the action of Γ on v given by $h \rightharpoonup v = \chi(h)v$ and the coaction by $\delta(v) = g \otimes v$. Since Γ is an abelian group, it is well-known that $V = \oplus_{\substack{g\in\Gamma\\ \chi\in\hat{\Gamma}}} V_g^{\chi}$.

Definition 2.1 *$V = \oplus_{i=1}^{t} K v_i \in_{\Gamma}^{\Gamma} \mathcal{YD}$ with $0 \neq v_i \in V_{g_i}^{\chi_i}$ is called a quantum linear space if $\chi_i(g_j)\chi_j(g_i) = 1$ for $i \neq j$. As well, $\chi_i(g_i)$ is a primitive r_ith root of unity with $1 < r_i < \infty$.*

Recall that for V a quantum linear space as above, $\mathcal{B}(V)$ is the Nichols algebra for V, and the Radford biproduct, $H = \mathcal{B}(V)\#K[\Gamma]$, is the coradically graded Hopf algebra generated by the $(1, g_i)$-primitives $v_i\#1$ (usually written just v_i) and the grouplike elements $h = 1\#h$, with $h \in \Gamma$. Multiplication is given by $hv_i = \chi_i(h)v_i h$ and $v_i v_j = \chi_j(g_i)v_j v_i$. Also if $\chi_i(g_i)$ is a primitive r_ith root of unity, $v_i^{r_i} = 0$, and then dim $\mathcal{B}(V) = \Pi_{i=1}^{t} r_i$.

Proposition 2.2 *(see [2] or [5]) For V a quantum linear space with $\chi_i(g_i)$ a primitive r_ith root of unity, all liftings A of $\mathcal{B}(V)\#K[\Gamma]$ are Hopf algebras generated by the grouplikes and by $(1, g_i)$-primitives $x_i, 1 \leq i \leq t$ where*

$$\begin{aligned} hx_i &= \chi_i(h)x_i h; \\ x_i^{r_i} &= \alpha_{ii}(g_i^{r_i} - 1); \\ x_i x_j &= \chi_j(g_i)x_j x_i + \alpha_{ij}(g_i g_j - 1). \end{aligned}$$

We may assume $\alpha_{ii} \in \{0, 1\}$ and then we have that

$$\begin{aligned} \alpha_{ii} &= 0 \textit{ if } g_i^{r_i} = 1 \textit{ or } \chi_i^{r_i} \neq \epsilon; \\ \alpha_{ij} &= 0 \textit{ if } g_i g_j = 1 \textit{ or } \chi_i\chi_j \neq \epsilon. \end{aligned}$$

Note that $\alpha_{ji} = -\chi_j(g_i)^{-1}\alpha_{ij} = -\chi_i(g_j)\alpha_{ij}$. Thus the lifting A is described by a matrix $\mathcal{A} = (\alpha_{ij})$ with 0's or 1's on the diagonal and with $\alpha_{ji} = -\chi_i(g_j)\alpha_{ij}$ for $i \neq j$. ∎

Throughout, we use the usual Sweedler notation for Hopf algebra computations. Also, we use the notation $\mathcal{M}^c(r, K)$ for an $r \times r$ matrix coalgebra over K and we call a basis $e(i, j), 1 \leq i, j \leq r$, a matrix coalgebra basis if $\Delta(e(i, j)) = \sum_{k=1}^{r} e(i, k) \otimes e(k, j)$ and $\epsilon(e(i, j)) = \delta_{i,j}$.

3 The Main Theorem

Let A be a lifting of $\mathcal{B}(V)\#K[\Gamma]$ with $V = \oplus_{i=1}^{t} K v_i \in_{\Gamma}^{\Gamma} \mathcal{YD}$ a quantum linear space and with $\chi_i(g_i)$ a primitive r_ith root of unity. Let $x_i \in A$ be the lifting of $v_i\#1$ to A, i.e. in A, x_i is $(1,g_i)$-primitive and $hx_i = \chi_i(h)x_ih$ for all $h \in \Gamma$. Then A has a vector space basis hz with $h \in \Gamma$ and $z \in \mathcal{Z} = \{x_1^{m_1}\dots x_t^{m_t} | 0 \leq m_i \leq r_i - 1\}$.

For $\gamma \in \hat{\Gamma}$, let $w(\gamma, x_1^{m_1}\dots x_t^{m_t}) \in A^*$ be the map which takes $hx_1^{m_1}\dots x_t^{m_t}$ to $\gamma(h)$ and all other basis elements to 0, so that $\mathcal{W} = \{w(\gamma,z) | \gamma \in \hat{\Gamma}, z \in \mathcal{Z}\}$ is a vector space basis for A^*. The map $w(\gamma, 1)$ is usually just written γ. Multiplication in A^* is given by the following lemma.

Lemma 3.1 *[3, Lemma 3.1] Let A be a lifting of $\mathcal{B}(V)\#K[\Gamma]$ with V a quantum linear space as above. Let $\gamma, \lambda \in \hat{\Gamma}$, $0 \leq m_i \leq r_i - 1$ and $1 \leq i < j \leq t$.*

(i) Then

$$w(\gamma, x_i^{m_i}) * w(\lambda, x_j^{m_j}) = (\gamma\chi_i^{-m_i})(g_j^{m_j})w(\gamma\lambda, x_i^{m_i}x_j^{m_j});$$

$$w(\lambda, x_j^{m_j}) * w(\gamma, x_i^{m_i}) = \lambda(g_i^{m_i})w(\gamma\lambda, x_i^{m_i}x_j^{m_j}).$$

In particular, $w(\chi_i, x_i)w(\chi_j, x_j) = w(\chi_i\chi_j, x_ix_j) = \chi_i(g_j)w(\chi_j, x_j)w(\chi_i, x_i)$ for $i < j$.

(ii)

$$\gamma * w(\lambda, x_1^{m_1}\dots x_t^{m_t}) = \gamma(g_1^{m_1}\dots g_t^{m_t})w(\lambda\gamma, x_1^{m_1}\dots x_t^{m_t});$$

$$w(\lambda, x_1^{m_1}\dots x_t^{m_t}) * \gamma = w(\gamma\lambda, x_1^{m_1}\dots x_t^{m_t}).$$

In particular $\gamma * w(\chi_i, x_i) = \gamma(g_i)w(\chi_i, x_i) * \gamma = \gamma(g_i)w(\chi_i\gamma, x_i)$.

(iii) For $0 \leq m_i \leq r_i - 1$ we have that

$$w(\lambda, x_i) * w(\gamma, x_i^{m_i}) = \lambda(g_i^{m_i})(q^{m_i} + q^{m_i-1} + \dots 1)w(\lambda\gamma, x_i^{m_i+1})$$

where $q = \chi_i(g_i)^{-1}$, a primitive r_ith root of 1, so that $w(\chi_i, x_i)^{r_i} = 0$ in A^*.

The multiplication formulas above show that $\gamma \in \hat{\Gamma}$ and the elements $w(\chi_i, x_i)$ generate A^* as an algebra with $\gamma * w(\chi_i, x_i) = \gamma(g_i)w(\chi_i, x_i) * \gamma$, $w(\chi_i, x_i) * w(\chi_j, x_j) = \chi_i(g_j)w(\chi_j, x_j) * w(\chi_i, x_i)$ for $i \neq j$, and $w(\chi_i, x_i)^{r_i} = 0$.

From now on, let $V \in {}^{\Gamma}_{\Gamma}\mathcal{YD}$, $V = Kv_1 \oplus Kv_2$ be a quantum linear space of dimension 2, with $v_i \in V_{g_i}^{\chi_i}$ and with $g_1 = g, \chi_1 = \chi, \chi_2 = \chi^m$. Also, for $i = 1, 2$,

1. $\chi_i(g_i)$ is a primitive rth root of unity for some odd integer $r > 1$.

2. χ_i has order r in the group $\hat{\Gamma}$, so that $(m, r) = 1$.

3. $g_i^r \neq 1$.

Let q denote the primitive rth root of unity $\chi(g)$. Let $H = \mathcal{B}(V)\#K\Gamma$ and let A be the nontrivial lifting of H with matrix $\mathcal{A} = \begin{bmatrix} 1 & \nu \\ -q^{-m}\nu & 1 \end{bmatrix}$. In A, let x be the lifting of v_1 and y the lifting of v_2. Then A is the pointed Hopf algebra generated by its group of grouplikes Γ and the skew-primitives x and y where for $h \in \Gamma$,

$$hx = \chi(h)xh, \text{ and } hy = \chi^m(h)yh, \tag{1}$$
$$x^r = g^r - 1, \text{ and } y^r = g_2^r - 1, \tag{2}$$
$$xy = \chi^m(g)yx + \nu(g_1g_2 - 1). \tag{3}$$

If $\nu \neq 0$, then $m = -1$, i.e., $\chi_2 = \chi^{-1}$.

As a K-space, A has basis $\{gz | g \in \Gamma, z \in \mathcal{Z}\}$ where $\mathcal{Z} = \{y^ix^j | 0 \leq i,j \leq r-1\}$. Then as a K-space, A^* has basis $\mathcal{W} = \{w(\gamma,z) | \gamma \in \hat{\Gamma}, z \in \mathcal{Z}\}$.

Definition 3.2 *$C(1)$ is defined to be the sub Hopf algebra of A^* of dimension r^3 with basis $\{w(\chi^i,z) | 0 \leq i \leq r-1, z \in \mathcal{Z}\}$.*

Note that Lemma 3.1 and the fact that $\Delta w(\gamma, y^jx^i)$ is a linear combination of terms of the form $w(\gamma, y^tx^s) \otimes w(\gamma\chi^{-mt-s}, y^nx^l)$ for some $0 \leq t,s,n,l \leq r-1$, guarantees that $C(1)$ is a sub Hopf algebra of A^*. Also we see that $\gamma C(1)$ is the subcoalgebra of A^* with basis $\mathcal{W}_\gamma = \{w(\gamma\chi^k, y^jx^l) | 0 \leq k,j,l \leq r-1\}$. From Lemma 3.1, we see that as a coalgebra, A^* is the direct sum of its subcoalgebras $\gamma C(1)$ where $\gamma \in \overline{\gamma} \in \hat{\Gamma}/<\chi>$.

Also we note that the only elements w in the basis $\mathcal{W}_\gamma$ for $\gamma C(1)$ such that a scalar multiple of $w \otimes w$ is a summand of Δv for some $v \in \mathcal{W}_\gamma$ are those of the form $w = w(\gamma\chi^k, y^jx^l)$, where $y^jx^l \in C_A(\Gamma) \cap \mathcal{Z}$, i.e., $y^jx^l \in \mathcal{Z}$ and commutes with all elements of Γ.

Now let $e(1,1)$ be a matrix coalgebra basis element for a matrix coalgebra in $\gamma C(1)$. Since $e(1,1) \otimes e(1,1)$ is a summand of $\Delta e(1,1)$, then $e(1,1)$ is a linear combination of basis elements $w(\gamma\chi^k, z)$, where $z \in C_A(\Gamma) \cap \mathcal{Z}$. Since terms of the form $\gamma\chi^k \otimes \gamma\chi^l, k \neq l$, do not occur in $\Delta(w)$ for any $w \in \mathcal{W}_\gamma$, then exactly one element of $\hat{\Gamma}$ is a summand of $e(1,1)$, say $\gamma\chi^k$. Then

$$e(1,1) = \gamma\chi^k + \sum_{1 \neq z \in C_A(\Gamma) \cap \mathcal{Z}} a_z w(\gamma\chi^{k_z}, z),$$

where the $a_z \in K$. Also since $\Delta a_z w(\gamma\chi^{k_z}, z)$ contains the summand $\gamma\chi^{k_z} \otimes a_z w(\gamma\chi^{k_z}, z)$ and does not contain the summand $\gamma\chi^k \otimes a_z w(\gamma\chi^{k_z}, z)$ for $k \neq k_z$, we must have $k_z = k$ for all z.

Theorem 3.3 *For $\mathcal{A}$ with lifting matrix* $\mathcal{A} = \begin{bmatrix} 1 & \nu \\ -q^{-m}\nu & 1 \end{bmatrix}$ *as above, let $\gamma \in \hat{\Gamma}$ with $\gamma(g_i^r - 1) \neq 0$ for $i = 1,2$. Then $\gamma C(1) \cong \oplus_{j=1}^{r} \mathcal{M}^c(r, K)$, the sum of r matrix coalgebras of dimension r^2, except possibly when ν belongs to a finite set of nonzero scalars.*

Proof. First we show that the matrix coalgebras in $\gamma C(1)$ have dimension at least r^2. This is essentially the argument from the proof of Theorem 3.6 in [3] and depends on the fact that $\gamma(g^r - 1) \neq 0$.

Suppose $e(1,1)$ as above is a matrix coalgebra basis element. Since $x^r = g^r - 1$ and $\gamma\chi^k(g^r - 1) = \gamma(g^r - 1) \neq 0$, then $\Delta(\gamma\chi^k)$ contains a summand $\xi w(\gamma\chi^k, x^j) \otimes w(\gamma\chi^k\chi^{-j}, x^{r-j})$ for some $\xi \in K^*$, all $1 \leq j \leq r-1$. Thus $\xi' w(\gamma\chi^k\chi^{-j}, x^{r-j})$ is a summand of $e(l,1)$ for some l, some $\xi' \in K^*$. But then $\gamma\chi^{k-j}$ is a summand of $e(l,l)$. Thus the matrix coalgebra contains at least r matrix coalgebra basis elements $e(l,l)$ and so has dimension at least r^2.

Now we find a family of right A-modules arising from matrix coalgebras in $\gamma C(1)$.

Let $E = \oplus_{i=1}^{r} K e_i$ and, with the understanding that subscripts on the e_i are taken modulo r, we define a right A-action on E by

$$\begin{aligned} e_i \leftharpoonup h &= (\gamma\chi^{i-1})(h)e_i; & (4)\\ e_i \leftharpoonup y &= e_{i-m} \text{ for } 1 \leq i \leq r-1 \text{ and } e_r \leftharpoonup y = \gamma(g_2^r - 1)e_{r-m}; & (5)\\ e_i \leftharpoonup x &= \eta_{i-1}e_{i-1} \text{ for } 1 \leq i \leq r, \eta_i \in K^*. & (6) \end{aligned}$$

Clearly $e_i \leftharpoonup y^r = e_i \leftharpoonup (g_2^r - 1)$, and also we have that

$$\begin{aligned} e_i \leftharpoonup (hx - \chi(h)xh) &= ((\gamma\chi^{i-1})(h) - \chi(h)(\gamma\chi^{i-2})(h))e_i \leftharpoonup x = 0;\\ e_i \leftharpoonup (hy - \chi^m(h)yh) &= ((\gamma\chi^{i-1})(h) - \chi^m(h)(\gamma\chi^{i-m-1})(h))e_i \leftharpoonup y = 0. \end{aligned}$$

Furthermore, we require that

$$\prod_{i=1}^{r} \eta_i = \gamma(g^r - 1) \neq 0, \qquad (7)$$

and this condition ensures that $e_i \leftharpoonup x^r = e_i \leftharpoonup (g^r - 1)$. Thus the action respects relations (1) and (2). As well, in order for the action to respect (3), we must have

$$\begin{aligned} \eta_r\gamma(g_2^r - 1) - q^m\eta_{r-m} &= \nu\gamma(g_1g_2 - 1), & (8)\\ \eta_{i-1} - q^m\eta_{i-1-m} &= \nu\gamma\chi^{i-1}(g_1g_2 - 1) \text{ for } 2 \leq i \leq r-1, & (9)\\ \eta_{r-1} - q^m\gamma(g_2^r - 1)\eta_{r-m-1} &= \nu\gamma\chi^{r-1}(g_1g_2 - 1). & (10) \end{aligned}$$

If $\nu = 0$, so that the right hand sides of equations (8),(9) and (10) are zero, then clearly this system of equations is consistent, each η_i, $2 \leq i \leq r$, is a nonzero multiple of η_1 and (7) becomes $\eta_1^r = \zeta$ for $\zeta \in K^*$, an equation with r distinct solutions in K (since K is algebraically closed).

If $\nu \neq 0$, so that $m = -1$, from equations (8) - (10), we obtain

$$\begin{aligned}\eta_r &= (q^{-1}\eta_1 + \nu\gamma(g_1g_2 - 1))/\gamma(g_2^r - 1)\\ \eta_{r-1} &= q^{-1}(q^{-1}\eta_1 + \nu\gamma(g_1g_2 - 1)) + \nu\gamma\chi^{-1}(g_1g_2 - 1)\\ &= q^{-2}\eta_1 + \nu\gamma(q^{-1} + \chi^{-1})(g_1g_2 - 1),\end{aligned}$$

and, in general, for $1 \leq t \leq r - 1$,

$$\eta_{r-t} = q^{-(t+1)}\eta_1 + \nu\gamma(q^{-t} + q^{-t+1}\chi^{-1} + \ldots + \chi^{-t})(g_1g_2 - 1).$$

Then equation (7) is again an equation of degree r in $\eta = \eta_1$ and thus, since K is algebraically closed, has a solution.

Thus E is a right A-module. Also, from the action of Γ on A, we see that any simple submodule of E must be isomorphic to a right A-module $\oplus_j Ke(j,k)$ where $e(i,l)$ are a basis for a matrix coalgebra in $\gamma C(1)$. But we have already shown that such a matrix coalgebra has dimension at least r^2 and so E is a simple right A-module.

Now we show that any two distinct solutions to equation (7) yield nonisomorphic modules. Let η and η' be two different solutions, and E, E' the corresponding modules.

Since Ke_i is the eigenspace for $\gamma\chi^{i-1}(g) = q^{i-1}\gamma(g)$ under the action of g, any isomorphism ϕ from E to E' must map e_i to $\alpha_i e_i'$ for some $\alpha_i \in K^*$. But then for $j < r$, $\phi(e_j \leftharpoonup y) = \phi(e_{j-m}) = \alpha_{j-m}e'_{j-m}$ and $\phi(e_j) \leftharpoonup y = \alpha_j e_j' \leftharpoonup y = \alpha_j e'_{j-m}$. Also $\phi(e_r \leftharpoonup y) = \phi(\gamma(g_2^r - 1)e_{r-m}) = \alpha_{r-m}\gamma(g_2^r - 1)e'_{r-m}$ while $\phi(e_r) \leftharpoonup y = \alpha_r e_r' \leftharpoonup y = \alpha_r\gamma(g_2^r - 1)e'_{r-m}$, so for all i, j, $\alpha_j = \alpha_i = \alpha$. Also $\phi(e_2 \leftharpoonup x) = \eta_1\alpha e_1'$ and $\phi(e_2) \leftharpoonup x = \alpha\eta_1' e_1'$, which is impossible unless $\eta_1 = \eta_1'$.

Finally we prove that for ν different from a finite set of nonzero scalars, equation (7) has r distinct solutions, and so $\gamma C(1)$ is the direct sum of r matrix coalgebras, each of dimension r^2.

Now suppose $\nu \neq 0$. Let $s_t = \gamma(q^{-t} + q^{-t+1}\chi^{-1} + \ldots + \chi^{-t})(g_1g_2 - 1)$, for $0 \leq t \leq r - 1$; note that $s_{r-1} = 0$.

Then writing (7) as a polynomial equation in the variable X, we have

$$\prod_{i=o}^{r-1}(q^i X + \nu s_{r-i-1}) - \gamma(g^r - 1)\gamma(g_2^r - 1) = 0.$$

Then, with $X = \nu Y$, (7) can be written as

$$g(Y) = Y^r + a_{r-1}Y^{r-1} + \ldots + a_1 Y + a_0 = 0,$$

where $a_0 = -\gamma(g^r - 1)\gamma(g_2^r - 1)/\nu^r$ and the scalar ν does not appear in any other coefficient a_i. If $g(Y)$ has a repeated root z then z is also a zero of $g'(Y) = 0$. Note that $g'(Y)$ is a polynomial whose coefficients do not depend upon ν. Let $\mathcal{Z}$ be the finite set of roots of $g'(Y)$. If $z \in \mathcal{Z}$ is also a zero of $g(Y)$ then $\nu^r = -\gamma(g^r - 1)\gamma(g_2^r - 1)/(z^r + a_{r-1}z^{r-1} + \ldots + a_1 z)$. If we choose ν so that this equation does not hold for any $z \in \mathcal{Z}$, then (7) has r distinct zeros. ∎

Proposition 3.4 *For V as above, suppose A is a lifting of $\mathcal{B}(V)\#K[\Gamma]$ with lifting matrix $\mathcal{A} = \begin{bmatrix} 1 & 1 \\ -q & 0 \end{bmatrix}$. Let $\gamma \in \hat{\Gamma}$ with $\gamma(g^r - 1) \neq 0$. Then $\gamma C(1) \cong \oplus_{j=1}^r \mathcal{M}^c(r, K)$, the sum of r matrix coalgebras of dimension r^2.*

Proof. Note that here $m = -1$. Also since $y^r = 0$, we may assume that if ν is nonzero then $\nu = 1$. The proof that any matrix coalgebra in $\gamma C(1)$ has dimension at least r^2 still holds and an A-module E may still be constructed using conditions (4) to (10) with $\gamma(g_2^r - 1)$ replaced by 0. However, here the equations (8) to (10) determine $\eta_1, \ldots, \eta_{r-1}$ and equation (7) is a linear equation in η_r, so that we obtain one simple A-module of dimension r in this way replacing γ in (8) - (10) by each of the r elements of $\gamma < \chi >$. Suppose two of these modules are isomorphic, say $E = \oplus_{i=1}^r Ke_i$, constructed using γ and $F = \oplus_{i=1}^r Kf_i$, constructed using $\gamma\chi^s$ with $s \neq 0$. Under the right action of g, Ke_i and Kf_{i-s} are the eigenspaces for $\gamma\chi^{i-1}(g)$. Thus any isomorphism from E to F must map e_i to $\alpha_i f_{i-s}$ for some nonzero scalar α_i. But $e_1 \leftharpoonup y^{r-1} = e_r$ while $f_{1-s} \leftharpoonup y^{r-1} = 0$. ∎

To describe the coradical of $C(1)$, note that the sub Hopf algebra $C(1)$ of A^* can be described in the following way. Let J be the Hopf ideal in A generated by $g_i^r - 1$ for $i = 1, 2$ and let $A' = A/J$ be the quotient Hopf algebra. In A', $x'^r = 0 = y'^r$ where x', y' are the images of x and y in A'. Then the dual of A' is isomorphic to a sub Hopf algebra of A^* and $C(1)$ is a sub Hopf algebra of A'^*. The structure of the sub Hopf algebra $C(1)$ of A'^* is different depending on whether ν is zero or nonzero.

If $\nu = 0$, then $C(1)$ is a Radford biproduct. Let $\overline{\Gamma}$ be the quotient of Γ by the subgroup generated by g_1^r, g_2^r and it was noted in [3] that $\hat{\overline{\Gamma}} \cong G(A^*) = \{\gamma \in \hat{\Gamma} | \gamma(g_1^r) = 1 = \gamma(g_2^r)\}$. The quotient Hopf algebra A' is isomorphic to the bicrossed product $\mathcal{B}(V)\#K[\overline{\Gamma}]$, so that A'^* is isomorphic to the bicrossed product $\mathcal{B}(W)\#K[G(A^*)]$ where $W = Kw_1 \oplus Kw_2 \in^{\hat{\Gamma}}_{\hat{\Gamma}} \mathcal{YD}$ with $w_i \in W^{g_i}_{\chi_i}$. Since $\chi \in G(A^*)$, then $\mathcal{B}(W)\#K[<\chi>]$ is a sub Hopf algebra of A'^* and this is $C(1)$. Thus $C(1)$ is pointed with coradical $K[<\chi>]$. More details can be found in [3].

If $\nu \neq 0$, then $C(1)$ is a Frobenius Lusztig kernel of dimension r^3. A' is isomorphic to the nontrivial lifting of $\mathcal{B}(V)\#K[\overline{\Gamma}]$ with lifting matrix $\mathcal{A} = \begin{bmatrix} 0 & 1 \\ -q & 0 \end{bmatrix}$. Again, $<\chi>\subseteq \hat{\overline{\Gamma}} = \{\gamma \in \hat{\Gamma} | \gamma(g_1^r) = 1 = \gamma(g_2^r)\}$ but here $G(C(1)) = \{\epsilon\}$ since for $i \neq 0$, $\chi^i(g_1g_2) = q^{2i} \neq 1$ since $(r,2) = 1$.

The representation theory for the Frobenius Lusztig kernels is known; however we make the following remark since the construction in the proof of Theorem 3.3 still is valid here.

Remark 3.5 Suppose that A has lifting matrix $\mathcal{A} = \begin{bmatrix} 1 & \nu \\ -q^{-m}\nu & 1 \end{bmatrix}$ but γ is such that $\gamma(g_i^r - 1) = 0$, for example, we could have $\gamma = \chi$. Then the argument that the simple modules must have dimension greater than or equal to r no longer holds. However, we still may define a right A-module E as in the proof of Theorem 3.3, where equations (4)-(10) hold with $\gamma(g_i^r - 1)$ replaced by 0. Suppose $\gamma \in <\chi>$ and $\nu \neq 0$. The various choices of γ will yield simple submodules of E of dimensions $1, \ldots, r$.
For example, if $\gamma = \epsilon$, then by (8), $\eta_1 = 0$ and then by (10), $\eta_{r-1} = \nu\chi^{r-1}(g_1g_2 - 1) = \nu(q^{-2} - 1)$, then by (9), $\eta_{r-2} = q^{-1}\eta_{r-1} + \nu\chi^{r-2}(g^2 - 1) = \nu(q^{-3} - q^{-1} + q^{-4} - 1)$, and then $\eta_{r-3} = \nu(q^{-4} - q^{-2} + q^{-5} - q^{-1} + q^{-6} - 1)$, etc. Thus $\eta_2, \ldots, \eta_{r-1}$ are nonzero and the submodule $Ke_2 \oplus \ldots \oplus Ke_r$ of E is a simple right $(r-1)$-dimensional A-module.
If $\eta_2 = 0$, and $\eta_3, \ldots, \eta_{r-1}$ are nonzero, then $Ke_3 \oplus \ldots \oplus Ke_r$ is a simple right $(r-2)$-dimensional A-module, etc. An example of this construction where $C(1)$ has dimension 27 and has coradical $\mathcal{M}^c(K,3) \oplus \mathcal{M}^c(K,2) \oplus K\epsilon$ appears in the next section. ∎

4 Examples of dimension 81

In the remainder of this note, we let $\Gamma = C_9 =< c >, \hat{\Gamma} =< c^* >, V = Kv_1 \oplus Kv_2$ with $v_1 \in V_c^{c^{*3}}, v_2 \in V_c^{c^{*6}}$. Let λ denote c^{*3} and let $q = \lambda(c)$, a primitive cube root of unity. Let ν be a nonzero scalar. Let A be the lifting of $\mathcal{B}(V)\#K[\Gamma]$ with lifting matrix $\mathcal{A} = \begin{bmatrix} 1 & \nu \\ -q\nu & \alpha_{2,2} \end{bmatrix}$ and we explicitly determine the matrix coalgebras in each of the subcoalgebras $c^{*i}C(1)$, $i = 1, 2$. Noice that here $G(A^*) = \{\epsilon\}$. D. Stefan has remarked that A cannot satisfy the Chevalley property. Otherwise the coradical of A^* would be a semisimple Hopf algebra of dimension 3^i, which, then, by a theorem of Masuoka, would have a nontrivial grouplike element. The irreducible representations of the sub Hopf algebra $C(1)$ are known [11], but the structure of the coradical of $C(1)$ also follows easily from our computations, and so we describe it as well.

By Theorem 3.3, or Proposition 3.4 the coalgebra $c^{*i}C(1)$, $1 \le i \le 2$, is the sum of three 9-dimensional matrix coalgebras except possibly when ν lies in a finite set of nonzero scalars. We find these coalgebras explicitly in terms of the basis $\mathcal{W}$ of A^*, $\mathcal{W} = \{w(\gamma, z) | \gamma \in \hat{\Gamma}, z \in \mathcal{Z}\}$, $\mathcal{Z} = \{y^i x^j | 0 \le i, j \le 2\}$.

4.1 Lifting matrix $\mathcal{A}$ with two nonzero diagonal elements

In this section, we let A be the lifting of $\mathcal{B}(V)\#K[\Gamma]$ with lifting matrix $\mathcal{A} = \begin{bmatrix} 1 & \nu \\ -q\nu & 1 \end{bmatrix}$.

Let $\gamma \in \hat{\Gamma}$. From the discussion in the previous section, some matrix coalgebra basis element say $e(1,1)$ in $\gamma C(1)$ has the form $e(1,1) = \gamma + aw(\gamma, yx) + bw(\gamma, y^2x^2)$ and noting that $\gamma(c^3 - 1) = \gamma\lambda^i(c^3 - 1)$ since $\lambda(c) = q$, we compute

$$
\begin{aligned}
& \Delta(\gamma + aw(\gamma, yx) + bw(\gamma, y^2x^2)) \qquad (11)\\
= \;& [\gamma \otimes \gamma + \nu\gamma(c^2 - 1)w(\gamma, x) \otimes w(\gamma\lambda^2, y)\\
+ \;& \gamma(c^3 - 1)w(\gamma, x) \otimes w(\gamma\lambda^2, x^2)\\
+ \;& \gamma(c^3 - 1)w(\gamma, y) \otimes w(\gamma\lambda, y^2) + \gamma(c^3 - 1)w(\gamma, x^2) \otimes w(\gamma\lambda, x)\\
+ \;& \nu^2\gamma(q - c^2)\gamma(c^2 - 1)w(\gamma, x^2) \otimes w(\gamma\lambda, y^2)\\
+ \;& \nu\gamma(q - c^2)\gamma(c^3 - 1)w(\gamma, x^2) \otimes w(\gamma\lambda, yx^2)\\
+ \;& q\gamma(c^3 - 1)^2 w(\gamma, yx) \otimes w(\gamma, y^2x^2) + \gamma(c^3 - 1)w(\gamma, y^2) \otimes w(\gamma\lambda^2, y)\\
+ \;& q^2\gamma(c^3 - 1)^2 w(\gamma, yx^2) \otimes w(\gamma\lambda^2, y^2x)\\
+ \;& \nu\gamma(c^3 - 1)\gamma(q - c^2)w(\gamma, y^2x) \otimes w(\gamma\lambda, y^2)\\
+ \;& q^2\gamma(c^3 - 1)^2 w(\gamma, y^2x) \otimes w(\gamma\lambda, yx^2)\\
+ \;& q\gamma(c^3 - 1)^2 w(\gamma, y^2x^2) \otimes w(\gamma, yx)\\
+ \;& q\nu\gamma(c^2q - 1)\gamma(c^3 - 1)^2 w(\gamma, y^2x^2) \otimes w(\gamma, y^2x^2)]\\
\\
+ \;& a[\gamma \otimes w(\gamma, yx) + w(\gamma, yx) \otimes \gamma + q^2 w(\gamma, x) \otimes w(\gamma\lambda^2, y)
\end{aligned}
$$

$$
\begin{aligned}
&+ \quad q\nu\gamma(1-c^2q)w(\gamma,x)\otimes w(\gamma\lambda^2,y^2x) + w(\gamma,y)\otimes w(\gamma\lambda,x)\\
&+ \quad q\nu\gamma(c^2-1)w(\gamma,x^2)\otimes w(\gamma\lambda,y^2) + q\gamma(c^3-1)w(\gamma,x^2)\otimes w(\gamma\lambda,yx^2)\\
&+ \quad \nu\gamma(c^2q^2-1)w(\gamma,yx)\otimes w(\gamma,yx) + \gamma(c^3-1)w(\gamma,y^2)\otimes w(\gamma\lambda^2,y^2x)\\
&+ \quad q\nu\gamma(1-c^2q)w(\gamma,yx^2)\otimes w(\gamma\lambda^2,y) + \gamma(c^3-1)w(\gamma,yx^2)\otimes w(\gamma\lambda^2,x^2)\\
&+ \quad \nu^2\gamma(1-c^2q)\gamma(c^2-q)w(\gamma,yx^2)\otimes w(\gamma\lambda^2,y^2x)\\
&+ \quad q\gamma(c^3-1)w(\gamma,y^2x)\otimes w(\gamma\lambda,y^2)\\
&+ \quad q^2\gamma(c^3-1)^2w(\gamma,y^2x^2)\otimes w(\gamma,y^2x^2)]
\end{aligned}
$$

$$
\begin{aligned}
&+ \quad b[\gamma\otimes w(\gamma,y^2x^2) + w(\gamma,y^2x^2)\otimes\gamma + qw(\gamma,x)\otimes w(\gamma\lambda^2,y^2x)\\
&+ \quad w(\gamma,y)\otimes w(\gamma\lambda,yx^2) + q^2w(\gamma,x^2)\otimes w(\gamma\lambda,y^2)\\
&+ \quad q^2w(\gamma,yx)\otimes w(\gamma,yx)\\
&+ \quad q\nu\gamma(1-c^2)w(\gamma,yx)\otimes w(\gamma,y^2x^2) + w(\gamma,y^2)\otimes w(\gamma\lambda^2,x^2)\\
&+ \quad qw(\gamma,yx^2)\otimes w(\gamma\lambda^2,y) + \nu\gamma(c^2-q)w(\gamma,yx^2)\otimes w(\gamma\lambda^2,y^2x)\\
&+ \quad w(\gamma,y^2x)\otimes w(\gamma\lambda,x) + \nu\gamma(c^2q-1)w(\gamma,y^2x)\otimes w(\gamma\lambda,yx^2)\\
&+ \quad q\nu\gamma(1-c^2)w(\gamma,y^2x^2)\otimes w(\gamma,yx)\\
&+ \quad q\nu^2\gamma(1-c^2)\gamma(c^2q-1)w(\gamma,y^2x^2)\otimes w(\gamma,y^2x^2)].
\end{aligned}
$$

From the first two paragraphs of the proof of Theorem 3.3, we recall that γ is a summand of the matrix coalgebra basis element $e(j,j)$ only when $j=1$. But then since for $i=1,2$, we have $\gamma\otimes aw(\gamma,y^iz^i)$ a summand of $\Delta aw(\gamma,y^iz^i)$, then $aw(\gamma,y^iz^i)$ cannot be a summand of $e(j,1), j\neq 1$. Then no scalar multiple of $w(\gamma,y^ix^i)\otimes w(\gamma,y^ix^i)$ can occur in $e(1,j)\otimes e(j,1)$ unless $j=1$. Thus:

$$
\begin{aligned}
a^2 &= a\nu\gamma(c^2q^2-1)+bq^2; && (12)\\
b^2 &= (q\nu\gamma(c^2q-1)+aq^2)\gamma(c^3-1)^2 + bq\nu^2\gamma(1-c^2)\gamma(c^2q-1), && (13)\\
ab &= q\gamma(c^3-1)^2 + bq\nu\gamma(1-c^2). && (14)
\end{aligned}
$$

Then from equation (12) and equation (14) we have

$$
\begin{aligned}
b &= (a^2-a\nu\gamma(c^2q^2-1))q && (15)\\
b(a-q\nu\gamma(1-c^2)) &= q\gamma(c^3-1)^2. && (16)
\end{aligned}
$$

so that

$$a(a-\nu\gamma(c^2q^2-1))(a-q\nu\gamma(1-c^2)) = \gamma(c^3-1)^2. \qquad (17)$$

Equation (17) is consistent with equation (13) so the system has solutions a and b. If $\gamma(c^3-1)\neq 0$ then both a and b are nonzero. (Note that equation (17) is just equation (7) with $r=3$, $a=\eta_1$, $\chi=\lambda$.)

4.1.1 Case: $\gamma = c^*$ or c^{*2}.

Now let $\gamma = c^{*i}$ with $i = 1$ or 2, let a be a solution to equation (17) and let b be defined by equation (15). Let $e(1,1) = \gamma + aw(\gamma, yx) + bw(\gamma, y^2x^2)$ as above.

Then define

$$e(2,1) = e(1,1) \leftharpoonup y = \gamma(c^3-1)w(\gamma\lambda, y^2) + aw(\gamma\lambda, x) + bw(\gamma\lambda, yx^2) \quad (18)$$

and

$$\begin{aligned} e(3,1) &= e(2,1) \leftharpoonup y = e(1,1) \leftharpoonup y^2 \quad (19) \\ &= \gamma(c^3-1)w(\gamma\lambda^2, y) + a\gamma(c^3-1)w(\gamma\lambda^2, y^2x) + bw(\gamma\lambda^2, x^2). \end{aligned}$$

Then

$$\begin{aligned} e(1,1) \leftharpoonup y^3 &= e(3,1) \leftharpoonup y \\ &= \gamma(c^3-1)\gamma + a\gamma(c^3-1)w(\gamma, yx) + b\gamma(c^3-1)w(\gamma, y^2x^2) \\ &= \gamma(c^3-1)e(1,1) = e(1,1) \leftharpoonup (c^3-1). \end{aligned}$$

Thus $e(i,1) \leftharpoonup y^3 = e(i,1) \leftharpoonup (c^3-1)$ for $1 \le i \le 3$.

Now from equation (11), we see that $\Delta e(1,1) = \sum_{i=1}^3 e(1,j) \otimes e(j,1)$ where

$$\begin{aligned} e(1,2) &= \alpha w(\gamma, x^2) + w(\gamma, y) + \frac{b}{a} w(\gamma, y^2x), \\ e(1,3) &= \beta w(\gamma, x) + w(\gamma, y^2) + a\beta w(\gamma, yx^2) \end{aligned}$$

with $\beta = \frac{\gamma(c^3-1)}{b}$, and $\alpha = \frac{\gamma(c^3-1)}{a}$. Now define

$$\begin{aligned} e(2,2) &= \gamma\lambda + \frac{b}{a} w(\gamma\lambda, yx) + \alpha\gamma(c^3-1)w(\gamma\lambda, y^2x^2), \\ e(2,3) &= w(\gamma\lambda, y) + a\beta w(\gamma\lambda, x^2) + \beta\gamma(c^3-1)w(\gamma\lambda, y^2x), \\ e(3,2) &= \frac{b}{a} w(\gamma\lambda^2, x) + \gamma(c^3-1)w(\gamma\lambda^2, y^2) + \alpha\gamma(c^3-1)w(\gamma\lambda^2, yx^2), \\ e(3,3) &= \gamma\lambda^2 + \beta\gamma(c^3-1)w(\gamma\lambda^2, yx) + a\beta\gamma(c^3-1)w(\gamma\lambda^2, y^2x^2). \end{aligned}$$

It is tedious but straightforward to check that the $e(i,j)$ are a matrix coalgebra basis. The 3-dimensional spaces $V_j = \oplus_{i=1}^3 Ke(i,j)$ are irreducible right A-modules. In general, the right action of $g \in \Gamma$ or y on V_j is given by

$$e(i,j) \leftharpoonup g = (\lambda^{i-1}\gamma)(g)e(i,j);$$

$$e(1,j) \leftharpoonup y = e(2,j),\ e(2,j) \leftharpoonup y = e(3,j),\ e(3,j) \leftharpoonup y = \gamma(c^3-1)e(1,j);$$

We compute the action of x on $e(1,1)$, $e(2,1)$ and $e(3,1)$.

From equation (11) we see that

$$\begin{aligned} e(1,1) \leftharpoonup x &= (\nu\gamma(c^2-1)w(\gamma\lambda^2,y) + \gamma(c^3-1)w(\gamma\lambda^2,x^2)) \\ &+ a(q^2 w(\gamma\lambda^2,y) + q\nu\gamma(1-c^2q)w(\gamma\lambda^2,y^2x)) + bqw(\gamma\lambda^2,y^2x) \\ &= (\nu\gamma(c^2-1) + aq^2)w(\gamma\lambda^2,y) \\ &+ (aq\nu\gamma(1-c^2q) + bq)w(\gamma\lambda^2,y^2x) + \gamma(c^3-1)w(\gamma\lambda^2,x^2). \end{aligned}$$

We see that $e(1,1) \leftharpoonup x = \beta e(3,1)$ if and only if

$$\nu\gamma(c^2-1) + aq^2 = \beta\gamma(c^3-1) = \frac{\gamma(c^3-1)^2}{b} \tag{20}$$

$$\text{and } aq\nu\gamma(1-c^2q) + bq = \beta a\gamma(c^3-1) = \frac{a\gamma(c^3-1)^2}{b}. \tag{21}$$

Now, (20) holds iff

$$\nu\gamma(c^2-1)b + abq^2 = \gamma(c^3-1)^2 \text{ iff}$$
$$\nu\gamma(c^2-1)b + \gamma(c^3-1)^2 + b\nu\gamma(1-c^2) = \gamma(c^3-1)^2 \text{ by equation (14),}$$

and this last equation clearly holds. Also, equation (21) holds if and only if

$$abq\nu\gamma(1-c^2q) + b^2q = a\gamma(c^3-1)^2$$

and using (13), we see that this equation reduces to equation (14). Thus $e(1,1) \leftharpoonup x = \beta e(3,1)$.

Similar computations now show that

$$e(1,1) \leftharpoonup x^2 = \beta e(3,1) \leftharpoonup x = \alpha e(2,1), \text{ and}$$

$$e(1,1) \leftharpoonup x^3 = \beta e(3,1) \leftharpoonup x^2 = \alpha e(2,1) \leftharpoonup x = \alpha a e(1,1) = \gamma(c^3-1)e(1,1),$$

and that, in general, the right action of x on the $e(i,j)$ is given by

$$e(1,j) \leftharpoonup x = \beta e(3,j),\ e(2,j) \leftharpoonup x = ae(1,j),\ e(3,j) \leftharpoonup x = \frac{b}{a}e(2,j).$$

This action preserves the relations (2) and (3) since,

$$\begin{aligned} e(i,j) \leftharpoonup y^3 &= \gamma(c^3-1)e(i,j) = e(i,j) \leftharpoonup (c^3-1); \\ e(i,j) \leftharpoonup x^3 &= \left(\frac{\gamma(c^3-1)}{b}\right)(a)(\frac{b}{a})e(i,j) = e(i,j) \leftharpoonup (c^3-1), \end{aligned}$$

and also denoting $xy - q^2yx$ by z and using (12),(13), (14), we have

$$\begin{aligned} e(1,j) \leftharpoondown z &= \left(\frac{\gamma(c^3-1)^2}{b} - q^2a\right) e(1,j) = \nu\gamma(c^2-1)e(1,j), \\ e(2,j) \leftharpoondown z &= \left(a - q^2\frac{b}{a}\right) e(2,j) = \nu(\gamma\lambda)(c^2-1)e(2,j), \\ e(3,j) \leftharpoondown z &= \left(\frac{b}{a} - q^2\frac{\gamma(c^3-1)^2}{b}\right) e(3,j) = \nu(\gamma\lambda^2)(c^2-1)e(3,j). \end{aligned}$$

We note that similarly, $U_1 = \oplus_{i=1}^3 Ke(1,i)$ is an irreducible left A-module. The left action of Γ is given by $h \rightharpoonup e(1,j) = \gamma\lambda^{j-1}(h)e(1,j)$. The left action of x is given by

$$x \rightharpoonup e(1,1) = ae(1,2);\; x \rightharpoonup e(1,2) = \frac{b}{a}e(1,3);\; x \rightharpoonup e(1,3) = \beta e(1,1),$$

and the left action of y is given by

$$y \rightharpoonup e(1,3) = e(1,2);\; y \rightharpoonup e(1,2) = e(1,1);\; y \rightharpoonup e(1,1) = \gamma(c^3-1)e(1,3).$$

Finally we consider the number of solutions to equation (17),

$$X(X - \nu\gamma(c^2q^2-1))(X - q\nu\gamma(1-c^2)) - \gamma(c^3-1)^2 = 0,$$

or equivalently,

$$Y^3 + \omega Y^2 + \theta Y = \gamma(c^3-1)^2/\nu^3$$

where $X = \nu Y$, $\omega = -\gamma(c^2q^2-1) - q\gamma(1-c^2)$,and $\theta = q\gamma(c^2q^2-1)\gamma(1-c^2)$. Then the solutions to $3Y^2 + 2\omega Y + \theta = 0$ are $\mathcal{Z} = \{(-\omega \pm \sqrt{\omega^2 - 3\theta})/3\}$, and in order to guarantee that $\gamma C(1)$ is $\oplus_{i=1}^3 \mathcal{M}^c(3,K)$, we should choose ν so that $\gamma(c^3-1)^2 \neq \nu^3(z^3 + \omega z^2 + \theta z)$ for either $z \in \mathcal{Z}$, i.e. 6 possible choices for ν are excluded.

Note that if $\gamma = c^*$, then $\omega^2 - 3\theta = -3\zeta^2q^2$ and if $\gamma = c^{*2}$, then $\omega^2 - 3\theta = -3\zeta$ where $\zeta = c^*(c)$ is a primitive 9th root of unity with $\zeta^3 = q$. Thus (17) always has at least 2 distinct solutions.

4.1.2 Case: $\gamma \in <\lambda>$.

Now consider the matrix coalgebras in $C(1)$, as described in Remark 3.5. For $\gamma \in <\lambda>$, $\gamma(c^3-1) = 0$ so that equations (13), (14) are different, and we have

$$\begin{aligned} a^2 &= a\nu\gamma(c^2q^2-1) + bq^2; \\ b^2 &= bq\nu^2\gamma(1-c^2)\gamma(c^2q-1); \\ ab &= b\nu q\gamma(1-c^2). \end{aligned}$$

If $\gamma = \epsilon$, then $b = 0$ and then either $a = 0$ or $a = \nu(q^2 - 1)$. If $\gamma = \lambda$, then $b = 0$ and then either $a = 0$ or $a = \nu(q - 1)$. If $\gamma = \lambda^2$, then $a^2 = bq^2$ and then either $a = b = 0$ or $a = \nu q(1 - q)$ and $b = \nu^2 q(1 - q)(q^2 - 1)$.

Then we obtain the 1 dimensional coalgebra $K \cdot \epsilon$, a 9-dimensional coalgebra E with basis $e(i,j)$, and a 4-dimensional F coalgebra with basis $f(i,j)$ where

$$\begin{aligned}
e(1,1) &= \lambda^2 + \nu q(1-q)w(\lambda^2, yx) + \nu^2(1-q)^2 w(\lambda^2, y^2x^2),\\
e(1,2) &= w(\lambda^2, y) + \nu(q^2-1)w(\lambda^2, y^2x);\\
e(1,3) &= w(\lambda^2, y^2);\\
e(2,1) &= \nu q(1-q)w(\epsilon, x) + \nu^2(1-q)^2 w(\epsilon, yx^2);\\
e(2,2) &= \epsilon + \nu(q^2-1)w(\epsilon, yx),\\
e(2,3) &= w(\epsilon, y);\\
e(3,1) &= \nu^2(1-q)^2 w(\lambda, x^2);\\
e(3,2) &= \nu(q^2-1)w(\lambda, x);\\
e(3,3) &= \lambda,
\end{aligned}$$

and

$$\begin{aligned}
f(1,1) &= \lambda^2,\\
f(1,2) &= \nu(q-1)w(\lambda^2, x);\\
f(2,1) &= w(\lambda, y);\\
f(2,2) &= \lambda + \nu(q-1)w(\lambda, yx).
\end{aligned}$$

4.2 Lifting matrix $\mathcal{A}$ with one nonzero diagonal element

Now let A have lifting matrix $\begin{bmatrix} 1 & \nu \\ -q\nu & 0 \end{bmatrix}$, i.e., $x^3 = c^3 - 1$, $y^3 = 0$ and we are in the situation of Proposition 3.4. Suppose $\gamma \notin <\lambda>$. Then equations (12),(13) and (14) become

$$a^2 = a\nu\gamma(c^2q^2 - 1) + bq^2; \tag{22}$$

$$b^2 = bq\nu^2\gamma(1 - c^2)\gamma(c^2q - 1), \tag{23}$$

$$ab = bq\nu\gamma(1 - c^2). \tag{24}$$

The three solutions to this set of equations are $a = b = 0$, $b = 0$ and $a = \nu\gamma(c^2q - 1)$ or $b = q\nu^2\gamma(1 - c^2)\gamma(c^2q - 1)$ and $a = q\nu\gamma(1 - c^2)$. Let us choose the solution where both a and b are nonzero and then we can construct a matrix coalgebra exactly as in the previous subsection. As before, let $\beta = \frac{\gamma(c^3-1)}{b}$, and $\alpha = \frac{\gamma(c^3-1)}{a}$. Then $Ke(1,1) \oplus Ke(2,1) \oplus Ke(3,1)$ is a simple right A-module and $e(i,j)$ with $1 \leq i,j \leq 3$ is a basis for a matrix coalgebra in $\gamma C(1)$ where

$$\begin{aligned}
e(1,1) &= \gamma + aw(\gamma, yx) + bw(\gamma, y^2x^2),\\
e(1,2) &= \alpha w(\gamma, x^2) + w(\gamma, y) + \frac{b}{a}w(\gamma, y^2x),\\
e(1,3) &= \beta w(\gamma, x) + w(\gamma, y^2) + a\beta w(\gamma, yx^2),\\
e(2,1) &= aw(\gamma\lambda, x) + bw(\gamma\lambda, yx^2),\\
e(2,2) &= \gamma\lambda + \frac{b}{a}w(\gamma\lambda, yx),\\
e(2,3) &= w(\gamma\lambda, y) + a\beta w(\gamma\lambda, x^2),\\
e(3,1) &= bw(\gamma\lambda^2, x^2),\\
e(3,2) &= \frac{b}{a}w(\gamma\lambda^2, x),\\
e(3,3) &= \gamma\lambda^2.
\end{aligned}$$

As in the previous section $e(1,j) \leftharpoonup y = e(2,j)$, $e(2,j) \leftharpoonup y = e(3,j)$ and $e(3,j) \leftharpoonup y = 0$. Had we not chosen a, b nonzero, then the action of y on $e(1,1)$ would not have yielded a 3-dimensional K-vector space.

To obtain the other two matrix coalgebras in $\gamma C(1)$, replace γ by $\gamma\lambda$ or $\gamma\lambda^2$ in equations (22)-(24) to obtain matrix coalgebras generated by elements

$$\begin{aligned}
e'(1,1) &= \gamma\lambda + a'w(\gamma\lambda, yx) + b'w(\gamma\lambda, y^2x^2),\\
e''(1,1) &= \gamma\lambda^2 + a''w(\gamma\lambda^2, yx) + b''w(\gamma\lambda^2, y^2x^2),
\end{aligned}$$

where $a' = q\nu\gamma(1 - c^2q^2)$, $b' = q\nu^2\gamma(1 - c^2q^2)\gamma(c^2 - 1)$, $a'' = q\nu\gamma(1 - c^2q)$, $b'' = q\nu^2\gamma(1 - c^2q)\gamma(c^2q^2 - 1)$.

Acknowledgement

Thanks to the referee for his/her careful reading of this paper and helpful comments.

References

[1] N. Andruskiewitsch and H.-J. Schneider, Finite quantum groups and Cartan matrices, Advances in Mathematics 154(2000), 1-45.

[2] N. Andruskiewitsch and H.-J. Schneider, Lifting of quantum linear spaces and pointed Hopf algebras of order p^3, J. Algebra 209 (1998), 659-691.

[3] M. Beattie, Duals of pointed Hopf algebras, J. Algebra, 262(2003), 54-76.

[4] M. Beattie, S. Dăscălescu, L. Grünenfelder, On the number of types of finite dimensional Hopf algebra, Invent. Math. 136 (1999), 1-7.

[5] M. Beattie, S. Dăscălescu, L. Grünenfelder, Constructing pointed Hopf algebras by Ore extensions, J. Algebra 225, (2000), 743-770.

[6] S. Caenepeel and S. Dăscălescu, Pointed Hopf algebras of dimension p^3, J. Algebra 209, (1998), 622-634.

[7] S. Caenepeel, S. Dăscălescu and S. Raianu, Classifying pointed Hopf algebras of dimension 16, Comm. Alg. 28(2)(2000), 541-568.

[8] S. Dăscălescu, Pointed Hopf algebras with large coradical, Comm. Alg. 27(10) (1999), 4827-4851.

[9] S. Gelaki, Pointed Hopf algebras and Kaplansky's 10th conjecture, J. Algebra 209 (1998), 635-657.

[10] M. Graña, A freeness theorem for Nichols algebras, J. Algebra 231 (2000), 235-257.

[11] J. C. Jantzen, Lectures on Quantum Groups, AMS Graduate Studies in Mathematics, Vol. 6, 1996.

[12] D. E. Radford, On the coradical of a finite-dimensional Hopf algebra, Proc. Amer. Math. Soc. 53 (1975), 8-15.

REPRESENTATIONS OF TWO-PARAMETER QUANTUM GROUPS AND SCHUR-WEYL DUALITY

GEORGIA BENKART[1]
DEPARTMENT OF MATHEMATICS
UNIVERSITY OF WISCONSIN
MADISON, WISCONSIN 53706
benkart@math.wisc.edu

SARAH WITHERSPOON[1]
DEPARTMENT OF MATHEMATICS AND COMPUTER SCIENCE
AMHERST COLLEGE
AMHERST, MASSACHUSETTS 01002
sjw@cs.amherst.edu

ABSTRACT. We determine the finite-dimensional simple modules for two-parameter quantum groups corresponding to the general linear and special linear Lie algebras $\mathfrak{gl}_n$ and $\mathfrak{sl}_n$ and present a complete reducibility result. These quantum groups have a natural n-dimensional module V. We prove an analogue of Schur-Weyl duality in this setting: the centralizer algebra of the quantum group action on the k-fold tensor power of V is a quotient of a Hecke algebra for all n and is isomorphic to the Hecke algebra in case $n \geq k$.

INTRODUCTION

Two-parameter general linear and special linear quantum groups were introduced by Takeuchi [T] in 1990. Our interest in these quantum groups

[1]The authors acknowledge with gratitude support from NSF Grant #DMS–9970119, NSA Grant #MDA904-01-1-0067, and the hospitality of the Mathematical Sciences Research Institute, Berkeley.

arose from our investigations [BW1] of down-up algebras and their embeddings into certain Hopf algebras. These Hopf algebras depend on two parameters r and s, and the Drinfel'd double of such a Hopf algebra is essentially the two-parameter quantum group $U_{r,s}(\mathfrak{sl}_3)$ of Takeuchi (defined below). More generally, as shown in [BW2], $U_{r,s}(\mathfrak{sl}_n)$ is a Drinfel'd double of a Borel-type subalgebra, and there is an R-matrix which comes from the double construction and which reduces to the standard R-matrix for the one-parameter quantum group $U_q(\mathfrak{sl}_n)$ (a quotient of $U_{q,q^{-1}}(\mathfrak{sl}_n)$). In the analogous quantum function algebra setting, allowing two parameters unifies the Drinfel'd-Jimbo quantum groups ($r = q, s = q^{-1}$) [D, Ji1] with the Dipper-Donkin quantum groups ($r = 1, s = q^{-1}$) [DD].

In this work we study the representations of the two-parameter quantum groups $\widetilde{U} = U_{r,s}(\mathfrak{gl}_n)$ and $U = U_{r,s}(\mathfrak{sl}_n)$, defined in Section 1. Our Hopf algebra $\widetilde{U}$ is isomorphic as an algebra to Takeuchi's quantum group $U_{r,s^{-1}}$ (see [T]), but as a Hopf algebra, it has the opposite coproduct. Our main goal is to prove a two-parameter analogue of Schur-Weyl duality, for which we need a result on complete reducibility of modules. To this end, in Sections 2 and 3 we adapt the methods of [Ja] and [L] to classify the finite-dimensional simple $\widetilde{U}$-modules when rs^{-1} is not a root of unity. We use a quantum Casimir operator to prove that all finite-dimensional $\widetilde{U}$-modules on which $\widetilde{U}^0$ (the subalgebra generated by the grouplike elements) acts semisimply are completely reducible. These results hold equally well for U. The hypothesis on $\widetilde{U}^0$ is necessary: In Remark 3.9, we present examples of finite-dimensional modules that are not completely reducible. These examples exist because the two-parameter quantum groups contain more grouplike elements than their one-parameter counterparts.

The construction of the R-matrix, which provides an isomorphism $R_{M',M} : M' \otimes M \to M \otimes M'$ for any two $\widetilde{U}$-modules M, M' in category $\mathcal{O}$ (defined in Section 3), is summarized in Section 4. On the tensor power $V^{\otimes k}$ of the natural module V (which is described in Section 1), the transformations $R_i = \mathrm{Id}^{\otimes(i-1)} \otimes R_{V,V} \otimes \mathrm{Id}^{\otimes(k-i-1)}$ $(1 \leq i < k)$ commute with the action of $\widetilde{U}$, and so they generate a subalgebra of $\mathrm{End}_{\widetilde{U}}(V^{\otimes k})$. This yields a map from the two-parameter Hecke algebra $H_k(r,s)$ to $\mathrm{End}_{\widetilde{U}}(V^{\otimes k})$. In the final section we prove a two-parameter analogue of Schur-Weyl duality: The transformations R_i generate the centralizer algebra $\mathrm{End}_{\widetilde{U}}(V^{\otimes k})$, and in case $n \geq k$, this centralizer algebra is isomorphic to $H_k(r,s)$. The proof of this result is elementary, relying only on basic facts about representations and explicit computations and is new in the one-parameter case as well (compare [DPS, Du, Ji2, KT, LR]). The proof in the $n \geq k$ case is similar to one for classical (nonquantum) Schur-Weyl duality given by De Concini and Procesi [DP]. It is a consequence of our result, Lemma 6.2 below, that $V^{\otimes k}$ is a cyclic $\widetilde{U}$-module in this case.

Throughout we will work over an algebraically closed field $\mathbb{K}$.

§1. Preliminaries

First we recall the definitions of the two-parameter quantum groups from [BW2], and some basics about their representations. Let $\epsilon_1, \ldots, \epsilon_n$ denote an orthonormal basis of a Euclidean space E with an inner product $\langle\ ,\ \rangle$. Let $\Pi = \{\alpha_j = \epsilon_j - \epsilon_{j+1} \mid j = 1, \ldots, n-1\}$ and $\Phi = \{\epsilon_i - \epsilon_j \mid 1 \leq i \neq j \leq n\}$. Then Φ is a finite root system of type A_{n-1} with Π a base of simple roots.

Fix nonzero elements r, s in $\mathbb{K}$ with $r \neq s$.

Let $\widetilde{U} = U_{r,s}(\mathfrak{gl}_n)$ be the unital associative algebra over $\mathbb{K}$ generated by elements $e_j,\ f_j, (1 \leq j < n)$, and $a_i^{\pm 1},\ b_i^{\pm 1}\quad (1 \leq i \leq n)$, which satisfy the following relations.

(R1) The $a_i^{\pm 1},\ b_j^{\pm 1}$ all commute with one another and $a_i a_i^{-1} = b_j b_j^{-1} = 1$,

(R2) $a_i e_j = r^{\langle \epsilon_i, \alpha_j \rangle} e_j a_i$ and $a_i f_j = r^{-\langle \epsilon_i, \alpha_j \rangle} f_j a_i$,

(R3) $b_i e_j = s^{\langle \epsilon_i, \alpha_j \rangle} e_j b_i$ and $b_i f_j = s^{-\langle \epsilon_i, \alpha_j \rangle} f_j b_i$,

(R4) $[e_i, f_j] = \dfrac{\delta_{i,j}}{r-s}(a_i b_{i+1} - a_{i+1} b_i)$,

(R5) $[e_i, e_j] = [f_i, f_j] = 0$ if $|i - j| > 1$,

(R6) $e_i^2 e_{i+1} - (r+s) e_i e_{i+1} e_i + rs e_{i+1} e_i^2 = 0$,
$e_i e_{i+1}^2 - (r+s) e_{i+1} e_i e_{i+1} + rs e_{i+1}^2 e_i = 0$,

(R7) $f_i^2 f_{i+1} - (r^{-1} + s^{-1}) f_i f_{i+1} f_i + r^{-1} s^{-1} f_{i+1} f_i^2 = 0$,
$f_i f_{i+1}^2 - (r^{-1} + s^{-1}) f_{i+1} f_i f_{i+1} + r^{-1} s^{-1} f_{i+1}^2 f_i = 0$.

We will be interested in the subalgebra $U = U_{r,s}(\mathfrak{sl}_n)$ of $\widetilde{U} = U_{r,s}(\mathfrak{gl}_n)$ generated by the elements e_j, f_j, ω_j, and ω_j' $(1 \leq j < n)$, where

$$\omega_j = a_j b_{j+1} \quad \text{and} \quad \omega_j' = a_{j+1} b_j. \tag{1.1}$$

These elements satisfy (R5)-(R7) along with the following relations:

(R1') The $\omega_i^{\pm 1},\ \omega_j'^{\pm 1}$ all commute with one another and
$\omega_i \omega_i^{-1} = \omega_j' (\omega_j')^{-1} = 1$,

(R2') $\omega_i e_j = r^{\langle \epsilon_i, \alpha_j \rangle} s^{\langle \epsilon_{i+1}, \alpha_j \rangle} e_j \omega_i$ and $\omega_i f_j = r^{-\langle \epsilon_i, \alpha_j \rangle} s^{-\langle \epsilon_{i+1}, \alpha_j \rangle} f_j \omega_i$,

(R3') $\omega_i' e_j = r^{\langle \epsilon_{i+1}, \alpha_j \rangle} s^{\langle \epsilon_i, \alpha_j \rangle} e_j \omega_i'$ and $\omega_i' f_j = r^{-\langle \epsilon_{i+1}, \alpha_j \rangle} s^{-\langle \epsilon_i, \alpha_j \rangle} f_j \omega_i'$,

(R4') $[e_i, f_j] = \dfrac{\delta_{i,j}}{r-s}(\omega_i - \omega_i')$.

When $r = q$ and $s = q^{-1}$, the algebra $U_{r,s}(\mathfrak{gl}_n)$ modulo the ideal generated by the elements $b_i - a_i^{-1}$, $1 \leq i \leq n$, is just the quantum general linear group $U_q(\mathfrak{gl}_n)$, and $U_{r,s}(\mathfrak{sl}_n)$ modulo the ideal generated by the elements $\omega_j' - \omega_j^{-1}$, $1 \leq j < n$, is $U_q(\mathfrak{sl}_n)$.

The algebras $\widetilde{U}$ and U are Hopf algebras, where the $a_i^{\pm 1}, b_i^{\pm 1}$ are group-like elements, and the remaining Hopf structure is given by

$$\begin{aligned} &\Delta(e_i) = e_i \otimes 1 + \omega_i \otimes e_i, \qquad \Delta(f_i) = 1 \otimes f_i + f_i \otimes \omega_i', \\ &\varepsilon(e_i) = \varepsilon(f_i) = 0, \quad S(e_i) = -\omega_i^{-1} e_i, \quad S(f_i) = -f_i(\omega_i')^{-1}. \end{aligned} \tag{1.2}$$

Let $\Lambda = \mathbb{Z}\epsilon_1 \oplus \cdots \oplus \mathbb{Z}\epsilon_n$, the weight lattice of $\mathfrak{gl}_n$, and $Q = \mathbb{Z}\Phi$ the root lattice. We assume Λ is equipped with the partial order in which $\nu \leq \lambda$ if and only if $\lambda - \nu \in \sum_{i=1}^{n-1} \mathbb{Z}_{\geq 0}\alpha_i$. Corresponding to $\lambda \in \Lambda$ is an algebra homomorphism $\hat{\lambda}$ from the subalgebra $\widetilde{U}^0$ of $\widetilde{U}$ generated by the elements $a_i^{\pm 1}, b_i^{\pm 1}$ $(1 \leq i \leq n)$ to $\mathbb{K}$ given by

$$\hat{\lambda}(a_i) = r^{\langle \epsilon_i, \lambda \rangle} \quad \text{and} \quad \hat{\lambda}(b_i) = s^{\langle \epsilon_i, \lambda \rangle}. \tag{1.3}$$

The restriction $\hat{\lambda} : U^0 \to \mathbb{K}$ of $\hat{\lambda}$ to the subalgebra U^0 of U generated by $\omega_j^{\pm 1}, (\omega_j')^{\pm 1}$ $(1 \leq j < n)$ satisfies

$$\hat{\lambda}(\omega_j) = r^{\langle \epsilon_j, \lambda \rangle} s^{\langle \epsilon_{j+1}, \lambda \rangle} \quad \text{and} \quad \hat{\lambda}(\omega_j') = r^{\langle \epsilon_{j+1}, \lambda \rangle} s^{\langle \epsilon_j, \lambda \rangle}. \tag{1.4}$$

Similarly for $U = U_{r,s}(\mathfrak{sl}_n)$, we let $\Lambda_{\mathfrak{sl}} = \mathbb{Z}\varpi_1 \oplus \cdots \oplus \mathbb{Z}\varpi_{n-1}$, the weight lattice of $\mathfrak{sl}_n$, where ϖ_i is the fundamental weight

$$\varpi_i = \epsilon_1 + \cdots + \epsilon_i - \frac{i}{n} \sum_{j=1}^{n} \epsilon_j.$$

If we fix nth roots $r^{1/n}$ and $s^{1/n}$ of r and s, respectively, then we may define algebra homomorphisms $\hat{\lambda} : U^0 \to \mathbb{K}$ by (1.4) for any $\lambda \in \Lambda_{\mathfrak{sl}}$.

Let M be a module for $\widetilde{U} = U_{r,s}(\mathfrak{gl}_n)$ of dimension $d < \infty$. As $\mathbb{K}$ is algebraically closed, we have

$$M = \bigoplus_{\chi} M_{\chi},$$

where each $\chi : \widetilde{U}^0 \to \mathbb{K}$ is an algebra homomorphism, and M_{χ} is the generalized eigenspace given by

(1.5) $M_\chi = \{m \in M \mid (a_i - \chi(a_i)\,1)^d m = 0 = (b_i - \chi(b_i)\,1)^d m \text{ for all } i\}.$

When $M_\chi \neq 0$ we say that χ is a *weight* and M_χ is the corresponding *weight space*. (If M decomposes into genuine eigenspaces relative to $\widetilde{U}^0$ (resp. U^0), then we say that $\widetilde{U}^0$ (resp. U^0) *acts semisimply on* M.)

From relations (R2) and (R3) we deduce that

$$\text{(1.6)}\qquad \begin{aligned} e_j M_\chi &\subseteq M_{\chi\cdot\widehat{\alpha_j}} \\ f_j M_\chi &\subseteq M_{\chi\cdot(\widehat{-\alpha_j})}, \end{aligned}$$

where $\widehat{\alpha_j}$ is as in (1.3), and $\chi\cdot\psi$ is the homomorphism with values $(\chi\cdot\psi)(a_i) = \chi(a_i)\psi(a_i)$ and $(\chi\cdot\psi)(b_i) = \chi(b_i)\psi(b_i)$. In fact, if $(a_i - \chi(a_i)\,1)^k m = 0$, then applying relation (R2) yields $(a_i - \chi(a_i) r^{\langle\epsilon_i,\alpha_j\rangle}\,1)^k e_j m = 0$, and similarly for b_i and for f_j. In particular, in the case $k = 1$, this implies that the sum of all eigenspaces of M is a submodule, and if M is simple, this sum must be M itself. Thus in (1.5), we may replace the power d by 1 whenever M is simple, and $\widetilde{U}^0$ must act semisimply in this case. We also can see from (1.6) that for each simple M there is a homomorphism χ so that all the weights of M are of the form $\chi\cdot\hat{\zeta}$, where $\zeta \in Q$.

It is shown in [BW2, Prop. 3.5] that if $\hat{\zeta} = \hat{\eta}$, then $\zeta = \eta$ $(\zeta, \eta \in \Lambda)$ provided rs^{-1} is not a root of unity. As a result, we have the following proposition.

Proposition 1.7. [BW2, Cor. 3.14] *Let M be a finite-dimensional module for $U_{r,s}(\mathfrak{sl}_n)$ or for $U_{r,s}(\mathfrak{gl}_n)$. If rs^{-1} is not a root of unity, then the elements e_i, f_i $(1 \le i < n)$ act nilpotently on M.*

When rs^{-1} is not a root of unity, a finite-dimensional simple module M is a *highest weight* module by Proposition 1.7 and (1.6). Thus there is some weight ψ and a nonzero vector $v_0 \in M_\psi$ such that $e_j v_0 = 0$ for all $j = 1, \dots, n-1$, and $M = \widetilde{U}.v_0$. It follows from the defining relations that $\widetilde{U}$ has a triangular decomposition: $\widetilde{U} = U^-\widetilde{U}^0 U^+$, where U^+ (resp., U^-) is the subalgebra generated by the elements e_i (resp., f_i). Applying this decomposition to v_0, we see that $M = \bigoplus_{\zeta\in Q^+} M_{\psi\cdot(\widehat{-\zeta})}$, where $Q^+ = \sum_{i=1}^{n-1} \mathbb{Z}_{\ge 0}\alpha_i$.

When all the weights of a module M are of the form $\hat{\lambda}$, where $\lambda \in \Lambda$, then for brevity we say that M has weights in Λ. Rather than writing $M_{\hat{\lambda}}$ for the weight space, we simplify the notation by writing M_λ. Then (1.6) can be rewritten as $e_j M_\lambda \subseteq M_{\lambda+\alpha_j}$ and $f_j M_\lambda \subseteq M_{\lambda-\alpha_j}$. Any simple $\widetilde{U}$-module having one weight in Λ has all its weights in Λ.

Next we give an example of a simple $\widetilde{U}$-module with weights in Λ, which is the analogue of the natural representation for $\mathfrak{gl}_n$.

The natural representation for $U_{r,s}(\mathfrak{gl}_n)$ and $U_{r,s}(\mathfrak{sl}_n)$.

Consider an n-dimensional vector space V over $\mathbb{K}$ with basis $\{v_j \mid 1 \leq j \leq n\}$. We define an action of the generators of $\widetilde{U} = U_{r,s}(\mathfrak{gl}_n)$ by specifying their matrices relative to this basis:

$$\begin{aligned} e_j &= E_{j,j+1}, \qquad f_j = E_{j+1,j}, && (1 \leq j < n) \\ a_i &= rE_{i,i} + \sum_{k\neq i} E_{k,k}, && (1 \leq i \leq n) \\ b_i &= sE_{i,i} + \sum_{k\neq i} E_{k,k} && (1 \leq i \leq n). \end{aligned}$$

It follows that $\omega_j = a_j b_{j+1} = rE_{j,j} + sE_{j+1,j+1} + \sum_{k\neq j,j+1} E_{k,k}$ and $\omega'_j = a_{j+1} b_j = sE_{j,j} + rE_{j+1,j+1} + \sum_{k\neq j,j+1} E_{k,k}$. It may be verified that this extends to an action of $\widetilde{U}$ (hence of $U = U_{r,s}(\mathfrak{sl}_n)$); that is, relations (R1)–(R7) hold.

It follows from the fact that $a_i v_j = r^{\langle \epsilon_i,\epsilon_j\rangle} v_j$ and $b_i v_j = s^{\langle \epsilon_i,\epsilon_j\rangle} v_j$ for all i, j that v_j corresponds to the weight $\epsilon_j = \epsilon_1 - (\alpha_1 + \cdots + \alpha_{j-1})$. Thus, $V = \bigoplus_{j=1}^n V_{\epsilon_j}$ is the natural analogue of the n-dimensional representation of $\mathfrak{gl}_n$ and $\mathfrak{sl}_n$, and it is a simple module for both $\widetilde{U}$ and U. When $r = q$ and $s = q^{-1}$, b_i acts as a_i^{-1} on V, and so V is a module for the quotient $U_q(\mathfrak{gl}_n)$ of $U_{q,q^{-1}}(\mathfrak{gl}_n)$ by the ideal generated by $b_i - a_i^{-1}$ $(1 \leq i \leq n)$. This is the natural module for the one-parameter quantum group $U_q(\mathfrak{gl}_n)$, and an analogous statement is true for $U_q(\mathfrak{sl}_n)$.

§2. Classification of finite-dimensional simple modules

Often results will be stated only for $\widetilde{U}$-modules, but generally everything holds as well for U-modules. We will indicate where there are differences in the theory.

Let $\widetilde{U}^{\geq 0}$ denote the subalgebra of $\widetilde{U}$ generated by a_i, b_i $(1 \leq i \leq n)$ and e_i $(1 \leq i < n)$. Let ψ be any algebra homomorphism from $\widetilde{U}^0$ to $\mathbb{K}$ and V^ψ be the one-dimensional $\widetilde{U}^{\geq 0}$-module on which e_i acts as multiplication by 0 $(1 \leq i < n)$, and $\widetilde{U}^0$ acts via ψ. We define the *Verma module* $M(\psi)$ with highest weight ψ to be the $\widetilde{U}$-module induced from V^ψ, that is

$$M(\psi) = \widetilde{U} \otimes_{\widetilde{U}^{\geq 0}} V^\psi.$$

Let $v_\psi = 1 \otimes v \in M(\psi)$, where v is any nonzero vector of V^ψ. Then $e_i.v_\psi = 0$ $(1 \le i < n)$ and $a.v_\psi = \psi(a)v_\psi$ for any $a \in \widetilde{U}^0$ by construction.

Notice that $\widetilde{U}^0$ acts semisimply on $M(\psi)$ by relations (R2) and (R3). If N is a $\widetilde{U}$-submodule of $M(\psi)$, then N is also a $\widetilde{U}^0$-submodule of the $\widetilde{U}^0$-module $M(\psi)$, and so $\widetilde{U}^0$ acts semisimply on N as well. If N is a *proper* submodule, it must be that $N \subset \sum_{\mu \in Q^+ \setminus \{0\}} M(\psi)_{\psi \cdot (\widehat{-\mu})}$ by (1.6), as $M(\psi)_\psi = \mathbb{K} v_\psi$ generates $M(\psi)$. Therefore $M(\psi)$ has a unique maximal submodule, namely the sum of all proper submodules, and a unique simple quotient, $L(\psi)$. In fact, all finite-dimensional simple $\widetilde{U}$-modules are of this form, as the following theorem demonstrates.

Theorem 2.1. *Let $\psi : \widetilde{U}^0 \to \mathbb{K}$ be an algebra homomorphism. Let M be a $\widetilde{U}$-module, on which $\widetilde{U}^0$ acts semisimply and which contains a nonzero element $m \in M_\psi$ such that $e_i.m = 0$ for all i $(1 \le i < n)$. Then there is a unique homomorphism of $\widetilde{U}$-modules $F : M(\psi) \to M$ with $F(v_\psi) = m$. In particular, if rs^{-1} is not a root of unity and M is a finite-dimensional simple $\widetilde{U}$-module, then $M \cong L(\psi)$ for some weight ψ.*

Proof. By the hypothesis on m, $\mathbb{K}m$ is a one-dimensional $\widetilde{U}^{\ge 0}$-submodule of M, considered as a $\widetilde{U}^{\ge 0}$-module by restriction. In fact, mapping v_ψ to m yields a $\widetilde{U}^{\ge 0}$-homomorphism from V^ψ to $\mathbb{K}m$. By the definition of $M(\psi)$, we have $\mathrm{Hom}_{\widetilde{U}}(M(\psi), M) \cong \mathrm{Hom}_{\widetilde{U}^{\ge 0}}(V^\psi, M)$, so there is a unique $\widetilde{U}$-module homomorphism $F : M(\psi) \to M$ with $F(v_\psi) = m$, namely $F(u \otimes v) = u.m$ for all $u \in \widetilde{U}$.

For the final assertion, note that $\widetilde{U}^0$ acts semisimply on any finite-dimensional simple module M, and by (1.6) and Proposition 1.7, there is some nonzero vector $m \in M_\psi$ such that $e_i.m = 0$ $(1 \le i < n)$. By the first part, M is a quotient of $M(\psi)$, and so $M \cong L(\psi)$, as $L(\psi)$ is the unique simple quotient of $M(\psi)$. □

As a special case, we will consider the modules $L(\lambda) = L(\hat{\lambda})$ where $\lambda \in \Lambda$. Let $\Lambda^+ \subset \Lambda$ be the subset of *dominant* weights, that is

$$\Lambda^+ = \{\lambda \in \Lambda \mid \langle \alpha_i, \lambda \rangle \ge 0 \text{ for } 1 \le i < n\}.$$

Similarly, the set of dominant weights for $\mathfrak{sl}_n$ is

$$\Lambda^+_{\mathfrak{sl}} = \{\lambda \in \Lambda_{\mathfrak{sl}} \mid \langle \alpha_i, \lambda \rangle \ge 0 \text{ for } 1 \le i < n\} = \left\{ \sum_{i=1}^{n-1} \ell_i \varpi_i \,\middle|\, \ell_i \in \mathbb{Z}_{\ge 0} \right\}.$$

We will show that if $L(\lambda)$ is finite-dimensional, then λ is dominant. This requires an identity for commuting e_i past powers of f_i. For $k \geq 1$, let

$$[k] = \frac{r^k - s^k}{r - s}. \tag{2.2}$$

Then the following lemma may be proven by induction.

Lemma 2.3. *If $k \geq 1$, then*

$$e_i f_i^k = f_i^k e_i + [k] f_i^{k-1} \frac{r^{1-k}\omega_i - s^{1-k}\omega_i'}{r - s}$$
$$e_i^k f_i = f_i e_i^k + [k] e_i^{k-1} \frac{s^{1-k}\omega_i - r^{1-k}\omega_i'}{r - s}.$$

Lemma 2.4. *Assume rs^{-1} is not a root of unity. Let M be a nonzero finite-dimensional module for $\widetilde{U} = U_{r,s}(\mathfrak{gl}_n)$ on which $\widetilde{U}^0$ acts semisimply, and let $\lambda \in \Lambda$. Suppose there is some nonzero vector $v \in M_\lambda$ with $e_i.v = 0$ for all i $(1 \leq i < n)$. Then $\lambda \in \Lambda^+$. A similar statement is true for $U = U_{r,s}(\mathfrak{sl}_n)$ with Λ replaced by $\Lambda_{\mathfrak{sl}}$ and Λ^+ by $\Lambda_{\mathfrak{sl}}^+$.*

Proof. Proposition 1.7 implies that for any given value of i there is some $k \geq 0$ such that $f_i^{k+1}.v = 0$ and $f_i^k.v \neq 0$. Applying e_i to $f_i^{k+1}.v = 0$ and using Lemma 2.3 and the fact that $e_i.v = 0$, we have

$$0 = [k+1] f_i^k \frac{r^{-k}\omega_i - s^{-k}\omega_i'}{r - s}.v = \frac{[k+1]}{r - s}(r^{-k}\hat{\lambda}(\omega_i) - s^{-k}\hat{\lambda}(\omega_i')) f_i^k.v.$$

Now $[k+1]/(r-s) \neq 0$ as rs^{-1} is not a root of unity. Therefore, since $f_i^k.v \neq 0$, we have $r^{-k}\hat{\lambda}(\omega_i) = s^{-k}\hat{\lambda}(\omega_i')$. Equivalently,

$$r^{-k} r^{\langle \epsilon_i, \lambda \rangle} s^{\langle \epsilon_{i+1}, \lambda \rangle} = s^{-k} r^{\langle \epsilon_{i+1}, \lambda \rangle} s^{\langle \epsilon_i, \lambda \rangle}, \quad \text{or} \quad r^{-k + \langle \alpha_i, \lambda \rangle} = s^{-k + \langle \alpha_i, \lambda \rangle}.$$

Again, because rs^{-1} is not a root of unity, this forces $\langle \alpha_i, \lambda \rangle = k \geq 0$, so $\lambda \in \Lambda^+$. □

Corollary 2.5. *When rs^{-1} is not a root of unity, any finite-dimensional simple $\widetilde{U}$-module with weights in Λ is isomorphic to $L(\lambda)$ for some $\lambda \in \Lambda^+$. An analogous result holds for U with Λ replaced by $\Lambda_{\mathfrak{sl}}$ and Λ^+ by $\Lambda_{\mathfrak{sl}}^+$.*

Next we will show that all modules $L(\lambda)$ with $\lambda \in \Lambda^+$ are indeed finite-dimensional, and that all other finite-dimensional simple $\widetilde{U}$-modules are shifts of these by one-dimensional modules. In doing this, it helps to consider first the special case of simple $U_{r,s}(\mathfrak{sl}_2)$-modules.

Highest weight modules for $U = U_{r,s}(\mathfrak{sl}_2)$.

For simplicity we drop the subscripts and just write e, f, ω, ω' for the generators of $U = U_{r,s}(\mathfrak{sl}_2)$. Any homomorphism $\phi : U^0 \to \mathbb{K}$ is determined by its values on ω and ω'. By abuse of notation, we adopt the shorthand $\phi = \phi(\omega)$ and $\phi' = \phi(\omega')$.

Corresponding to each such ϕ, there is a Verma module $M(\phi) = U \otimes_{U^{\geq 0}} \mathbb{K}v$ with basis $v_j = f^j \otimes v \quad (0 \leq j < \infty)$ such that the U-action is given by

$$
\begin{aligned}
&f.v_j = v_{j+1} \\
&e.v_j = [j]\frac{\phi r^{-j+1} - \phi' s^{-j+1}}{r-s} v_{j-1} \qquad (v_{-1} := 0) \\
&\omega.v_j = \phi r^{-j\langle \epsilon_1, \alpha_1\rangle} s^{-j\langle \epsilon_2, \alpha_1\rangle} v_j = \phi r^{-j} s^j v_j \\
&\omega'.v_j = \phi' r^{-j\langle \epsilon_2, \alpha_1\rangle} s^{-j\langle \epsilon_1, \alpha_1\rangle} v_j = \phi' r^j s^{-j} v_j.
\end{aligned} \tag{2.6}
$$

Note that $M(\phi)$ is a simple U-module if and only if $[j]\frac{\phi r^{-j+1} - \phi' s^{-j+1}}{r-s} \neq 0$ for any $j \geq 1$.

Suppose $[\ell+1]\frac{\phi r^{-\ell} - \phi' s^{-\ell}}{r-s} = 0$ for some $\ell \geq 0$. Then either $r^{\ell+1} = s^{\ell+1}$, which implies rs^{-1} is a root of unity, or $\phi' = \phi r^{-\ell} s^{\ell}$. Assuming that rs^{-1} is not a root of unity and $\phi' = \phi r^{-\ell} s^{\ell}$, we see that the elements v_i, $i \geq \ell+1$, span a maximal submodule. The quotient is the $(\ell+1)$-dimensional simple module $L(\phi)$, which we can suppose is spanned by $v_0, v_1, \ldots, v_\ell$ and has U-action given by

$$
\begin{aligned}
&f.v_j = v_{j+1}, \qquad (v_{\ell+1} = 0) \\
&e.v_j = \phi r^{-\ell} [j]\,[\ell+1-j]\, v_{j-1} \qquad (v_{-1} = 0) \\
&\omega.v_j = \phi r^{-j} s^j v_j \\
&\omega'.v_j = \phi r^{-\ell+j} s^{\ell-j} v_j.
\end{aligned} \tag{2.7}
$$

When $M(\phi)$ is not simple and rs^{-1} is not a root of unity, $j = \ell+1$ is the unique value such that $[j]\frac{\phi r^{-j+1} - \phi' s^{-j+1}}{r-s} = 0$. In this case, $M(\phi)$ has a unique proper submodule, namely the maximal submodule generated by $v_{\ell+1}$ as above.

We now have the following classification of simple modules for $U_{r,s}(\mathfrak{sl}_2)$.

Proposition 2.8.

(i) *Assume* $U = U_{r,s}(\mathfrak{sl}_2)$, *where* rs^{-1} *is not a root of unity. Let* $\phi : U^0 \to \mathbb{K}$ *be an algebra homomorphism such that* $\phi(\omega') = \phi(\omega)r^{-\ell}s^{\ell}$ *for some* $\ell \geq 0$. *Then there is an* $(\ell+1)$*-dimensional simple* U*-module* $L(\phi)$ *spanned by vectors* $v_0, v_1, \dots, v_\ell$ *and having* U*-action given by (2.7). Any* $(\ell+1)$*-dimensional simple* U*-module is isomorphic to some such* $L(\phi)$.

(ii) *If* $\nu = \nu_1\epsilon_1 + \nu_2\epsilon_2 \in \Lambda^+$, *then* $\nu_1 - \nu_2 = \ell$ *for some* $\ell \in \mathbb{Z}_{\geq 0}$, *and* $\nu(\omega') = r^{\nu_2}s^{\nu_1} = r^{\nu_1-\ell}s^{\nu_2+\ell} = \nu(\omega)r^{-\ell}s^{\ell}$ *in this case. Thus, the module* $L(\nu)$ *is* $(\ell+1)$*-dimensional and has* U*-action given by (2.7) with* $\phi = r^{\nu_1}s^{\nu_2} = r^{\nu_1}s^{\nu_1-\ell}$.

Finite-dimensionality of $L(\lambda)$ for $\lambda \in \Lambda^+$.

We show below that the simple modules $L(\lambda)$ for $\widetilde{U} = U_{r,s}(\mathfrak{gl}_n)$ with $\lambda \in \Lambda^+$ are finite-dimensional. For this it suffices to prove that $M(\lambda)$ has a $\widetilde{U}$-submodule of finite codimension, as $L(\lambda)$ is the quotient of $M(\lambda)$ by its unique maximal submodule.

As λ is dominant, $k_i = \langle \alpha_i, \lambda \rangle$ for $i = 1, \dots, n-1$, are nonnegative integers. Define a $\widetilde{U}$-submodule $M'(\lambda)$ of $M(\lambda)$ by

$$M'(\lambda) = \sum_{i=1}^{n-1} \widetilde{U} f_i^{k_i+1}.v_\lambda. \tag{2.9}$$

Our goal is to prove that the module $L'(\lambda) := M(\lambda)/M'(\lambda)$ is nonzero and finite-dimensional.

By Lemma 2.3 we have $e_i f_i^{k_i+1}.v_\lambda = 0$. If $j \neq i$, $e_j f_i^{k_i+1}.v_\lambda = f_i^{k_i+1}e_j.v_\lambda = 0$ by the defining relations. Consequently, by Theorem 2.1, $\widetilde{U} f_i^{k_i+1}.v_\lambda$ is a homomorphic image of $M(\lambda - (k_i+1)\alpha_i)$, and so all its weights are less than or equal to $\lambda - (k_i+1)\alpha_i$. This implies that $v_\lambda \notin M'(\lambda)$, hence $L'(\lambda) \neq 0$.

Lemma 2.10. *The elements* e_j, f_j $(1 \leq j < n)$ *act locally nilpotently on* $L'(\lambda)$.

Proof. As the Verma module $M(\lambda)$ is spanned over $\mathbb{K}$ by all elements $x_1 \cdots x_t.v_\lambda$ where $x_1, \dots, x_t \in \{f_1, \dots, f_{n-1}\}$, $t \in \mathbb{Z}_{\geq 0}$, it is enough to argue by induction on t that a sufficiently high power of e_j (resp., f_j) takes such an element to $M'(\lambda)$. If $t = 0$, then $e_j.v_\lambda = 0 \in M'(\lambda)$, and $f_j^{k_j+1}.v_\lambda \in M'(\lambda)$ by construction. Now assume that there are positive integers N_j such that

$$e_j^{N_j} x_2 \cdots x_t.v_\lambda \in M'(\lambda) \quad \text{and} \quad f_j^{N_j} x_2 \cdots x_t.v_\lambda \in M'(\lambda).$$

Suppose that $x_1 = f_i$. If $j \neq i$, then $e_j^{N_j} x_1 \cdots x_t.v_\lambda = f_i e_j^{N_j} x_2 \cdots x_t.v_\lambda \in M'(\lambda)$. Otherwise by Lemma 2.3,

$$e_i^{N_i+1} x_1 \cdots x_t.v_\lambda = f_i e_i^{N_i+1} x_2 \cdots x_t.v_\lambda + [N_i+1] e_i^{N_i} \frac{s^{-N_i}\omega_i - r^{-N_i}\omega_i'}{r-s} x_2 \cdots x_t.v_\lambda.$$

Applying relation (R2') and the induction hypothesis, we see that these terms are both in $M'(\lambda)$.

Now $f_i^{N_i-1} x_1 \cdots x_t.v_\lambda = f_i^{N_i} x_2 \cdots x_t.v_\lambda \in M'(\lambda)$, and if $|i-j| > 1$, we also have $f_j^{N_j} x_1 \cdots x_t.v_\lambda = f_i f_j^{N_j} x_2 \cdots x_t.v_\lambda \in M'(\lambda)$. Finally, we need to show that if $|i-j| = 1$, then $f_j^{N_j+1} x_1 \cdots x_t.v_\lambda \in M'(\lambda)$. This will follow from the induction hypothesis once we know that $f_j^{N_j+1} f_i \in \mathbb{K} f_j f_i f_j^{N_j} + \mathbb{K} f_i f_j^{N_j+1}$.

We argue by induction on $m \geq 1$ that

$$f_j^{m+1} f_i \in \mathbb{K} f_j f_i f_j^m + \mathbb{K} f_i f_j^{m+1}.$$

Indeed if $m = 1$, this follows from relation (R7); but if $m > 1$, then by induction and (R7),

$$f_j^{m+1} f_i \in f_j(\mathbb{K} f_j f_i f_j^{m-1} + \mathbb{K} f_i f_j^m) \subseteq \mathbb{K} f_j f_i f_j^m + \mathbb{K} f_i f_j^{m+1}. \quad \square$$

Lemma 2.11. *Assume rs^{-1} is not a root of unity, and let V be a module for $U = U_{r,s}(\mathfrak{sl}_2)$ on which U^0 acts semisimply. Suppose $V = \bigoplus_{j \in \mathbb{Z}_{\geq 0}} V_{\lambda - j\alpha}$ for some weight $\lambda \in \Lambda$; each weight space of V is finite-dimensional; and e and f act locally nilpotently on V. Then V is finite-dimensional, and the weights of V are preserved under the simple reflection taking α to $-\alpha$.*

Proof. Let $\mu = \mu_1\epsilon_1 + \mu_2\epsilon_2$ be a weight of V, and $v \in V_\mu \setminus \{0\}$. As e acts locally nilpotently on V, there is a nonnegative integer k such that $e^{k+1}.v = 0$ and $e^k.v \neq 0$. By Theorem 2.1, $Ue^k.v$ is a homomorphic image of $M(\mu+k\alpha)$. But since f acts locally nilpotently on $Ue^k.v$, this image cannot be isomorphic to $M(\mu+k\alpha)$. Thus because $M(\mu+k\alpha)$ has a unique proper submodule, $Ue^k.v \cong L(\mu+k\alpha)$, and so it is finite-dimensional. Corollary 2.5 implies that $\mu + k\alpha$ is dominant. As there are only finitely many dominant weights less than or equal to the given weight λ (under the partial order $\nu \leq \lambda$ if and only if $\lambda - \nu \in \mathbb{Z}_{\geq 0}\alpha$), and each weight space is finite-dimensional, it must be that V itself is finite-dimensional.

In particular, V has a composition series with factors isomorphic to $L(\nu)$ for some $\nu \in \Lambda^+$. Any weight μ of V is a weight of some such $L(\nu)$ with

$\nu = \nu_1\epsilon_1 + \nu_2\epsilon_2 \in \Lambda^+$. By (ii) of Proposition 2.8, $L(\nu)$ has weights $\nu, \nu - \alpha, \ldots, \nu - \ell\alpha$ where $\ell = \nu_1 - \nu_2$. Thus, $\mu = \nu - j\alpha$ for some $j \in \{0, 1, \ldots, \ell\}$. But then $\mu - \langle \mu, \alpha \rangle \alpha = \nu - (\ell - j)\alpha$ is a weight of $L(\nu)$ since $\ell - j \in \{0, 1, \ldots, \ell\}$, hence it is a weight of V. Thus, the weights of V are preserved under the simple reflection taking α to $-\alpha$. □

Lemma 2.12. *Assume rs^{-1} is not a root of unity, and let $\lambda \in \Lambda^+$. Then the module $L(\lambda)$ for $\widetilde{U} = U_{r,s}(\mathfrak{gl}_n)$ is finite-dimensional. A similar statement holds for $U = U_{r,s}(\mathfrak{sl}_n)$, where Λ^+ is replaced by $\Lambda^+_{\mathfrak{sl}}$.*

Proof. This follows once we show that $L'(\lambda) = M(\lambda)/M'(\lambda)$, where $M'(\lambda)$ is as in (2.9), is finite-dimensional. We will prove that the set of weights of $L'(\lambda)$ is preserved under the action of the symmetric group S_n (the Weyl group of $\mathfrak{gl}_n$) on Λ which is generated by the simple reflections $s_i : \mu \to \mu - \langle \mu, \alpha_i \rangle \alpha_i$ $(1 \le i < n)$. Each S_n-orbit contains a dominant weight, and there are only finitely many dominant weights less than or equal to λ. As the weights in $M(\lambda)$ are all less than or equal to λ, and the weight spaces are finite-dimensional, the same is true of $L'(\lambda)$. Therefore $L'(\lambda)$ is finite-dimensional.

To see that s_i preserves the set of weights of $L'(\lambda)$, let $\mu = \mu_1\epsilon_1 + \cdots + \mu_n\epsilon_n$ be a weight of $L'(\lambda)$. Consider $L'(\lambda)$ as a module for the copy U_i of $U_{r,s}(\mathfrak{sl}_2)$ generated by $e_i, f_i, \omega_i, \omega_i'$, and let $L_i'(\mu)$ be the U_i-submodule of $L'(\lambda)$ generated by $L'(\lambda)_\mu$. As all weights of $L'(\lambda)$ are less than or equal to λ, we have

$$L_i'(\mu) = \bigoplus_{j \in \mathbb{Z}_{\ge 0}} L_i'(\mu)_{\lambda' - j\alpha_i}$$

for some weight $\lambda' \le \lambda$. By Lemmas 2.10 and 2.11, the simple reflection s_i preserves the weights of $L_i'(\mu)$, so in particular, $s_i(\mu)$ is also a weight of $L'(\lambda)$. □

Remark 2.13. It will follow from Lemma 3.7 in the next section that $L(\lambda) \cong L'(\lambda)$, since $L(\lambda)$ is the unique simple quotient of $M(\lambda)$, $L'(\lambda)$ is a finite-dimensional quotient of $M(\lambda)$, and by that lemma, every finite-dimensional quotient is simple. Thus, we have

Corollary 2.14. *When rs^{-1} is not a root of unity, the finite-dimensional simple $\widetilde{U}$-modules having weights in Λ are precisely the modules $L(\lambda)$ where $\lambda \in \Lambda^+$. Moreover, $L(\lambda) \cong L(\mu)$ if and only if $\lambda = \mu$. Similar statements hold for $U = U_{r,s}(\mathfrak{sl}_n)$, where Λ is replaced by $\Lambda_{\mathfrak{sl}}$ and Λ^+ by $\Lambda^+_{\mathfrak{sl}}$.*

Proof. The first statement is a consequence of Corollary 2.5 and Lemma 2.12 (see also Remark 2.13). Assume there is an isomorphism of $\widetilde{U}$-modules from $L(\lambda)$ to $L(\mu)$. The highest weight vector of $L(\lambda)$ must be sent to a weight vector of $L(\mu)$, so $\lambda \leq \mu$. As a similar argument shows that $\mu \leq \lambda$, we have $\lambda = \mu$. □

Shifts by one-dimensional modules.

Suppose now that we have a one-dimensional module L for $\widetilde{U} = U_{r,s}(\mathfrak{gl}_n)$. Then by Theorem 2.1, $L = L(\chi)$ for some algebra homomorphism $\chi : \widetilde{U}^0 \to \mathbb{K}$, with the elements e_i, f_i $(1 \leq i < n)$ acting as multiplication by 0. Relation (R4) yields

$$\text{(2.15)} \qquad \chi(\omega_i) = \chi(a_i b_{i+1}) = \chi(a_{i+1} b_i) = \chi(\omega_i') \qquad (1 \leq i < n).$$

Conversely, if an algebra homomorphism χ satisfies this equation, then $L(\chi)$ is one-dimensional by relation (R4). We will write $L_\chi = L(\chi)$ to emphasize that the module is one-dimensional.

Proposition 2.16. *Assume rs^{-1} is not a root of unity and $L(\psi)$ is the finite-dimensional simple module for $\widetilde{U} = U_{r,s}(\mathfrak{gl}_n)$ with highest weight ψ. Then there exists a homomorphism $\chi : \widetilde{U}^0 \to \mathbb{K}$ such that (2.15) holds and an element $\lambda \in \Lambda^+$ so that $\psi = \chi \cdot \hat{\lambda}$. Thus, the weights of $L(\psi)$ belong to $\chi \cdot \hat{\Lambda}$. A similar statement holds for $U = U_{r,s}(\mathfrak{sl}_n)$ with Λ replaced by $\Lambda_{\mathfrak{sl}}$ and Λ^+ by $\Lambda^+_{\mathfrak{sl}}$.*

Proof. When $L(\psi)$ is viewed as a module for the copy U_i of $U_{r,s}(\mathfrak{sl}_2)$ generated by $e_i, f_i, \omega_i, \omega_i'$, it has a composition series whose factors are simple U_i-modules as described by Proposition 2.8. As the highest weight vector of $L(\psi)$ gives a highest weight vector of some composition factor, there is a weight ϕ_i of U_i and a nonnegative integer ℓ_i so that $\psi(\omega_i) = \phi_i(\omega_i)$ and $\psi(\omega_i') = \phi_i(\omega_i') = \phi_i(\omega_i) r^{-\ell_i} s^{\ell_i} = \psi(\omega_i) r^{-\ell_i} s^{\ell_i}$.

For the case $\widetilde{U} = U_{r,s}(\mathfrak{gl}_n)$, set $\ell_n = 0$ and define $\lambda_i = \ell_i + \cdots + \ell_n$ for $i = 1, \ldots, n$. Let $\lambda = \sum_{i=1}^{n} \lambda_i \epsilon_i$, which belongs to Λ^+. Now define $\chi : \widetilde{U}^0 \to \mathbb{K}$ by the formulas

$$\chi(a_i) = \psi(a_i) r^{-\langle \epsilon_i, \lambda \rangle} = \psi(a_i) r^{-(\ell_i + \cdots + \ell_n)}$$
$$\chi(b_i) = \psi(b_i) s^{-\langle \epsilon_i, \lambda \rangle} = \psi(b_i) s^{-(\ell_i + \cdots + \ell_n)}.$$

Then it follows that $\chi(\omega_i') = \chi(\omega_i)$ for $i = 1, \ldots, n-1$, and $\psi = \chi \cdot \hat{\lambda}$ as desired.

For the case $U = U_{r,s}(\mathfrak{sl}_n)$, fix an nth root $(rs)^{1/n}$ of rs in $\mathbb{K}$. As above, there exist nonnegative integers ℓ_i $(1 \leq i < n)$ so that $\psi(\omega_i') = \psi(\omega_i) r^{-\ell_i} s^{\ell_i}$. Let $\lambda = \sum_{i=1}^{n-1} \ell_i \varpi_i \in \Lambda^+_{\mathfrak{sl}}$. Define $\chi : U^0 \to \mathbb{K}$ by

$$\chi(\omega_i) = \psi(\omega_i) r^{-(\ell_i + \cdots + \ell_{n-1})} s^{-(\ell_{i+1} + \cdots + \ell_{n-1})} (rs)^{(c/n)}$$
$$\chi(\omega_i') = \psi(\omega_i') r^{-(\ell_{i+1} + \cdots + \ell_{n-1})} s^{-(\ell_i + \cdots + \ell_{n-1})} (rs)^{(c/n)}$$

where $c = \sum_{j=1}^{n-1} j\ell_j$. Then (2.15) holds, and $\psi = \chi \cdot \hat{\lambda}$. □

Remark 2.17. If M is any finite-dimensional module, then $M = \bigoplus_{i=1}^{m} \bigoplus_{\lambda\in\Lambda} M_{\psi_i\cdot\hat{\lambda}}$ for some weights ψ_i such that $\psi_i \cdot \hat{\Lambda}$ $(1 \le i \le m)$ are distinct cosets in $\mathrm{Hom}(\widetilde{U}^0, \mathbb{K})/\hat{\Lambda}$ (viewed as a $\mathbb{Z}$-module under the action $k \cdot \psi = \psi^k$). Then $M_i := \bigoplus_{\lambda\in\Lambda} M_{\psi_i\cdot\hat{\lambda}}$ is a submodule, and $M = \bigoplus_{i=1}^{m} M_i$. Therefore, if M is an indecomposable $\widetilde{U}$-module, $M = \bigoplus_{\lambda\in\Lambda} M_{\psi\cdot\hat{\lambda}}$ for some $\psi \in \mathrm{Hom}(\widetilde{U}^0, \mathbb{K})$. A simple submodule S of M has weights in $\psi \cdot \hat{\Lambda}$. By replacing ψ with the homomorphism χ for S given by Proposition 2.16, we may assume that for any indecomposable module M, there is a χ satisfying (2.15) so that $M = \bigoplus_{\lambda\in\Lambda} M_{\chi\cdot\hat{\lambda}}$.

Lemma 2.18. *Let $\chi : \widetilde{U}^0 \to \mathbb{K}$ be an algebra homomorphism with $\chi(\omega_i) = \chi(\omega_i')$ $(1 \le i < n)$. Let M be a finite-dimensional $\widetilde{U}$-module whose weights are all in $\chi \cdot \hat{\Lambda}$. If $\widetilde{U}^0$ acts semisimply on M, then*

$$M \cong L_\chi \otimes N$$

for some $\widetilde{U}$-module N whose weights are all in Λ. A similar statement holds for $U = U_{r,s}(\mathfrak{sl}_n)$ with Λ replaced by $\Lambda_{\mathfrak{sl}}$.

Proof. Let $\chi^{-1} : \widetilde{U}^0 \to \mathbb{K}$ be the algebra homomorphism defined by $\chi^{-1}(a_i) = \chi(a_i^{-1}) = (\chi(a_i))^{-1}$ and $\chi^{-1}(b_i) = \chi(b_i^{-1}) = (\chi(b_i))^{-1}$ for $1 \le i \le n$. Note that $L_\chi \otimes L_{\chi^{-1}}$ is isomorphic to the trivial module L_ε corresponding to the counit. Let

$$N = L_{\chi^{-1}} \otimes M.$$

Then $M \cong L_\chi \otimes N$ as L_ε is a multiplicative identity (up to isomorphism) for $\widetilde{U}$-modules. The weights of N are all in $\chi^{-1} \cdot \chi \cdot \hat{\Lambda} = \hat{\Lambda}$. □

We now have a classification of finite-dimensional simple modules for $\widetilde{U} = U_{r,s}(\mathfrak{gl}_n)$ and for $U = U_{r,s}(\mathfrak{sl}_n)$.

Theorem 2.19. *Assume rs^{-1} is not a root of unity. The finite-dimensional simple $\widetilde{U}$-modules are precisely the modules*

$$L_\chi \otimes L(\lambda),$$

where $\chi : \widetilde{U}^0 \to \mathbb{K}$ is an algebra homomorphism with $\chi(\omega_i) = \chi(\omega_i')$ $(1 \le i < n)$, and $\lambda \in \Lambda^+$. An analogous statement holds for U with Λ^+ replaced by $\Lambda^+_{\mathfrak{sl}}$.

Proof. Let M be a finite-dimensional simple $\widetilde{U}$-module. By Theorem 2.1, Proposition 2.16, and Lemma 2.18, $M \cong L_\chi \otimes N$ for some χ satisfying

(2.15) and some simple module N with weights in Λ. By Corollary 2.5, $N \cong L(\lambda)$ for some $\lambda \in \Lambda^+$. Conversely, any $\widetilde{U}$-module of this form is finite-dimensional by Lemma 2.12 and simple by its construction. □

Remark 2.20. If $r = q$ and $s = q^{-1}$ for some $q \in \mathbb{K}$, the classification of finite-dimensional simple $U_q(\mathfrak{sl}_n)$-modules is a consequence of Theorem 2.19 applied to $U_{q,q^{-1}}(\mathfrak{sl}_n)$: The simple $U_q(\mathfrak{sl}_n)$-modules are precisely those simple $U_{q,q^{-1}}(\mathfrak{sl}_n)$-modules on which ω_i' acts as ω_i^{-1}, so that

$$\chi(\omega_i) = \chi(\omega_i') = \chi(\omega_i^{-1}).$$

This implies $\chi(\omega_i) = \pm 1$ $(1 \le i < n)$. Each choice of algebra homomorphism $\chi : U^0 \to \mathbb{K}$ with $\chi(\omega_i) = \chi(\omega_i') = \pm 1$ yields a one-dimensional $U_{q,q^{-1}}(\mathfrak{sl}_n)$-module L_χ, and so the simple $U_q(\mathfrak{sl}_n)$-modules are the $L_\chi \otimes L(\lambda)$ with $\lambda \in \Lambda_{\mathfrak{sl}}^+$ and χ as above. We have

$$\hat{\lambda}(\omega_i) = q^{\langle \epsilon_i, \lambda \rangle} q^{-\langle \epsilon_{i+1}, \lambda \rangle} = q^{\langle \alpha_i, \lambda \rangle}.$$

Thus, we recover the results of [Ja, 5.2 and 5,10].

Remark 2.21. We can interpret Proposition 2.8 in light of Theorem 2.19: Let $L(\phi)$ be the simple $U_{r,s}(\mathfrak{sl}_2)$-module described in the proposition. Let $\lambda = (\ell/2)\alpha \in \Lambda_{\mathfrak{sl}}^+$ and define $\chi : U^0 \to \mathbb{K}$ by $\chi(\omega) = \phi(\omega) r^{-\ell/2} s^{\ell/2}$, $\chi(\omega') = \phi(\omega') r^{\ell/2} s^{-\ell/2} = \phi(\omega) r^{-\ell/2} s^{\ell/2} = \chi(\omega)$. Then $\phi = \chi \cdot \hat{\lambda}$ and $L(\phi) \cong L_\chi \otimes L(\lambda)$.

§3. Complete reducibility

In this section we will establish complete reducibility of all finite-dimensional modules for $\widetilde{U}$ (resp., U) on which $\widetilde{U}^0$ (resp., U^0) acts semisimply. Statements made for $\widetilde{U}$ hold equally well for U, and we will point out where there are differences. It is helpful to work in a more general context.

Let $\mathcal{O}$ denote the category of modules M for $\widetilde{U} = U_{r,s}(\mathfrak{gl}_n)$ which satisfy the conditions:

- ($\mathcal{O}$1) $\widetilde{U}^0$ acts semisimply on M, and the set $\mathrm{wt}(M)$ of weights of M belongs to Λ: $M = \bigoplus_{\lambda \in \mathrm{wt}(M)} M_\lambda$, where $M_\lambda = \{m \in M \mid a_i.m = r^{\langle \epsilon_i, \lambda \rangle},\ b_i.m = s^{\langle \epsilon_i, \lambda \rangle} \text{ for all } i\}$;
- ($\mathcal{O}$2) $\dim_{\mathbb{K}} M_\lambda < \infty$ for all $\lambda \in \mathrm{wt}(M)$;
- ($\mathcal{O}$3) $\mathrm{wt}(M) \subseteq \bigcup_{\mu \in F} (\mu - Q^+)$ for some finite set $F \subset \Lambda$.

The morphisms in $\mathcal{O}$ are $\widetilde{U}$-module homomorphisms. In defining category $\mathcal{O}$ for $U = U_{r,s}(\mathfrak{sl}_n)$, we replace Λ by the weight lattice $\Lambda_{\mathfrak{sl}}$ of $\mathfrak{sl}_n$.

All finite-dimensional $\widetilde{U}$-modules which satisfy ($\mathcal{O}$1) belong to category $\mathcal{O}$, as do all highest weight modules with weights in Λ such as the Verma modules $M(\lambda)$.

We recall the definition of the quantum Casimir operator [BW2, Sec. 4]. It is a consequence of (R2) and (R3) that the subalgebra U^+ of $\widetilde{U}$ (or of $U = U_{r,s}(\mathfrak{sl}_n)$) generated by 1 and e_i $(1 \le i < n)$ has the decomposition $U^+ = \bigoplus_{\zeta \in Q^+} U_\zeta^+$ where

$$U_\zeta^+ = \{z \in U^+ \mid a_i z = r^{\langle \epsilon_i, \zeta \rangle} z a_i,\ b_i z = s^{\langle \epsilon_i, \zeta \rangle} z b_i \ (1 \le i < n)\}.$$

The weight space U_ζ^+ is spanned by all the monomials $e_{i_1} \cdots e_{i_\ell}$ such that $\alpha_{i_1} + \cdots + \alpha_{i_\ell} = \zeta$. Similarly, the subalgebra U^- generated by 1 and the f_i has the decomposition $U^- = \bigoplus_{\zeta \in Q^+} U_{-\zeta}^-$. The spaces U_ζ^+ and $U_{-\zeta}^-$ are nondegenerately paired by the Hopf pairing specified by

$$
\begin{aligned}
&(f_i, e_j) = \frac{\delta_{i,j}}{s - r} \\
(3.1)\qquad &(\omega_i', \omega_j) = r^{\langle \epsilon_j, \alpha_i \rangle} s^{\langle \epsilon_{j+1}, \alpha_i \rangle} \\
&(b_n, a_n) = 1, \quad (b_n, \omega_j) = s^{-\langle \epsilon_n, \alpha_j \rangle}, \quad (\omega_i', a_n) = r^{\langle \epsilon_n, \alpha_i \rangle}.
\end{aligned}
$$

(See [BW2, Sec. 2].) The Hopf algebras $\widetilde{U}$ and U are Drinfel'd doubles of certain Hopf subalgebras with respect to this pairing [BW2, Thm. 2.7]. Let $d_\zeta = \dim_{\mathbb{K}} U_\zeta^+$. Assume $\{u_k^\zeta\}_{k=1}^{d_\zeta}$ is a basis for U_ζ^+, and $\{v_k^\zeta\}_{k=1}^{d_\zeta}$ is the dual basis for $U_{-\zeta}^-$ with respect to the pairing.

Now let

$$(3.2)\qquad \Omega = \sum_{\zeta \in Q^+} \sum_{k=1}^{d_\zeta} S(v_k^\zeta) u_k^\zeta,$$

where S denotes the antipode. All but finitely many terms in this sum will act as multiplication by 0 on any weight space M_λ of $M \in \mathcal{O}$. Therefore Ω is a well-defined operator on such a module M.

The second part of the Casimir operator involves a function $g : \Lambda \to \mathbb{K}^\#$ defined as follows. If ρ denotes half the sum of the positive roots, then $2\rho = \sum_{j=1}^n (n + 1 - 2j)\epsilon_j \in \Lambda$. For $\lambda \in \Lambda$, set

$$(3.3)\qquad g(\lambda) = (rs^{-1})^{\frac{1}{2}\langle \lambda + 2\rho, \lambda \rangle}.$$

When M is a $\widetilde{U}$-module in $\mathcal{O}$, we define the linear operator $\Xi : M \to M$ by

$$\Xi(m) = g(\lambda)m$$

for all $m \in M_\lambda$, $\lambda \in \Lambda$. Then Ξ is well-defined, as $\hat{\lambda} = \hat{\mu}$ if and only if $\lambda = \mu$ $(\lambda, \mu \in \Lambda)$ [BW2, Prop. 3.5]. (When $U = U_{r,s}(\mathfrak{sl}_n)$, it is necessary to first fix roots $r^{1/2n}$ and $s^{1/2n}$ of r and s in $\mathbb{K}$.) Then Ξ, as given above, is well-defined. We have the following result from [BW2].

Proposition 3.4. [BW2, Thm. 4.20] *The operator $\Omega\Xi : M \to M$ commutes with the action of $\widetilde{U}$ on any $\widetilde{U}$-module $M \in \mathcal{O}$.*

We require the next lemma in order to prove complete reducibility.

Lemma 3.5. *Assume rs^{-1} is not a root of unity, and let $\lambda, \mu \in \Lambda^+$. If $\lambda \geq \mu$ and $g(\lambda) = g(\mu)$, then $\lambda = \mu$.*

Proof. Because $\lambda \geq \mu$, we may suppose $\lambda = \mu + \beta$ where $\beta = \sum_{i=1}^{n-1} k_i\alpha_i$ and $k_i \in \mathbb{Z}_{\geq 0}$. By assumption we have

$$(rs^{-1})^{\frac{1}{2}\langle\lambda+2\rho,\lambda\rangle} = g(\lambda) = g(\mu) = (rs^{-1})^{\frac{1}{2}\langle\mu+2\rho,\mu\rangle},$$

and as rs^{-1} is not a root of unity, it must be that $\langle\lambda + 2\rho, \lambda\rangle = \langle\mu + 2\rho, \mu\rangle$, or equivalently, $2\langle\mu + \rho, \beta\rangle + \langle\beta, \beta\rangle = 0$. Since $\mu \in \Lambda^+$, $\mu = \mu_1\epsilon_1 + \mu_2\epsilon_2 + \cdots + \mu_n\epsilon_n$ where $\mu_i \in \mathbb{Z}$ for all i and $\mu_1 \geq \mu_2 \geq \cdots \geq \mu_n$. A calculation yields

$$\begin{aligned} 0 &= \langle 2\mu + 2\rho, \beta\rangle + \langle\beta, \beta\rangle \\ &= \sum_{i=1}^{n-1} 2k_i(\mu_i - \mu_{i+1} + 1) + \sum_{i=1}^{n}(k_i - k_{i-1})^2, \quad (k_0 = 0 = k_n). \end{aligned}$$

The only way this can happen is if $k_i = 0$ for all i and $\lambda = \mu$. $\square$

Lemma 3.6. *Assume rs^{-1} is not a root of unity.*

(i) *$\Omega\Xi$ acts as multiplication by the scalar $g(\lambda) = (rs^{-1})^{\frac{1}{2}\langle\lambda+2\rho,\lambda\rangle}$ on the Verma module $M(\lambda)$ with $\lambda \in \Lambda$, hence on any submodule or quotient of $M(\lambda)$.*

(ii) *The eigenvalues of the operator $\Omega\Xi : M \to M$ are integral powers of $(rs^{-1})^{\frac{1}{2}}$ on any finite-dimensional $M \in \mathcal{O}$. (For $U = U_{r,s}(\mathfrak{sl}_n)$, the eigenvalues are integral powers of $(rs^{-1})^{1/2n^2}$.)*

Proof. By its construction, $\Omega\Xi$ acts by multiplication by $g(\lambda) = (rs^{-1})^{\frac{1}{2}\langle\lambda+2\rho,\lambda\rangle}$ on the maximal vector v_λ of $M(\lambda)$. But since $M(\lambda) = \widetilde{U}.v_\lambda$ and $\Omega\Xi$ commutes with $\widetilde{U}$ on modules in $\mathcal{O}$, $\Omega\Xi$ acts as multiplication by $(rs^{-1})^{\frac{1}{2}\langle\lambda+2\rho,\lambda\rangle}$ on all of $M(\lambda)$.

If $M \in \mathcal{O}$ is finite-dimensional, it has a composition series. Each factor is a finite-dimensional simple $\widetilde{U}$-module with weights in Λ, and in particular, is a quotient of $M(\lambda)$ for some $\lambda \in \Lambda$. On such a factor, $\Omega\Xi$ acts as multiplication by $g(\lambda)$. Therefore the action of $\Omega\Xi$ on M may be expressed by an upper triangular matrix with each diagonal entry equal to $g(\lambda)$ for some $\lambda \in \Lambda$. □

Lemma 3.7. *Assume rs^{-1} is not a root of unity. Let $\lambda \in \Lambda$ and M be a nonzero finite-dimensional quotient of the Verma module $M(\lambda)$. Then M is simple.*

Proof. First observe that by Lemma 2.4, $\lambda \in \Lambda^+$. Assume M' is a proper submodule of M. As M is generated by its one-dimensional subspace M_λ, we must have $M'_\lambda = 0$. Let $\mu \in \Lambda$ be maximal such that $M'_\mu \neq 0$, and note that $\mu < \lambda$. Let m' be a nonzero vector of M'_μ. It follows from the maximality of μ that $e_i.m' = 0$ for all i $(1 \leq i < n)$. Letting $M'' = \widetilde{U}.m'$, a nonzero finite-dimensional quotient of $M(\mu)$, we see that $\mu \in \Lambda^+$ as well. By Lemma 3.6 (i), $\Omega\Xi$ acts as multiplication by $g(\lambda)$ on M, and by $g(\mu)$ on M''. This forces $g(\lambda) = g(\mu)$, which contradicts Lemma 3.5 as $\mu < \lambda$. □

Finally we state the needed complete reducibility result, whose proof parallels that of [L, Thm. 6.2.2].

Theorem 3.8. *Assume rs^{-1} is not a root of unity. Let M be a nonzero finite-dimensional $\widetilde{U}$-module on which $\widetilde{U}^0$ acts semisimply. Then M is completely reducible.*

Proof. Suppose first that M has weights in Λ. Then M is a direct sum of generalized eigenspaces for $\Omega\Xi$, which by Proposition 3.4, is a direct sum decomposition of M as a $\widetilde{U}$-module. Therefore, we may assume M is itself a generalized eigenspace of $\Omega\Xi$, so that $(\Omega\Xi - (rs^{-1})^c)^d(M) = 0$ for some $c \in \frac{1}{2}\mathbb{Z}$, $d = \dim_{\mathbb{K}} M$, by Lemma 3.6 (ii).

Let $P = \{m \in M \mid e_i.m = 0\ (1 \leq i < n)\}$, and note that $P = \bigoplus_{\lambda\in\Lambda} P_\lambda$, $P_\lambda = P \cap M_\lambda$. If $m \in P_\lambda - \{0\}$, the $\widetilde{U}$-submodule $\widetilde{U}.m$ of M is a nonzero quotient of $M(\lambda)$ by Theorem 2.1. By Lemma 3.7, each such $\widetilde{U}.m$ is a simple $\widetilde{U}$-module, and so the $\widetilde{U}$-submodule M' of M generated by P is a sum of simple $\widetilde{U}$-modules. That is, M' is completely reducible. Let $M'' = M/M'$.

Assuming $M'' \neq 0$, there is a weight $\mu \in \Lambda$ maximal such that $M''_\mu \neq 0$. It follows from the maximality of μ that $e_i.m'' = 0$ for $m'' \in M''_\mu - \{0\}$ and for all i $(1 \leq i < n)$. By Lemma 2.4, $\mu \in \Lambda^+$, and by Theorem 2.1 and Lemma 3.6, $\Omega\Xi$ acts as multiplication by $g(\mu)$ on the U-module $U.m''$ generated by m''. This implies $g(\mu) = (rs^{-1})^c$.

Let $m \in M_\mu$ be a representative for $m'' \in (M/M')_\mu$, and $M_1 = \widetilde{U}.m$. Then there is a weight $\eta \in \Lambda$ maximal such that $M_1 \cap M_\eta \neq 0$. Let $m_1 \in M_1 \cap M_\eta - \{0\}$, so that $e_i.m_1 = 0$ for all i $(1 \leq i < n)$. Again applying Theorem 2.1 and Lemmas 2.4 and 3.6, we have $\eta \in \Lambda^+$ and $\Omega\Xi(m_1) = g(\eta)m_1$. Therefore $g(\eta) = (rs^{-1})^c$.

We now have $g(\mu) = g(\eta)$, where $\eta, \mu \in \Lambda^+$, and $\eta \geq \mu$ by construction. By Lemma 3.5, $\eta = \mu$, so M_1 is the one-dimensional space spanned by m, and $e_i.m = 0$ $(1 \leq i < n)$, that is $m \in P$. This implies $m'' = 0$, a contradiction to the assumption that $M'' \neq 0$. Therefore $M'' = 0$, and $M = M'$ is completely reducible.

Finally, when M does not have weights in Λ, we may assume that M is indecomposable. By Remark 2.17, M has all its weights in $\chi \cdot \hat{\Lambda}$ for some χ satisfying (2.15). By Lemma 2.18, $M \cong L_\chi \otimes N$ for some $\widetilde{U}$-module N whose weights are all in Λ. Note that $\widetilde{U}^0$ acts semisimply on N as well $(N = L_{\chi^{-1}} \otimes M)$, and so N is completely reducible by the above argument. This implies that M itself is completely reducible. □

Remark 3.9. It is necessary to include the hypothesis that $\widetilde{U}^0$ acts semisimply in Theorem 3.8, as the next examples illustrate. (Recall that $\widetilde{U}^0$ does indeed act semisimply on any simple $\widetilde{U}$-module, as remarked in the text following (1.6).) Let $V = \mathbb{K}^m$ for $m \geq 2$ and $\xi, \xi' \in \mathbb{K} \setminus \{0\}$. We define a $\widetilde{U}$-module structure on V by requiring that e_i, f_i act as multiplication by 0 and a_i, b_i act via the $m \times m$ Jordan blocks with diagonal entries ξ, ξ', respectively. Then relations (R1)-(R7) of $\widetilde{U}$ hold on V. The scalars ξ, ξ' may be chosen so that V has weights in Λ, for example choose an integer c, let $\lambda = c(\epsilon_1 + \cdots + \epsilon_n)$, and set $\xi = r^c = \hat{\lambda}(a_i)$, $\xi' = s^c = \hat{\lambda}(b_i)$. Clearly V is not completely reducible. A related example for $U_{r,s}(\mathfrak{sl}_n)$ is given by sending each ω_i, ω'_i to the Jordan block with 1s on the diagonal, thus corresponding to the weight 0.

§4. The R-matrix

In this section we recall the definition of the R-matrix from [BW2, Sec. 4] and use it to prove a more general result on the commutativity of the tensor product of finite-dimensional modules than was given there (compare [BW2,

Thm. 4.11] with Theorem 4.2 below). Let M, M' be $\widetilde{U}$-modules in category $\mathcal{O}$. We define an isomorphism of $\widetilde{U}$-modules $R_{M',M} : M' \otimes M \to M \otimes M'$ as follows. If $\lambda = \sum_{i=1}^{n} \lambda_i \alpha_i \in \Lambda$, where $\alpha_n = \epsilon_n$, set

$$\omega_\lambda = \omega_1^{\lambda_1} \cdots \omega_{n-1}^{\lambda_{n-1}} a_n^{\lambda_n} \quad \text{and} \quad \omega'_\lambda = (\omega'_1)^{\lambda_1} \cdots (\omega'_{n-1})^{\lambda_{n-1}} b_n^{\lambda_n}.$$

Also let

$$\Theta = \sum_{\zeta \in Q^+} \sum_{k=1}^{d_\zeta} v_k^\zeta \otimes u_k^\zeta,$$

where the notation is as in the paragraph following (3.1). Define

$$R_{M',M} = \Theta \circ \widetilde{f} \circ P$$

where $P(m' \otimes m) = m \otimes m'$, $\widetilde{f}(m \otimes m') = (\omega'_\mu, \omega_\lambda)^{-1}(m \otimes m')$ when $m \in M_\lambda$ and $m' \in M'_\mu$, and the Hopf pairing (,) is defined in (3.1). (There is an equivalent definition of $\widetilde{f}$ that works in the case $U = U_{r,s}(\mathfrak{sl}_n)$, given in [BW2] after (4.3).) Then $R_{M',M}$ is an isomorphism of $\widetilde{U}$-modules that satisfies the quantum Yang-Baxter equation and the hexagon identities [BW2, Thms. 4.11, 5.4, and 5.7].

We will show that the tensor product of *any* two finite-dimensional $\widetilde{U}$-modules in $\mathcal{O}$ is commutative (up to module isomorphism), starting first with the special case that one of the modules is a one-dimensional module $L_\chi = L(\chi)$, as defined in Section 2.

Lemma 4.1. *Let M be a $\widetilde{U}$-module in category $\mathcal{O}$, and let L_χ be a one-dimensional $\widetilde{U}$-module. Then*

$$L_\chi \otimes M \cong M \otimes L_\chi.$$

Proof. Fix a basis element v of L_χ. Define a linear function $F : L_\chi \otimes M \to M \otimes L_\chi$ as follows. If $m \in M_\lambda$ and $\lambda = -\sum_{i=1}^{n} c_i \alpha_i$, then

$$F(v \otimes m) = \chi_1^{c_1} \cdots \chi_n^{c_n} m \otimes v,$$

where $\chi_i = \chi(\omega_i) = \chi(\omega'_i)$ $(1 \le i < n)$ and $\chi_n = \chi(a_n)$. Clearly F is bijective, and we check that F is a $\widetilde{U}$-homomorphism:

$$e_i.F(v \otimes m) = \chi_1^{c_1} \cdots \chi_n^{c_n}(e_i \otimes 1 + \omega_i \otimes e_i)(m \otimes v) = \chi_1^{c_1} \cdots \chi_n^{c_n} e_i.m \otimes v.$$

On the other hand, as $e_i.m \in M_{\lambda+\alpha_i}$, we have

$$\begin{aligned} F(e_i.(v \otimes m)) &= \chi_i F(v \otimes e_i m) \\ &= \chi_i(\chi_1^{c_1} \cdots \chi_i^{c_i - 1} \cdots \chi_n^{c_n}) e_i.m \otimes v = e_i.F(v \otimes m). \end{aligned}$$

Similarly, F commutes with f_i. As the action by a_i, b_i preserves the weight spaces, F commutes with a_i, b_i $(1 \le i \le n)$ as well. Therefore F is an isomorphism of $\widetilde{U}$-modules. □

Theorem 4.2. *Let M and M' be finite-dimensional modules for $\widetilde{U}$ (resp., U) with $\widetilde{U}^0$ (resp., U^0) acting semisimply. Then*

$$M \otimes M' \cong M' \otimes M.$$

Proof. As the tensor product distributes over direct sums, we may assume that M and M' are indecomposable. Therefore the weights of M are all in $\chi \cdot \hat{\Lambda}$ for some algebra homomorphism $\chi : \widetilde{U}^0 \to \mathbb{K}$ with $\chi(\omega_i) = \chi(\omega_i')$. (See Remark 2.17.) By Lemma 2.18, $M \cong L_\chi \otimes N$ for some module N with weights in Λ. Similarly, $M' \cong L_{\chi'} \otimes N'$ for some χ'. By Lemma 4.1 and [BW2, Thm. 4.11],

$$\begin{aligned} M \otimes M' \cong (L_\chi \otimes N) \otimes (L_{\chi'} \otimes N') &\cong (L_\chi \otimes L_{\chi'}) \otimes (N \otimes N') \\ &\cong (L_{\chi'} \otimes L_\chi) \otimes (N' \otimes N) \\ &\cong (L_{\chi'} \otimes N') \otimes (L_\chi \otimes N) \\ &\cong M' \otimes M. \quad \square \end{aligned}$$

§5. Tensor powers of the natural module

In this section we consider tensor powers $V^{\otimes k} = V \otimes V \otimes \cdots \otimes V$ (k factors) of the natural module V for $\widetilde{U}$ (defined in Section 1). Set $R = R_{V,V}$, and for $1 \leq i < k$, let R_i be the $\widetilde{U}$-module isomorphism on $V^{\otimes k}$ defined by

$$R_i(z_1 \otimes z_2 \otimes \cdots \otimes z_k) = z_1 \otimes \cdots \otimes z_{i-1} \otimes R(z_i \otimes z_{i+1}) \otimes z_{i+2} \otimes \cdots \otimes z_k.$$

Then it is a consequence of the quantum Yang-Baxter equation that the braid relations hold:

$$\begin{aligned} R_i \circ R_{i+1} \circ R_i &= R_{i+1} \circ R_i \circ R_{i+1} \qquad \text{for} \quad 1 \leq i < k \\ R_i \circ R_j &= R_j \circ R_i \qquad \text{for} \quad |i - j| \geq 2. \end{aligned} \tag{5.1}$$

We would like to argue that

$$R_i^2 = (1 - rs^{-1})R_i + rs^{-1}\mathrm{Id} \tag{5.2}$$

for all $i = 1, \ldots, k-1$. For this it suffices to work with the 2-fold tensor product $V \otimes V$.

Proposition 5.3. *Whenever $s \neq -r$, the $\widetilde{U}$-module $V \otimes V$ decomposes into two simple $\widetilde{U}$-submodules, $S^2_{r,s}(V)$ (the (r,s)-symmetric tensors) and $\Lambda^2_{r,s}(V)$ (the (r,s)-antisymmetric tensors). These modules are defined as follows:*

(i) *$S^2_{r,s}(V)$ is the span of*
$\{v_i \otimes v_i \mid 1 \leq i \leq n\} \cup \{v_i \otimes v_j + s v_j \otimes v_i \mid 1 \leq i < j \leq n\}$.

(ii) *$\Lambda^2_{r,s}(V)$ is the span of $\{v_i \otimes v_j - r v_j \otimes v_i \mid 1 \leq i < j \leq n\}$.*

Proof. Direct computation shows that $S^2_{r,s}(V)$ and $\Lambda^2_{r,s}(V)$ are $\widetilde{U}$-submodules of $V \otimes V$. Note that each weight space of $S^2_{r,s}(V)$ is one-dimensional and is spanned by one of the weight vectors listed in (i). Therefore any submodule of $S^2_{r,s}(V)$ must contain one of these vectors. It may be checked that any of these vectors generates all of $S^2_{r,s}(V)$ in case $s \neq -r$. In particular, $v_1 \otimes v_1$ is a highest weight vector, and given any other vector in (i), there is an element of U taking it to $v_1 \otimes v_1$. Therefore $S^2_{r,s}(V)$ is simple. A similar argument proves that $\Lambda^2_{r,s}(V)$ is simple, with highest weight vector $v_1 \otimes v_2 - r v_2 \otimes v_1$. □

Remark 5.4. The $s = -r$ case is "nongeneric," and in this exceptional case, $V \otimes V$ need not be completely reducible. For example, when $n = 2$ what happens is that $v_1 \otimes v_2 - r v_2 \otimes v_1$ spans a one-dimensional module (as it does for $n = 2$ generic) that is not complemented in $V \otimes V$. Modulo that submodule, $v_1 \otimes v_1$ spans a one-dimensional module. Modulo the resulting two-dimensional module, $v_1 \otimes v_2 + r v_2 \otimes v_1$ and $v_2 \otimes v_2$ span a two-dimensional module.

Proposition 5.5. *The minimum polynomial of $R = R_{V,V}$ on $V \otimes V$ is $(t-1)(t + rs^{-1})$ if $s \neq -r$.*

Proof. It follows from the definition of R that $R(v_1 \otimes v_1) = v_1 \otimes v_1$ and $R(v_1 \otimes v_2 - r v_2 \otimes v_1) = -rs^{-1}(v_1 \otimes v_2 - r v_2 \otimes v_1)$. By Proposition 5.3, $S^2_{r,s}(V)$ and $\Lambda^2_{r,s}(V)$ are simple, and in fact, $v_1 \otimes v_1$ and $v_1 \otimes v_2 - r v_2 \otimes v_1$ are the highest weight vectors. In particular, each is a cyclic module generated by its highest weight vector. As $Ra(v_1 \otimes v_1) = aR(v_1 \otimes v_1) = a(v_1 \otimes v_1)$ for all $a \in \widetilde{U}$, this implies that $S^2_{r,s}(V)$ is in the eigenspace of R corresponding to eigenvalue 1. Analogously, $\Lambda^2_{r,s}(V)$ corresponds to the eigenvalue $-rs^{-1}$, and since $V \otimes V$ is the direct sum of those submodules, we have the desired result. □

From Proposition 5.5 it follows that R acts as

$$r\sum_{i<j} E_{j,i}\otimes E_{i,j} + s^{-1}\sum_{i<j} E_{i,j}\otimes E_{j,i} \tag{5.6}$$
$$+ (1-rs^{-1})\sum_{i<j} E_{j,j}\otimes E_{i,i} + \sum_{i} E_{i,i}\otimes E_{i,i}$$

on $V\otimes V$. Indeed, (5.6) is a linear operator that acts on $S^2_{r,s}(V)$ as multiplication by 1, and on $\Lambda^2_{r,s}(V)$ as multiplication by $-rs^{-1}$. By Proposition 5.5, R has the same properties, and so R is equal to this sum on $V\otimes V$.

§6. Quantum Schur-Weyl duality

Assume $r,s\in\mathbb{K}$. Let $H_k(r,s)$ be the unital associative algebra over $\mathbb{K}$ with generators T_i, $1\le i<k$, subject to the relations:

(H1) $T_iT_{i+1}T_i = T_{i+1}T_iT_{i+1}$, $\quad 1\le i<k$
(H2) $T_iT_j = T_jT_i$, $\quad |i-j|\ge 2$
(H3) $T_i^2 = (s-r)T_i + rs1$.

When $r\neq 0$, the elements $t_i = r^{-1}T_i$ satisfy the braid relations (H1), (H2), along with the relation

$$t_i^2 = (q-1)t_i + q1, \tag{H3'}$$

where $q=r^{-1}s$. The *Hecke algebra* $H_k(q)$ (of type A_{k-1}) is generated by elements t_i, $1\le i<k$, which satisfy (H1), (H2), (H3'). It has dimension $k!$ and is semisimple whenever q is not a root of unity. At $q=1$, the Hecke algebra $H_k(q)$ is isomorphic to $\mathbb{K}S_k$, the group algebra of the symmetric group S_k, where we may identify t_i with the transposition $(i\ i+1)$.

The two-parameter Hecke algebra $H_k(r,s)$ defined above is isomorphic to $H_k(r^{-1}s)$ whenever $r\neq 0$. Thus, it is semisimple whenever $r^{-1}s$ is not a root of unity. For any $\sigma\in S_k$, we may define $T_\sigma = T_{i_1}\cdots T_{i_\ell}$ where $\sigma = t_{i_1}\ldots t_{i_\ell}$ is a reduced expression for σ as a product of transpositions. It follows from (H1) and (H2) that T_σ is independent of the reduced expression and these elements give a basis.

The results of Section 5 show that the $\widetilde{U}$-module $V^{\otimes k}$ affords a representation of the Hecke algebra $H_k(r,s)$:

$$\begin{aligned} H_k(r,s) &\to \mathrm{End}_{\widetilde{U}}(V^{\otimes k}) \\ T_i &\mapsto sR_i \qquad (1\le i<k). \end{aligned} \tag{6.1}$$

When $k = 2$ and $s \neq -r$, $V^{\otimes 2} = S^2_{r,s}(V) \oplus \wedge^2_{r,s}(V)$ is a decomposition of $V^{\otimes 2}$ into simple $\widetilde{U}$-modules by Proposition 5.3. The maps $p_1 = (sR_1 + r)/(s + r)$ and $p_2 = (s - sR_1)/(s + r)$, $(R_1 = R_{V,V})$, are the corresponding projections onto the simple summands. Thus, the map in (6.1) is an isomorphism for $k = 2$. More generally, we will show next that it is surjective whenever rs^{-1} is not a root of unity, and it is an isomorphism when $n \geq k$. This is the two-parameter version of the well-known result of Jimbo [Ji] that $H_k(q) \cong \mathrm{End}_{U_q(\mathfrak{gl}_n)}(V^{\otimes k})$ and is the analogue of classical Schur-Weyl duality, $\mathbb{K}S_k \cong \mathrm{End}_{\mathfrak{gl}_n}(V^{\otimes k})$ for $n \geq k$. We will apply the following lemma. The case $n < k$ is dealt with separately, and it uses the isomorphism $H_k(r, s) \cong \mathrm{End}_{\widetilde{U}}(V^{\otimes k})$ in the $n = k$ case.

Lemma 6.2. *If $n \geq k$ and V is the natural representation of $\widetilde{U}$, then $V^{\otimes k}$ is a cyclic $\widetilde{U}$-module generated by $v_1 \otimes \cdots \otimes v_k$.*

Proof. Let $\underline{v} = v_1 \otimes \cdots \otimes v_k$. We begin by showing that if $\sigma \in S_k$, then $v_{\sigma(1)} \otimes \cdots \otimes v_{\sigma(k)} \in \widetilde{U}.\underline{v}$.

Suppose we have an arbitrary permutation $x_1 \otimes \cdots \otimes x_k$ ($x_i \in \{v_1, \ldots, v_k\}$ for all i) of the factors of $\underline{v}$. For some $\ell < m$, assume that $x_\ell = v_j$ and $x_m = v_{j+1}$. Then because of the formulas

(6.3)
$$\Delta^{k-1}(e_j) = \sum_{i=1}^{k} \underbrace{\omega_j \otimes \cdots \otimes \omega_j}_{i-1} \otimes e_j \otimes \underbrace{1 \otimes \cdots \otimes 1}_{k-i}$$
$$\Delta^{k-1}(f_j) = \sum_{i=1}^{k} \underbrace{1 \otimes \cdots \otimes 1}_{k-i} \otimes f_j \otimes \underbrace{\omega_j' \otimes \cdots \otimes \omega_j'}_{i-1},$$

there are nonzero scalars c and c' such that

$$(ce_j f_j + c').(x_1 \otimes \cdots \otimes x_k) = x_1 \otimes \cdots \otimes x_m \otimes \cdots \otimes x_\ell \otimes \cdots \otimes x_k.$$

Similarly, there are nonzero scalars d and d' such that

$$(de_j f_j + d').(x_1 \otimes \cdots \otimes x_m \otimes \cdots \otimes x_\ell \otimes \cdots \otimes x_k) = x_1 \otimes \cdots \otimes x_k.$$

As the transpositions $(j\ j+1)$ generate S_k, $v_{\sigma(1)} \otimes \cdots \otimes v_{\sigma(k)} \in \widetilde{U}.\underline{v}$ for all $\sigma \in S_k$.

Next we will use induction on k to establish the following. For any k elements $i_1, \ldots, i_k \in \{1, \ldots, n\}$ satisfying $i_1 \leq i_2 \leq \cdots \leq i_k$, there is a $u \in \widetilde{U}$ such that $u.\underline{v} = v_{i_1} \otimes \cdots \otimes v_{i_k}$ and u does not contain any terms with factors of $e_m, e_{m+1}, \ldots, e_{n-1}, f_{m+1}, f_{m+2}, \ldots, f_{n-2}$, or f_{n-1} where $m = \max\{i_k, k\}$. If $k = 1$, we may apply $f_{m-1} \cdots f_1$ to $\underline{v} = v_1$ to obtain v_m for any $m \in \{1, \ldots, n\}$. If $k > 1$, let ℓ be such that $i_\ell < i_k$, $i_{\ell+1} = i_{\ell+2} = \cdots = i_k$. (If no such ℓ exists, that is if $i_1 = \cdots = i_k$, then set $\ell = 0$ and apply u' from (6.5) below to $v_1 \otimes \cdots \otimes v_k$ to obtain a nonzero scalar multiple of $v_{i_1} \otimes \cdots \otimes v_{i_k}$.) By induction, there is an element $u \in \widetilde{U}$ such that

$$u.(v_1 \otimes \cdots \otimes v_\ell) = v_{i_1} \otimes \cdots \otimes v_{i_\ell}, \tag{6.4}$$

where u has no terms with factors of $e_{m'}, e_{m'+1}, \ldots, e_{n-1}, f_{m'+1}, \ldots, f_{n-1}$ ($m' = \max\{i_\ell, \ell\}$).

Suppose initially that $i_\ell \leq \ell$. Then $m' = \ell$, and so $u.(v_1 \otimes \cdots \otimes v_k)$ is a nonzero scalar multiple of $(v_{i_1} \otimes \cdots \otimes v_{i_\ell}) \otimes (v_{\ell+1} \otimes \cdots \otimes v_k)$. Now apply

$$u' = \begin{cases} (f_{i_k-1} f_{i_k-2} \cdots f_{\ell+1}) \cdots (f_{i_k-1} f_{i_k-2})(f_{i_k-1}) \times \\ \qquad (e_{i_k} e_{i_k+1} \cdots e_{k-1}) \cdots (e_{i_k} e_{i_k+1})(e_{i_k}) & \text{if } i_k < k \\ (f_{i_k-1} f_{i_k-2} \cdots f_{\ell+1}) \cdots (f_{i_k-1} f_{i_k-2} \cdots f_{k-1})(f_{i_k-1} f_{i_k-2} \cdots f_k) \\ & \text{if } i_k \geq k \end{cases} \tag{6.5}$$

to obtain a nonzero scalar multiple of $v_{i_1} \otimes \cdots \otimes v_{i_k}$, as desired. (Note that we did not use any factors of $e_m, e_{m+1}, \ldots, e_{n-1}, f_{m+1}, \ldots, f_{n-1}$ for $m = \max\{i_k, k\}$.)

If on the other hand, $i_\ell > \ell$ (so that $m' = i_\ell$ and $i_k > \ell + 1$), first apply u' from (6.5) to $v_1 \otimes \cdots \otimes v_k$ to obtain a nonzero scalar multiple of

$$(v_1 \otimes \cdots \otimes v_\ell) \otimes (v_{i_k} \otimes \cdots \otimes v_{i_k}),$$

and then apply u from (6.4) to obtain a nonzero scalar multiple of $v_{i_1} \otimes \cdots \otimes v_{i_k}$, as desired.

Finally, if $i_1, \ldots, i_k \in \{1, \ldots, n\}$ are *any* k elements (not necessarily in nondecreasing numerical order), let $\sigma \in S_k$ be a permutation such that

$$i_{\sigma(1)} \leq i_{\sigma(2)} \leq \cdots \leq i_{\sigma(k)}.$$

By the first paragraph of the proof, there is an element of $\widetilde{U}$ taking $\underline{v}$ to $v_{\sigma^{-1}(1)} \otimes \cdots \otimes v_{\sigma^{-1}(k)}$. Now we may apply u from (6.4) and u' from (6.5) in the appropriate order (as above) to $v_{\sigma^{-1}(1)} \otimes \cdots \otimes v_{\sigma^{-1}(k)}$ to obtain a nonzero scalar multiple of $v_{i_1} \otimes \cdots \otimes v_{i_k}$. $\square$

This leads to the two-parameter analogue of Schur-Weyl duality.

Theorem 6.6. *Assume rs^{-1} is not a root of unity. Then:*

(i) *$H_k(r,s)$ maps surjectively onto* $\mathrm{End}_{\widetilde{U}}(V^{\otimes k})$.

(ii) *If $n \geq k$, then $H_k(r,s)$ is isomorphic to* $\mathrm{End}_{\widetilde{U}}(V^{\otimes k})$.

Proof. We establish part (ii) first. Assume $F \in \mathrm{End}_{\widetilde{U}}(V^{\otimes k})$ and $\underline{v} = v_1 \otimes \cdots \otimes v_k$. As F commutes with the action of $\widetilde{U}$, $F(\underline{v})$ must have the same weight as $\underline{v}$, that is, $\epsilon_1 + \cdots + \epsilon_k$. The only vectors of $V^{\otimes k}$ with this weight are the linear combinations of the permutations of $v_1 \otimes \cdots \otimes v_k$, so that

$$F(\underline{v}) = \sum_{\sigma \in S_k} c_\sigma v_{\sigma(1)} \otimes \cdots \otimes v_{\sigma(k)}, \tag{6.7}$$

for some scalars $c_\sigma \in \mathbb{K}$. We will show that there is an element R^σ in the image of $H_k(r,s)$ in $\mathrm{End}_{\widetilde{U}}(V^{\otimes k})$ such that $R^\sigma(\underline{v}) = v_{\sigma(1)} \otimes \cdots \otimes v_{\sigma(k)}$. (Previously we constructed an element $u \in \widetilde{U}$ with this property.)

We begin with the transposition $\tau = t_j = (j\ j+1)$. For any tensor product $v_{i_1} \otimes \cdots \otimes v_{i_k}$ of distinct basis vectors, we have by (5.6) that

$$v_{i_{\tau(1)}} \otimes \cdots \otimes v_{i_{\tau(k)}} = \begin{cases} r^{-1} R_j(v_{i_1} \otimes \cdots \otimes v_{i_k}) & \text{if } i_j < i_{j+1} \\ (sR_j + (r-s)\mathrm{Id})(v_{i_1} \otimes \cdots \otimes v_{i_k}) & \text{if } i_j > i_{j+1}. \end{cases}$$

Therefore, if $\sigma = t_{j_1} \cdots t_{j_m}$, a product of such transpositions, we can set $R^{t_{j_\ell}} := r^{-1} R_{j_\ell}$ or $R^{t_{j_\ell}} := sR_{j_\ell} + (r-s)\mathrm{Id}$, depending on the numerical order of the appropriate indices i_{j_ℓ} and $i_{j_\ell+1}$ in $R^{t_{j_{\ell-1}}} \circ \cdots \circ R^{t_{j_1}} \underline{v}$. Then defining $R^\sigma = R^{t_{j_m}} \circ \cdots \circ R^{t_{j_1}} \in \mathrm{End}_{\widetilde{U}}(V^{\otimes k})$, we have the desired map such that $R^\sigma(\underline{v}) = v_{\sigma(1)} \otimes \cdots \otimes v_{\sigma(k)}$.

Now let $F_0 = F - \sum_{\sigma \in S_k} c_\sigma R^\sigma \in \mathrm{End}_{\widetilde{U}}(V^{\otimes k})$ (with the c_σ coming from (6.7)). By Lemma 6.2, $F_0(V^{\otimes k}) = F_0(\widetilde{U}.\underline{v}) = \widetilde{U}.F_0(\underline{v}) = 0$. Therefore $F = \sum_{\sigma \in S_k} c_\sigma R^\sigma$ is in the image of $H_k(r,s)$. Consequently, the map $H_k(r,s) \to \mathrm{End}_{\widetilde{U}}(V^{\otimes k})$ in (6.1) is a surjection, and $\mathrm{End}_{\widetilde{U}}(V^{\otimes k})$ is the $\mathbb{K}$-linear span of $\{R^\sigma \mid \sigma \in S_k\}$. Now suppose that $\sum_{\sigma \in S_k} c_\sigma R^\sigma = 0$ for some scalars $c_\sigma \in \mathbb{K}$. Then in particular, $0 = \sum_{\sigma \in S_k} c_\sigma R^\sigma(\underline{v}) = \sum_{\sigma \in S_k} c_\sigma v_{\sigma(1)} \otimes \cdots \otimes v_{\sigma(k)}$, which implies $c_\sigma = 0$ for all $\sigma \in S_k$. Therefore $\{R^\sigma \mid \sigma \in S_k\}$ is a basis for the vector space $\mathrm{End}_{\widetilde{U}}(V^{\otimes k})$ and $\dim_{\mathbb{K}} \mathrm{End}_{\widetilde{U}}(V^{\otimes k}) = k! = \dim_{\mathbb{K}} H_k(r,s)$. It follows that $H_k(r,s)$ is isomorphic to $\mathrm{End}_{\widetilde{U}}(V^{\otimes k})$ for $n \geq k$, as asserted.

Next we turn to the proof of (i) and assume here that $n < k$. For $i = n, k$, let $\widetilde{U}_i = U_{r,s}(\mathfrak{gl}_i)$, let Λ_i be the weight lattice of $\mathfrak{gl}_i$, and let V_i be the natural $\widetilde{U}_i$-module. By (ii), we may identify $H_k(r,s)$ with $\mathrm{End}_{\widetilde{U}_k}(V_k^{\otimes k})$. We will show that $H_k(r,s)$ maps surjectively onto $\mathrm{End}_{\widetilde{U}_n}(V_n^{\otimes k})$.

Consider $V_k^{\otimes k}$ as a $\widetilde{U}_n$-module via the inclusion of $\widetilde{U}_n$ into $\widetilde{U}_k$, and regard $V_n^{\otimes k}$ as a $\widetilde{U}_n$-submodule of $V_k^{\otimes k}$ in the obvious way. Now $V_n^{\otimes k}$ is a finite-dimensional $\widetilde{U}_n$-module on which $\widetilde{U}_n^0$ acts semisimply, so by Theorem 3.8, it is completely reducible. Therefore,

$$V_n^{\otimes k} = L_1 \oplus \cdots \oplus L_t \tag{6.8}$$

for simple $\widetilde{U}_n$-modules L_i. It suffices to show that the projections onto the simple summands L_i can be obtained from $H_k(r,s)$.

Consider

$$\widetilde{U}_k.V_n^{\otimes k} = \widetilde{U}_k.L_1 + \cdots + \widetilde{U}_k.L_t, \tag{6.9}$$

the $\widetilde{U}_k$-submodule of $V_k^{\otimes k}$ generated by $V_n^{\otimes k}$. By Corollary 2.5, each L_i is isomorphic to some $L(\lambda_i)$, $\lambda_i \in \Lambda_n^+$, and in particular is generated by a highest weight vector m_i with $e_j.m_i = 0$ for all j $(1 \le j < n)$. We claim that $e_j.m_i = 0$ as well when $n \le j < k$. This follows from the expression for $\Delta^{k-1}(e_j)$ in (6.3) and the action of e_j on the natural module V_k for $\widetilde{U}_k$ given by $e_j.v_i = \delta_{i,j+1}v_j$, because m_i must be some linear combination of vectors $v_{i_1} \otimes \cdots \otimes v_{i_k}$ with $i_1, \ldots, i_k \in \{1, \ldots, n\}$. Therefore m_i is also a highest weight vector for the finite-dimensional $\widetilde{U}_k$-module $\widetilde{U}_k.L_i$. By Theorem 2.1 and Lemma 3.7, $\widetilde{U}_k.L_i = \widetilde{U}_k.m_i$ is a simple $\widetilde{U}_k$-module. Therefore (6.9) must be a direct sum:

$$\widetilde{U}_k.V_n^{\otimes k} = \widetilde{U}_k.L_1 \oplus \cdots \oplus \widetilde{U}_k.L_t.$$

Because $V_k^{\otimes k}$ is a completely reducible $\widetilde{U}_k$-module, there is some complementary $\widetilde{U}_k$-submodule W such that

$$V_k^{\otimes k} = \widetilde{U}_k.L_1 \oplus \cdots \oplus \widetilde{U}_k.L_t \oplus W. \tag{6.10}$$

Let $\pi_i \in H_k(r,s)$ be the projection of $V_k^{\otimes k}$ onto $\widetilde{U}_k.L_i$. Then, π_i commutes with the $\widetilde{U}_k$-action, and acts as the identity map on $\widetilde{U}_k.L_i$ and as 0 on the other summands in (6.10). Since $L_j \subseteq \widetilde{U}_k.L_j$ for all j, the map π_i restricted to $V_n^{\otimes k}$ commutes with the $\widetilde{U}_n$-action and is the projection onto L_i. Thus, $H_k(r,s) \to \mathrm{End}_{\widetilde{U}_n}(V_n^{\otimes k})$ is onto. $\square$

Finally, we note that because $V^{\otimes k}$ is a semisimple $\widetilde{U}$-module, it follows from the double commutant theorem (see for example, [GW, Thm. 3.3.7]) that $\mathrm{End}_{H_k(r,s)}(V^{\otimes k})$ is isomorphic to the image of $\widetilde{U}$ in $\mathrm{End}_{\mathbb{K}}(V^{\otimes k})$.

References

[BW1] G. Benkart and S. Witherspoon, *A Hopf structure for down-up algebras*, Math. Z. **238** (2001), 523-553.

[BW2] G. Benkart and S. Witherspoon, *Two-parameter quantum groups and Drinfel'd doubles*, Algebr. Represent. Theory (to appear).

[D] V.G. Drinfel'd, *Quantum groups*, in Proceedings of the International Congress of Mathematicians, Berkeley, 1986, A.M. Gleason (ed.), pp. 798-820, Amer. Math. Soc. 1987.

[DD] R. Dipper and S. Donkin, *Quantum* GL_n, Proc. London Math. Soc. **63** (1991), 165-211.

[DP] C. De Concini and C. Procesi, *A characteristic free approach to invariant theory*, Adv. Math. **21** (1976), 330-354.

[DPS] J. Du, B. Parshall, and L. Scott, *Quantum Weyl reciprocity and tilting modules*, Comm. Math. Phys. **195** (1998), 321-352.

[Du] J. Du, *A note on quantized Weyl reciprocity at roots of unity*, Algebra Colloq. **2:4** (1995), 363-372.

[GW] R. Goodman and N.R. Wallach, *Representations and Invariants of the Classical Groups, Encyclopedia of Mathematics and Its Applications*, vol. 68, Cambridge University Press, 1998.

[Ja] J. C. Jantzen, *Lectures on Quantum Groups*, vol. 6, Graduate Studies in Math., Amer. Math. Soc., 1993.

[Ji1] M. Jimbo, *A q-difference analog of* $U(\mathfrak{g})$ *and the Yang-Baxter equation*, Lett. Math. Phys. **10** (1985), 63-69.

[Ji2] M. Jimbo, *A q-analogue of* $U(\mathfrak{gl}_n(N+1))$, *Hecke algebra, and the Yang-Baxter equation*, Lett. Math. Phys. **11** (1986), 247-252.

[Jo] A. Joseph, *Quantum Groups and Their Primitive Ideals*, Ergebnisse der Mathematik und ihrer Grenzgebiete, Springer-Verlag, Berlin, 1995.

[KT] D. Krob and J.-Y. Thibon, *Noncommutative symmetric functions V: A degenerate version of* $U_q(\mathfrak{gl}_n)$, Internat. J. Algebra Comput. **9** (1999), 405-430.

[LR] R. Leduc and A. Ram, *A ribbon Hopf algebra approach to the irreducible representations of centralizer algebras: the Brauer, Birman-Wenzl, and type A Iwahori-Hecke algebras*, Adv. Math. **125** (1997), 1-94.

[L] G. Lusztig, *Introduction to Quantum Groups*, Birkhäuser, 1993.

[T] M. Takeuchi, *A two-parameter quantization of* $GL(n)$, Proc. Japan Acad. Ser. A Math. Sci. **66** (1990), 112-114.

A New Proof of the Skolem-Noether Theorem

JEFFREY BERGEN

Department of Mathematics
DePaul University
Chicago, Illinois 60614
jbergen@condor.depaul.edu

In this paper, we provide a new proof of the following version of the Skolem-Noether Theorem:

THEOREM. If R is a simple algebra which is finite dimensional over its center F, then all F-linear skew derivations δ and automorphisms σ are inner.

Previous proofs of the Skolem-Noether Theorem are applications of the Wedderburn-Artin Theorem on the structure of simple Artinian rings. However, the proof in this paper is almost entirely self-contained. It only makes use of the fact that if a simple algebra R is finite dimensional over its center F, then the ring $R \otimes R^{op}$ is a simple ring which is isomorphic to $End_F(R)$, the ring of F-linear maps from R to R. In this case, R^{op} represents the opposite ring to R and the tensor product is taken over F. Furthermore, the action of $R \otimes R^{op}$ on R is defined as $a \otimes b \rightharpoonup r = arb$ for all $a, b, r \in R$.

If σ is an automorphism of R, then a function $\delta : R \to R$, is called a skew-derivation or a σ-derivation if δ is an additive map such that $\delta(rs) = \delta(r)s + \sigma(r)\delta(s)$, for all $r, s \in R$. We say that σ is inner if there exists some $a \in R$ such that $\sigma(r) = ara^{-1}$, for all $r \in R$. We say that δ is inner if there exists some $a \in R$ such that $\delta(r) = ar - \sigma(r)a$, for all $r \in R$.

The argument used in our proof has the flavor of the shortest length arguments frequently used in the study of group rings, enveloping algebras, and smash products of Hopf algebras.

Proof of the Theorem. Let x and y be variables and then let M and N be the sets

$$M = (R \otimes R^{op}) \oplus (R \otimes R^{op})x \quad \text{and} \quad N = (R \otimes R^{op}) \oplus (R \otimes R^{op})y.$$

We can give M and N the structure of (R, R)-bimodules. First, the left R-module structures of M and N are defined as

$$r \cdot (a \otimes b) = ra \otimes b, \quad r \cdot (c \otimes d)x = (rc \otimes d)x, \quad \text{and} \quad r \cdot (e \otimes f)y = (re \otimes f)y,$$

for all $a, b, c, d, e, f, r \in R$.

Using the automorphism σ and σ-derivation δ, we define the right R-module structures of M and N as

$$(a \otimes b) \cdot r = ar \otimes b,$$

$$(c \otimes d)x \cdot r = (c\sigma(r) \otimes d)x + (c\delta(r) \otimes d),$$

and

$$(e \otimes f)y \cdot r = (e\sigma(r) \otimes f)y,$$

for all $a, b, c, d, e, f, r \in R$. For convenience, we can express the essential aspects of the last two equations as

$$(1 \otimes 1)x \cdot r = (\sigma(r) \otimes 1)x + (\delta(r) \otimes 1)$$

and

$$(1 \otimes 1)y \cdot r = (\sigma(r) \otimes 1)y.$$

We can also consider the sets M and N to be acting on R. In particular,

$$a \otimes b \rightharpoonup r = arb, \quad (c \otimes d)x \rightharpoonup r = c\delta(r)d, \quad \text{and} \quad (e \otimes f)y \rightharpoonup r = e\sigma(r)f,$$

for all $a, b, c, d, e, f, r \in R$. It is important to note that the (R, R)-bimodule structures of M and N are compatible with their actions on R. By this we mean that

$$(1) \qquad (r \cdot m \cdot s) \rightharpoonup t = r(m \rightharpoonup st),$$

for all $r, s, t \in R$ and $m \in M \cup N$. To see this, observe that

$$(r \cdot (a \otimes b) \cdot s) \rightharpoonup t = (ras \otimes b) \rightharpoonup t = rastb = r(astb) = r(a \otimes b \rightharpoonup st),$$

$$(r \cdot (c \otimes d)x \cdot s) \rightharpoonup t = ((rc\sigma(s) \otimes d)x + (rc\delta(s) \otimes d)) \rightharpoonup t =$$

$$rc\sigma(s)\delta(t)d + rc\delta(s)td = rc\delta(st)d = r((c \otimes d)x \rightharpoonup st),$$

and

$$(r \cdot (e \otimes f)y \cdot s) \rightharpoonup t = (re\sigma(s) \otimes f)y \rightharpoonup t = re\sigma(s)\sigma(t)f =$$

$$re\sigma(st)f = r((e \otimes f)y \rightharpoonup st),$$

for all $a, b, c, d, e, f, r, s, t \in R$.

Next, let $V = Ann_M(R)$ and $W = Ann_N(R)$ be the annihilators of R in, respectively, M and N. It follows from (1) that V and W are (R, R)-subbimodules of, respectively, M and N. Since $R \otimes R^{op}$ is isomorphic to

$End_F(R)$, it follows that if δ is F-linear then there exists $\sum_i a_i \otimes b_i \in R \otimes R^{op}$ such that $\delta(r) = \sum_i a_i r b_i$, for all $r \in R$. Therefore

$$v = (1 \otimes 1)x - \sum_i a_i \otimes b_i \in V.$$

Observe that we can choose the b_i to be part of an F-basis of R with $b_1 = 1$. Also note that since no nonzero element of $R \otimes R^{op}$ annihilates R, we have $(R \otimes R^{op}) \cap V = (R \otimes R^{op}) \cap W = 0$. Since $v \in V$, we now have $v \cdot \sigma(r) - r \cdot v \in V$, for all $r \in R$. Thus

$$v \cdot r - \sigma(r) \cdot v = ((1 \otimes 1)x - \sum_i a_i \otimes b_i) \cdot r - \sigma(r) \cdot ((1 \otimes 1)x - \sum_i a_i \otimes b_i) =$$

$$(\sigma(r) \otimes 1)x + (\delta(r) \otimes 1) - \sum_i a_i r \otimes b_i - (\sigma(r) \otimes 1)x + \sum_i \sigma(r) a_i \otimes b_i =$$

$$(2) \quad (\delta(r) \otimes 1) - \sum_i a_i r \otimes b_i + \sum_i \sigma(r) a_i \otimes b_i \in (R \otimes R^{op}) \cap V = 0.$$

Since $b_1 = 1$ and the b_i are linearly independent, it follows from (2) that

$$(\delta(r) - a_1 r + \sigma(r) a_1 \otimes 1) + \sum_{i \geq 2} (\sigma(r) a_i - a_i r) \otimes b_i = 0$$

and so,

$$\delta(r) - a_1 r + \sigma(r) a_1 = 0,$$

for all $r \in R$. This immediately implies that

$$\delta(r) = a_1 r - \sigma(r) a_1,$$

for all $r \in R$. Thus δ is inner.

Now, if σ is F-linear, then there also exists $\sum_i c_i \otimes d_i \in R \otimes R^{op}$ such that $\sigma(r) = \sum_i c_i r d_i$, for all $r \in R$. Therefore

$$w = (1 \otimes 1)y - \sum_i c_i \otimes d_i \in W.$$

Since $w \in W$, we also have $w \cdot r - \sigma(r) \cdot w \in W$, for all $r \in R$. Thus

$$w \cdot r - \sigma(r) \cdot w = ((1 \otimes 1)y - \sum_i c_i \otimes d_i) r - \sigma(r)((1 \otimes 1)y - \sum_i c_i \otimes d_i) =$$

$$(\sigma(r) \otimes 1)y - \sum_i c_i r \otimes d_i - (\sigma(r) \otimes 1)y + \sum_i \sigma(r) c_i \otimes d_i =$$

$$-\sum_i c_i r \otimes d_i + \sum_i \sigma(r) c_i \otimes d_i = \sum_i (\sigma(r) c_i - c_i r) \otimes d_i \in$$

$$(R \otimes R^{op}) \cap W = 0. \tag{3}$$

Since we may assume that the d_i are linearly independent, equation (3) indicates that $\sigma(r)c_i - c_i r = 0$, for all i and all $r \in R$. As a result,

$$\sigma(r)c_i = c_i r, \tag{4}$$

for all i and all $r \in R$. Since $\sum_i c_i \otimes d_i \neq 0$, there is some i such that $c_i \neq 0$. Therefore, (4) implies that $Rc_i = c_i R$ is a nonzero, two-sided ideal of R. Since R is simple, this tells us that $Rc_i = R$ and so, c_i is invertible. It now follows from (4) that

$$\sigma(r) = c_i r c_i^{-1},$$

for all $r \in R$. Thus σ is inner. □

ACKNOWLEDGMENT

The author was supported by a grant from the University Research Council at DePaul University.

Projectivity of a relative Hopf module over the subring of coinvariants

S. CAENEPEEL AND T. GUEDENON Faculty of Applied Sciences, Vrije Universiteit Brussel, VUB, B-1050 Brussels, Belgium
e-mail address: scaenepe@vub.ac.be, guedenon@caramail.com

ABSTRACT. Let k be a commutative ring, H a faithfully flat Hopf algebra with bijective antipode, A a k-flat right H-comodule algebra. We investigate when a relative Hopf module is projective over the subring of coinvariants $B = A^{\mathrm{co}H}$, and we study the semisimplicity of the category of relative Hopf modules.

Introduction

Let H be a Hopf algebra with bijective antipode, and A a left H-module algebra. We can then form the smash product $A\#H$. In [8], the second author investigated necessary and sufficient conditions for the projectivity of an $A\#H$-module M over the subring of invariants A^H.

Our starting point is the following: if H is finitely generated and projective, then H^* is also a Hopf algebra, A is a right H^*-comodule algebra, and $A^H = A^{\mathrm{co}H^*}$ The category of left $A\#H$-modules is isomorphic to ${}_A\mathcal{M}^{H^*}$, the category of relative (A, H^*)-Hopf modules, i.e. k-modules together with a left A-action and a right H^*-coaction, satisfying an appropriate compatibility condition. Thus [8] brings us necessary and sufficient conditions for a relative Hopf module to be projective as a module over the ring of coinvariants. In this paper, we will generalize these results to relative (A, H)-Hopf modules, where H is an arbitrary Hopf algebra with bijective antipode over a commutative ring k, and A is a (k-flat) right H-comodule algebra.

Our main result is Theorem 2.1, where we give necessary and sufficient conditions for projectivity of a relative Hopf module over the subring of coinvariants $B = A^{\mathrm{co}H}$. The main tool - based on the methods developed in [7] and [8] - is the basic fact that the canonical structure of A as left-right (A, H)-Hopf module is such that ${}_A\mathrm{Hom}^{\mathrm{H}}(\mathrm{A}, \mathrm{A})$, the k-module consisting of A-linear and H-colinear maps, is isomorphic to B (see Section 1). The result can be improved if we assume that there exists a total integral $\phi : H \to A$, see Proposition 2.2. In Proposition 2.5, we look at relative modules that are coinvariantly generated, and we will present some conditions that are sufficient, but in general not necessary for projectivity. These conditions have the advantage that they are easier to verify than the ones from

Research supported by the project G.0278.01 "Construction and applications of non-commutative geometry: from algebra to physics" from FWO Vlaanderen.

Theorem 2.1; they turn out to be necessary if the coinvariants functor is exact. In Sections 3 and 4, we will work over a field k. Our methods will be applied to discuss properties of injective and projective dimension in the category of relative Hopf modules (Section 3), and semisimplicity of the the category of relative Hopf modules. Our main result is Corollary 4.5, where we give a sufficient condition for the category of relative Hopf modules to be semisimple.
Let us finally mention that our results may be applied in the following particular cases:

- $A = H$ with comodule structure map Δ. In this case, B is isomorphic to k as k-algebra and trivial H-comodule.
- A crossed product $A = R\#_\sigma H$ where R is an algebra with H-action and σ is an invertible map in $Hom_k(H \otimes_k H, R)$ [10, Definition 7.1.1] is a right H-comodule algebra such that $A^{coH} = R$. The comodule map is given by $a\#_\sigma h :\rightarrow a\#_\sigma h_1 \otimes h_2$.
- $A = \oplus_{g\in G} A_g$ is a G-graded k-algebra for a group G and $H = kG$ is the group algebra of G. Then A is a right H-comodule algebra, $A^{\mathrm{co}H} = A_1$ and ${}_A\mathcal{M}^H$ is the category of G-graded left A-modules.
- $H = k[G]$, the affine coordinate ring of an affine k-group scheme G and A is a right $k[G]$-comodule algebra. A right $k[G]$-comodule is also called a left G-module (see [5], [9]). If G is linearly reductive, meaning that every G-module is completely reducible, then the functor $(-)^{\mathrm{co}k[G]} : \mathcal{M}^{k[G]} \rightarrow {}_k\mathcal{M}$ is exact. In the particular case of an affine algebraic group G over an algebraically closed field k acting rationally on A, ${}_A\mathcal{M}^{k[G]}$ is the category of rational (A, G)-modules (see [6], [10]).

For more detail on Hopf algebras and the category of relative Hopf modules, we refer to the literature, see for example [1], [2], [3], [4], [11], [13].

1. Preliminary Results

Troughout this paper, k will be a commutative ring, and H will be a Hopf algebra with bijective antipode S. We freely use the Sweedler-Heyneman notation for the comultiplication:

$$\Delta(h) = h_1 \otimes h_2$$

We use a similar notation for the right H-coaction ρ on a right H-comodule M:

$$\rho(m) = m_0 \otimes m_1$$

$\mathcal{M}^H$ will be the category of right H-comodules and right H-colinear maps. The k-module consisting of all right H-comodule maps between two right H-comodules M and N will be denoted by $\mathrm{Hom}^{\mathrm{H}}(\mathrm{M}, \mathrm{N})$.
The tensor product of two right H-comodules M and N is again a right H-comodule. The right H-coaction on $M \otimes N$ is given by

$$\rho(m \otimes n) = m_0 \otimes n_0 \otimes m_1 n_1$$

For any right H-comodule M,

$$M^{\mathrm{co}H} = \{m \in M \mid \rho(m) = m \otimes 1\}$$

is called the submodule of coinvariants of M.
A right H-comodule algebra A is an algebra in the category $\mathcal{M}^H$. This means that A is a right H-comodule and a k-algebra such that the unit and multiplication maps are right H-colinear:

$$\rho_A(ab) = a_0b_0 \otimes a_1b_1 \quad \text{and} \quad \rho_A(1_A) = 1_A \otimes 1_H$$

A (left-right) relative (A, H)-Hopf module is a k-module M together with a left A-action and a right H-coaction such that

$$\rho(am) = a_0m_0 \otimes a_1m_1$$

for all $a \in A$ and $m \in M$. ${}_A\mathcal{M}^H$ is the category of relative (A, H)-Hopf modules and left A-linear and right H-colinear maps. For two relative Hopf modules M and N, we write ${}_A\mathrm{Hom}^{\mathrm{H}}(\mathrm{M},\mathrm{N})$ for the k-module consisting of all A-linear H-colinear maps from M to N.
$B = A^{\mathrm{co}H}$ is a k-algebra. The coinvariants $M^{\mathrm{co}H}$ of a relative Hopf module M form a B-module.

LEMMA 1.1. 1. *Take $M \in \mathcal{M}^H$ and $N \in {}_A\mathcal{M}^H$.*
a) *$N \otimes M \in {}_A\mathcal{M}^H$, with left A-action given by*

$$a(n \otimes m) = (an) \otimes m$$

b) *If H is commutative, then $M \otimes N \in {}_A\mathcal{M}^H$, with left A-action given by*

$$a(m \otimes n) = m \otimes an$$

2. *Let A be commutative, and take $M, N \in {}_A\mathcal{M}^H$. Then $M \otimes_A N \in {}_A\mathcal{M}^H$, under the coaction*

$$\rho(m \otimes n) = m_0 \otimes n_0 \otimes m_1n_1$$

3. *Take $M, N \in {}_A\mathcal{M}^H$, and assume that M is finitely generated projective as a left A-module.*
a) *${}_A\mathrm{Hom}(\mathrm{M},\mathrm{N}) \in \mathcal{M}^{\mathrm{H}}$ and*

$${}_A\mathrm{Hom}^{\mathrm{H}}(\mathrm{M},\mathrm{N}) = {}_{\mathrm{A}}\mathrm{Hom}(\mathrm{M},\mathrm{N})^{\mathrm{coH}}$$

b) *$N \cong {}_A\mathrm{Hom}(\mathrm{A},\mathrm{N}) \in {}_{\mathrm{A}}\mathcal{M}^{\mathrm{H}}$, with left A-action*

$$(af)(u) = f(ua)$$

c) *If A is commutative, then ${}_A\mathrm{Hom}(\mathrm{M},\mathrm{N}) \in {}_{\mathrm{A}}\mathcal{M}^{\mathrm{H}}$.*

Proof. 1a) The only nontrivial thing that we have to show is the compatibility of the right H-coaction with the left A-action. For any $a \in A$, $m \in M$ and $n \in N$, we have

$$\rho_{N\otimes M}((an) \otimes m) = a_0n_0 \otimes m_0 \otimes a_1n_1m_1 = a_0(n \otimes m)_0 \otimes a_1(n \otimes m)_1$$

1b) As in 1a), we have to prove compatibility. For $a \in A$, $m \in M$ and $n \in N$, we have

$$\begin{aligned}\rho_{M\otimes N}(m \otimes an) &= m_0 \otimes a_0n_0 \otimes m_1a_1n_1 \\ &= m_0 \otimes a_0n_0 \otimes a_1m_1n_1 = a_0(m \otimes n)_0 \otimes a_1(m \otimes n)_1\end{aligned}$$

2) The right H-coaction on $M \otimes_A N$ is well-defined since

$$\rho(ma \otimes n) = m_0 a_0 \otimes n_0 \otimes m_1 a_1 n_1 = \rho(m \otimes an)$$

and it is straightforward to verify that $M \otimes_A N$ is a right H-comodule. Finally

$$\rho_{M\otimes_A N}(am \otimes n) = a_0 m_0 \otimes n_0 \otimes a_1 m_1 n_1 = a_0(m \otimes n)_0 \otimes a_1(m \otimes n)_1$$

3a) The fact that M is finitely generated projective as a left A-module implies that we have a natural isomorphism

$$_A\mathrm{Hom}(\mathrm{M},\mathrm{N}) \otimes \mathrm{H} \cong {}_A\mathrm{Hom}(\mathrm{M},\mathrm{N} \otimes \mathrm{H})$$

Using this isomorphism, we define a map

$$\pi : {}_A\mathrm{Hom}(\mathrm{M},\mathrm{N}) \to {}_A\mathrm{Hom}(\mathrm{M},\mathrm{N}) \otimes \mathrm{H}; \quad \pi(\mathrm{f}) = \mathrm{f}_0 \otimes \mathrm{f}_1$$

by

$$(\pi(f))(m) = f_0(m) \otimes f_1 = f(m_0)_0 \otimes S^{-1}(m_1) f(m_0)_1$$

defining a right H-coaction on $_A\mathrm{Hom}(\mathrm{M},\mathrm{N})$.
Take $f \in {}_A\mathrm{Hom}^{\mathrm{H}}(\mathrm{M},\mathrm{N})$ and $m \in M$. Then

$$f(m)_0 \otimes f(m)_1 = f(m_0) \otimes m_1$$

so

$$\begin{aligned}\pi(f)(m) &= f_0(m) \otimes f_1 = f(m_0)_0 \otimes S^{-1}(m_1) f(m_0)_1 \\ &= f(m_0) \otimes S^{-1}(m_2) m_1 = f(m_0) \otimes \varepsilon(m_1) 1 = (f \otimes 1)(m)\end{aligned}$$

proving that f is coinvariant. Conversely, take a coinvariant $f \in {}_A\mathrm{Hom}(\mathrm{M},\mathrm{N})^{\mathrm{coH}}$. For every $m \in M$ we have

$$f_0(m) \otimes f_1 = f(m_0)_0 \otimes S^{-1}(m_1) f(m_0)_1 = f(m) \otimes 1$$

and we deduce that

$$\begin{aligned}f(m_0) \otimes m_1 &= f(m_0) \otimes m_1 1 \\ &= f(m_0)_0 \otimes m_2 S^{-1}(m_1) f(m_0)_1 \\ &= f(m)_0 \otimes f(m)_1\end{aligned}$$

so f is H-colinear, as needed.
3b) By 3a), $_A\mathrm{Hom}(\mathrm{A},\mathrm{N})$ is a right H-comodule. The map

$$\psi : {}_A\mathrm{Hom}(\mathrm{A},\mathrm{N}) \to \mathrm{N}, \ \psi(\mathrm{f}) = \mathrm{f}(1)$$

is clearly an isomorphism of A-modules. ψ is also right H-colinear, since

$$\begin{aligned}\psi(f)_o \otimes \psi(f)_1 &= f(1)_0 \otimes f(1)_1 = f(1_0)_0 \otimes S^{-1}(1) f(1_0)_1 \\ &= f_0(1) \otimes f_1 = \psi(f_0) \otimes f_1\end{aligned}$$

3c) $_A\mathrm{Hom}(\mathrm{M},\mathrm{N})$ is an A-module, and, by 3a), it is a right H-comodule. The A-action and H-coaction are compatible since

$$\begin{aligned}((af)_0 \otimes (af)_1)(m) &= ((af)(m_0))_0 \otimes S^{-1}(m_1)((af)(m_0))_1 \\ &= a_0(f(m_0)_0) \otimes S^{-1}(m_1) a_1(f(m_0)_1) \\ &= a_0(f(m_0)_0) \otimes a_1 S^{-1}(m_1)(f(m_0)_1) = a_0(f_0(m)) \otimes a_1 f_1 \\ &= (a_0 f_0)(m) \otimes a_1 f_1 = (a_0 f_0 \otimes a_1 f_1)(m)\end{aligned}$$

and this proves that ${}_A\mathrm{Hom}(\mathrm{M},\mathrm{N}) \in {}_A\mathcal{M}^{\mathrm{H}}$. □

Consider the functor

$$(-)^{\mathrm{co}H} : {}_A\mathcal{M}^H \longrightarrow {}_B\mathcal{M}$$

$(-)^{coH}$ commutes with direct sums, and has a left adjoint

$$T = A \otimes_B - : {}_B\mathcal{M} \longrightarrow {}_A\mathcal{M}^H$$

The unit and counit of the adjunction are the following: for $N \in {}_B\mathcal{M}$ and $M \in {}_A\mathcal{M}^H$:

$$u_N : N \longrightarrow (A \otimes {}_B N)^{\mathrm{co}H},\ u_N(n) = 1 \otimes n$$

$$c_M : A \otimes_B M^{\mathrm{co}H} \longrightarrow M,\ c_M(a \otimes m) = am$$

LEMMA 1.2. 1. *The functor* $(-)^{\mathrm{co}H}$ *is naturally isomorphic to*

$${}_A\mathrm{Hom}^{\mathrm{H}}(\mathrm{A}, -) : {}_A\mathcal{M}^{\mathrm{H}} \longrightarrow {}_B\mathcal{M}$$

2. $c_{A^{(I)}}$ *is an isomorphism for any set* I.
3. $u_{B^{(I)}}$ *is an isomorphism for any set* I.

Proof. 1) By Lemma 1.1 (3), ${}_A\mathrm{Hom}^{\mathrm{H}}(\mathrm{A},\mathrm{M}) = {}_A\mathrm{Hom}(\mathrm{A},\mathrm{M})^{\mathrm{coH}}$, and ${}_A\mathrm{Hom}(\mathrm{A},\mathrm{M})$ is isomorphic to M in ${}_A\mathcal{M}^H$. So ${}_A\mathrm{Hom}^{\mathrm{H}}(\mathrm{A},\mathrm{M})$ is B-isomorphic to $M^{\mathrm{co}H}$.
2) It suffices to observe that

$$(A^{(I)})^{\mathrm{co}H} = (A^{\mathrm{co}H})^{(I)} = B^{(I)}$$

and

$$c_{A^{(I)}} : A \otimes_B (A^{\mathrm{co}H})^{(I)} \longrightarrow A^{(I)}$$

is the canonical isomorphism.
3) From the fact that u and c are the unit and the counit of the adjunction, we derive that

$$(c_{A^{(I)}})^{\mathrm{co}H} \circ u_{(A^{(I)})^{\mathrm{co}H}} = 1_{(A^{(I)})^{\mathrm{co}H}}$$

and it follows that $u_{(A^{(I)})^{\mathrm{co}H}} = u_{(A^{\mathrm{co}H})^{(I)}}$ is an isomorphism. □

LEMMA 1.3. *Take* $M \in {}_A\mathcal{M}^H$. *Then*

1. $M \otimes H \in {}_A\mathcal{M}^H$ *and* $(M \otimes H)^{\mathrm{co}H} \cong M$ *as a B-module.*
2. u_M *is an injection of B-modules.*

Proof. 1) The structure maps on $M \otimes H$ are the following:

$$a(m \otimes h) = am \otimes h\ ;\ \rho(m \otimes h) = m_0 \otimes h_1 \otimes m_1 h_2$$

It is clear that the map

$$f : M \to M \otimes H;\ f(m) = m_0 \otimes S(m_1)$$

is left B-linear. Also

$$\rho(f(m)) = m_0 \otimes S(m_3) \otimes m_1 S(m_2) = m_0 \otimes S(m_1) \otimes 1$$

so $f(m) \in (M \otimes H)^{\mathrm{co}H}$, and we have a B-linear map

$$f : M \to (M \otimes H)^{\mathrm{co}H}$$

Its inverse is

$$g : (M \otimes H)^{\mathrm{co}H} \to M;\ g(m \otimes h) = \varepsilon(h)m$$

It is easy to see that g is a left inverse of f. Take $\sum_i m_i \otimes h_i \in (M \otimes H)^{\mathrm{co}H}$. Then

$$f(g(\sum_i m_i \otimes h_i)) = \sum_i \varepsilon(h_i)f(m_i) = \sum_i \varepsilon(h_i)m_{i0} \otimes S(m_{i1})$$

We also have that

$$\sum_i m_{i0} \otimes h_{i1} \otimes m_{i1}h_{i2} = \sum_i m_i \otimes h_i \otimes 1$$

applying S to the third tensor factor, and then multiplying the second and third factor, we obtain

$$\sum_i \varepsilon(h_i)m_{i0} \otimes S(m_{i1}) = \sum_i m_i \otimes h_i$$

and it follows that g is also a right inverse of f.

2) Set $W = M \otimes H$. By Lemma 1.2,

$$(c_W)^{coH} \circ u_{W^{coH}} = 1_{W^{coH}}$$

By 1), $W^{\mathrm{co}H} \cong M$ as B-modules, so u_M is an injection of B-modules. □

2. Projectivity of relative Hopf modules

We keep the notation of Section 1. We will now discuss when a relative Hopf module is a projective object as a B-module.

THEOREM 2.1. *For a relative Hopf module $P \in {}_A\mathcal{M}^H$, the following conditions are equivalent:*

1. *P is projective in ${}_B\mathcal{M}$;*
2. *$A \otimes_B P$ is isomorphic in ${}_A\mathcal{M}^H$ to a direct summand of a direct sum of copies of A, and u_P is surjective (bijective);*
3. *There is a direct summand M in ${}_A\mathcal{M}^H$ of a direct sum of copies of A such that $M^{\mathrm{co}H} \cong P$ as a B-module.*

Proof. 1) ⇒ 2). Let $p: B^{(I)} \longrightarrow P$ be a split epimorphism in ${}_B\mathcal{M}$. Then

$$1 \otimes p: A \otimes_B B^{(I)} \longrightarrow A \otimes_B P$$

is a split epimorphism in ${}_A\mathcal{M}^H$ and $A \otimes_B B^{(I)} \cong A^{(I)}$ in ${}_A\mathcal{M}^H$. Furthermore

$$u_P \circ p = (1 \otimes p)^{\mathrm{co}H} \circ u_{B^{(I)}}$$

and, by Lemma 1.2, $u_{B^{(I)}}$ is an isomorphism. Thus u_P is an epimorphism and it follows from Lemma 1.3 that u_P is bijective.
2) ⇒ 3). If $u_P : P \longrightarrow (A \otimes_B P)^{\mathrm{co}H}$ is an epimorphism (and therefore an isomorphism by Lemma 1.3) and if M is a direct summand in ${}_A\mathcal{M}^H$ of a direct sum of copies of A isomorphic to $A \otimes_B P$, then M satisfies the required condition.
3) ⇒ 1) Let M be a direct summand in ${}_A\mathcal{M}^H$ of a direct sum of copies of A, such that $M^{\mathrm{co}H} \cong P$ as a B-module. Then there exists a split epimorphism $f: A^{(I)} \longrightarrow M$ in ${}_A\mathcal{M}^H$. Thus

$$f^{\mathrm{co}H} : (A^{(I)})^{\mathrm{co}H} \longrightarrow M^{\mathrm{co}H} \cong P$$

is a split epimorphism in ${}_B\mathcal{M}$ and $(A^{(I)})^{\mathrm{co}H} \cong B^{(I)}$. So P is projective in ${}_B\mathcal{M}$. □

Recall that a total integral is an H-colinear map $\phi : H \to A$ such that $\phi(1_H) = 1_A$. If there exists a total integral, then u_N is an isomorphism of B-modules, for every $N \in {}_B\mathcal{M}$ (see e.g. [3, Lemma 23]). From Theorem 2.1, we easily obtain the following result:

PROPOSITION 2.2. *Assume that there exists a total integral $H \to A$, and take $P \in {}_A\mathcal{M}^H$. P is projective as a B-module if and only if $A \otimes_B P$ is isomorphic in ${}_A\mathcal{M}^H$ to a direct summand of a direct sum of copies of A.*

An H-ideal I of A is an H-subcomodule of A which is also an ideal of A. We will say that A is H-simple if A has no nontrivial H-ideals.

LEMMA 2.3. *If A is commutative H-simple, then $A^{\mathrm{co}H}$ is a field.*

Proof. Let a be a nonzero element in $A^{\mathrm{co}H}$. Then Aa is a nonzero H-ideal of A. But A is H-simple and $Aa \neq 0$, so $Aa = A$. Hence, we can find an element b in A such that $ba = 1$. □

Theorem 2.1 gives necessary and sufficient conditions for the projectivity of $M \in {}_A\mathcal{M}^H$ as a B-module. These conditions might be difficult to check, and this is why we look for sufficient conditions that are easier.
$M \in {}_A\mathcal{M}^H$ is called coinvariantly generated if $M = AM^{\mathrm{co}H}$, or, equivalently, if the canonical map c_M is surjective.

LEMMA 2.4. *For any B-module M, $A \otimes_B M$ is coinvariantly generated. In particular, A is coinvariantly generated.*

Proof. For every $m \in M$, $1 \otimes_B m \in (A \otimes_B M)^{\mathrm{co}H}$, and $c_{A\otimes_B M}(a \otimes_B 1 \otimes_B m = a \otimes_B m$, hence $c_{A\otimes_B M}$ is surjective. □

PROPOSITION 2.5. *Take $P \in {}_A\mathcal{M}^H$ and consider the following conditions:*

1. *$A \otimes_B P$ is projective in ${}_A\mathcal{M}^H$;*
2. *there exists a projective coinvariantly generated object $M \in {}_A\mathcal{M}^H$ such that $M^{\mathrm{co}H} \cong P$ as a B-module;*
3. *P is projective as a B-module.*

Then 1) ⇒ 2) ⇒ 3). If the functor $(-)^{\mathrm{co}H} : {}_A\mathcal{M}^H \to {}_B\mathcal{M}$ is exact, then conditions 1), 2) and 3) are equivalent.

Note that the functor $(-)^{\mathrm{co}H} : {}_A\mathcal{M}^H \to {}_B\mathcal{M}$ is exact in the following situations:

- $(-)^{\mathrm{co}H} : {}_A\mathcal{M}^H \to {}_k\mathcal{M}$ is exact; this is the case if k is a field and H is a cosemisimple Hopf algebra [11, Lemma 2.4.3].
- A is right H-coflat (see e.g. [3, Lemma 22]); if k is a field, then this condition is equivalent to A to being injective in $\mathcal{M}^H$ (see e.g.[3, Theorem 1]).

Proof. 1) ⇒ 2). Let $f : B^{(I)} \to P$ be an epimorphism in ${}_B\mathcal{M}$. Then

$$1 \otimes f : A \otimes_B B^{(I)} \to A \otimes_B P$$

is an epimorphism in ${}_A\mathcal{M}^H$. Therefore $1 \otimes f$ splits and we have the following commutative diagram with exact rows:

$$\begin{array}{ccccc} B^{(I)} & \xrightarrow{\quad f \quad} & P & \longrightarrow & 0 \\ \downarrow u_{B^{(I)}} & & \downarrow u_P & & \\ (A \otimes_B B^{(I)})^{\mathrm{co}H} & \xrightarrow{(1\otimes f)^{\mathrm{co}H}} & (A \otimes_B P)^{\mathrm{co}H} & \longrightarrow & 0 \end{array}$$

By Lemma 1.2, $u_{B^{(I)}}$ is an isomorphism. Therefore, u_P is surjective and, by Lemma 1.3, it is an isomorphism. Thus $M = A \otimes_B P$ satisfies condition 2).
2) ⇒3) Consider the map

$$f:\ A^{(M^{\mathrm{co}H})} \to M;\ f((a_m)_{m\in M^{\mathrm{co}H}}) = \sum_{m\in M^{\mathrm{co}H}} a_m m$$

For each $m \in M^{\mathrm{co}H}$, we have $\rho(m) = m \otimes 1$, hence

$$\rho_M(\sum_{m\in M^{\mathrm{co}H}} (a_m m)) = \sum_{m\in M^{\mathrm{co}H}} (a_m)_0 m \otimes (a_m)_1$$

On the other hand

$$(a_m)_{m\in M^{\mathrm{co}H}} = \sum_{m\in M^{\mathrm{co}H}} (0, 0, \cdots, a_m, 0, \cdots, 0)$$

The linearity of ρ implies that

$$\begin{aligned} &((a_m)_{m\in M^{\mathrm{co}H}})_0 \otimes ((a_m)_{m\in M^{\mathrm{co}H}})_1 \\ &= \sum_{m\in M^{\mathrm{co}H}} (0, \cdots 0, a_m, 0, \cdots 0)_0 \otimes (0, \cdots 0, a_m, 0, \cdots 0)_1 \\ &= \sum_{m\in M^{\mathrm{co}H}} (0, \cdots 0, (a_m)_0, 0, \cdots 0) \otimes (a_m)_1 \end{aligned}$$

Thus

$$\begin{aligned} &(f \otimes id_H)(((a_m)_{m\in M^{\mathrm{co}H}})_0 \otimes ((a_m)_{m\in M^{\mathrm{co}H}})_1) \\ &= (f \otimes id_H)(\sum_{m\in M^{\mathrm{co}H}} (0, \cdots 0, (a_m)_0, 0, \cdots 0) \otimes (a_m)_1) \\ &= \sum (a_m)_0 m \otimes (a_m)_1 \end{aligned}$$

and it follows that f is H-colinear. It is clear that f is A-linear, so it is a morphism in ${}_A\mathcal{M}^H$. f is an epimorphism by the assumption $M = AM^{\mathrm{co}H}$. f splits since M is a projective object in ${}_A\mathcal{M}^H$, so M is a direct summand in ${}_A\mathcal{M}^H$ of a direct sum of copies of A. By Theorem 2.1, P is projective as a B-module.
3) ⇒1) P is projective in ${}_B\mathcal{M}$, so the functor $\mathrm{Hom}_B(P, -) :\ {}_B\mathcal{M} \to {}_k\mathcal{M}$ is exact. On the other hand, we have an adjoint pair of functors $(A \otimes_B (-), (-)^{\mathrm{co}H})$ and a natural isomorphism

$${}_A\mathrm{Hom}^H(A \otimes_B P, -) \cong \mathrm{Hom}_B(P, (-)^{\mathrm{co}H})$$

By assumption, the functor $(-)^{\mathrm{co}H}$ is exact, and it follows that

$$ {}_A\mathrm{Hom}^H(A \otimes_B P, -) : {}_A\mathcal{M}^H \to {}_k\mathcal{M} $$

is exact, since it is isomorphic to the composition of two exact functors. This means that $A \otimes_B P$ is a projective object in ${}_A\mathcal{M}^H$. □

Remark 2.6. It was pointed out to us by the referee that the proof of 2) ⇒3) becomes much simpler if we work over a field k; indeed, $A \otimes M^{\mathrm{co}H}$ is a direct sum in ${}_A\mathcal{M}^H$ of copies of A (indexed by a basis of $M^{\mathrm{co}H}$, and the map $g = p \circ c_M : A \otimes M^{\mathrm{co}H} \to M$ is surjective ($p : A \otimes M^{\mathrm{co}H} \to A \otimes_B M^{\mathrm{co}H}$ being the canonical projection). g splits since M is projective in ${}_A\mathcal{M}^H$, and it follows from Theorem 2.1 that P is projective as a B-module.

3. Projective and injective dimension in ${}_A\mathcal{M}^H$

LEMMA 3.1. *Assume that H and A are commutative, and take $M, N, P \in {}_A\mathcal{M}^H$, with N finitely generated projective as an A-module.*

1. *We have a k-isomorphism*

$$ {}_A\mathrm{Hom}^H(M, {}_A\mathrm{Hom}(N, P)) \cong {}_A\mathrm{Hom}^H(M \otimes_A N, P) $$

2. *The functor ${}_A\mathrm{Hom}(N, -) : {}_A\mathcal{M}^H \to {}_A\mathcal{M}^H$ preserves injectives.*

Proof. 1) We have a natural isomorphism

$$ \phi : {}_A\mathrm{Hom}(M, {}_A\mathrm{Hom}(N, P)) \to {}_A\mathrm{Hom}(M \otimes_A N, P) $$

$$ \phi(f)(m \otimes n) = f(m)(n) $$

A standard computation shows that f is H-colinear if and only if $\phi(f)$ is H-colinear.
2) Let I be an injective object in ${}_A\mathcal{M}^H$. Then the functor

$$ {}_A\mathrm{Hom}^H(-, I) : {}_A\mathcal{M}^H \to {}_k\mathcal{M} $$

is exact. N is projective as an A-module, so $(-) \otimes_A N$ is also exact, and it follows from 1) that

$$ {}_A\mathrm{Hom}^H(-, {}_A\mathrm{Hom}(N, I)) : {}_A\mathcal{M}^H \to {}_k\mathcal{M} $$

is exact. □

Let k be a field. Then ${}_A\mathcal{M}^H$ is a Grothendieck category with enough injective objects, and for any $M \in {}_A\mathcal{M}^H$ we can consider the right derived functors ${}_A\mathrm{Ext}^{H^i}(M, -)$ of ${}_A\mathrm{Hom}^H(M, -) : {}_A\mathcal{M}^H \to {}_k\mathcal{M}$.

PROPOSITION 3.2. *Let k be a field, and assume that H and A are commutative. Take $M, N, P \in {}_A\mathcal{M}^H$, with N finitely generated projective as an A-module. Then*

$$ {}_A\mathrm{Ext}^{H^i}(M, {}_A\mathrm{Hom}(N, P)) \cong {}_A\mathrm{Ext}^{H^i}(M \otimes_A N, P) $$

Proof. By Lemma 3.1, the functors ${}_A\mathrm{Hom}^H(M, {}_A\mathrm{Hom}(N, -))$ and ${}_A\mathrm{Hom}^H(M \otimes_A N, -)$ are isomorphic. ${}_A\mathrm{Hom}(N, -)$ is exact, since N is projective, and it preserves injectives, by Lemma 3.1. Therefore it preserves injective resolutions. □

Let ${}_A\mathrm{pdim}^H(-)$ and ${}_A\mathrm{injdim}^H(-)$ denote respectively the projective and injective dimension in ${}_A\mathcal{M}^H$.

COROLLARY 3.3. *With assumptions as in Proposition 3.2, we have*

$$ {}_A\mathrm{pdim}^H(M \otimes_A N) \leq {}_A\mathrm{pdim}^H(M) \quad (1)$$

$$ {}_A\mathrm{injdim}^H({}_A\mathrm{Hom}(N,P)) \leq {}_A\mathrm{injdim}^H(P) \quad (2)$$

Remarks 3.4. 1) The conclusions of Lemma 3.1, Proposition 3.2 and Corollary 3.3 remain valid without the assumption that N is projective, if we assume that A is semisimple.
2) As a consequence of the Fundamental Theorem for Hopf modules ([11, 1.9.4], the results of Proposition 3.2 and Corollary 3.3 also hold without the assumption that N is projective, in the case where $A = H$.

4. Semisimplicity of the category of relative Hopf modules

Throughout this Section, k will be a field, and A a k-algebra. If V is a finite dimensional vector space, then $A \otimes V$ is finitely generated projective as an A-module. If V is an H-comodule, then $A \otimes V \in {}_A\mathcal{M}^H$. So if $N \in {}_A\mathcal{M}^H$, then, by Lemma 1.1, ${}_A\mathrm{Hom}(A \otimes V, N) \in \mathcal{M}^H$, and ${}_A\mathrm{Hom}(A \otimes V, N)$ and $\mathrm{Hom}(V, N)$ are isomorphic as H-comodules.
Let V be a finite-dimensional projective object of $\mathcal{M}^H$. Then $A \otimes V$ is a projective object of ${}_A\mathcal{M}^H$; this follows from the fact that

$$\begin{aligned} {}_A\mathrm{Hom}^H(A \otimes V, N) &\cong {}_A\mathrm{Hom}(A \otimes V, N)^{coH} \\ &\cong \mathrm{Hom}(V, N)^{coH} \cong \mathrm{Hom}^H(V, N) \end{aligned}$$

for all $N \in {}_A\mathcal{M}^H$.

PROPOSITION 4.1. *$M \in {}_A\mathcal{M}^H$ is finitely generated as an A-module if and only if there exist a finite-dimensional H-comodule V and an epimorphism of A-modules and H-comodules $\pi : A \otimes V \longrightarrow M$.*

Proof. Assume that V and π exist. Then $A \otimes V$ is finitely generated as A-module and M is a quotient of $A \otimes V$ in ${}_A\mathcal{M}$, and therefore finitely generated as a left A-module.
Conversely, let M be finitely generated as an A-module by $\{m_1, \cdots, m_n\}$. For each i, we can find a finite dimensional H-subcomodule W_i of M containing m_i, see [11, 5.1.1]. $V = \sum W_i$ is a finite dimensional H-subcomodule of M containing the m_i and the k-linear map

$$\pi : A \otimes V \to M; \ \pi(a \otimes v) = av$$

is an epimorphism of A-modules and H-comodules. □

Let H^* be the linear dual of H, and let M and N be two right H-comodules. Then $\mathrm{Hom}_k(M, N)$ is a left H^*-module, under the action

$$(h^* f)(m) = h^*(S^{-1}(m_1) f(m_0)_1) f(m_0)$$

for all $h^* \in H^*$, $f \in \mathrm{Hom}_k(M, N)$ and $m \in M$. If M and N are relative Hopf modules, then ${}_A\mathrm{Hom}(M, N)$ is a left H^*-submodule of $\mathrm{Hom}_k(M, N)$ (see the proof of [13, Propositions 1.1 and 1.5].
A left H^*-module M is called rational if the H^*-action comes from a right H-coaction.

PROPOSITION 4.2. *Take $M, N \in {}_A\mathcal{M}^H$, and assume that M is finitely generated as an A-module. Then ${}_A\mathrm{Hom}(M,N)$ is a right H-comodule.*

Proof. Let V and π be as in Proposition 4.1. The map

$$ {}_A\mathrm{Hom}_A(\pi, N) : \ {}_A\mathrm{Hom}(M,N) \to {}_A\mathrm{Hom}(A \otimes V, N) $$

is injective, and it is easy to show that it is H^*-linear. ${}_A\mathrm{Hom}(A \otimes V, N)$ is an H-comodule, by Lemma 1.1, so it is a rational H^*-module, and - being a submodule - ${}_A\mathrm{Hom}(M,N)$ is also a rational H^*-module. The H^*-action on it is induced by an H-coaction, making ${}_A\mathrm{Hom}(M,N)$ a right H-comodule. □

It follows from Proposition 4.2 that if A is commutative and if $M, N \in {}_A\mathcal{M}^H$, with M finitely generated as an A-module, then ${}_A\mathrm{Hom}(M,N) \in {}_A\mathcal{M}^H$.
Recall that $M \in {}_A\mathcal{M}^H$ is called simple if it has no proper subobjects; a direct sum of simple objects is called semisimple. ${}_A\mathcal{M}^H$ is termed semisimple if every $M \in {}_A\mathcal{M}^H$ is semisimple.
We say that ${}_A\mathcal{M}^H$ satisfies condition (†) if the functor

$$ {}_A\mathrm{Hom}(M, -) : \ {}_A\mathcal{M}^H \to \mathcal{M}^H $$

is exact, for every $M \in {}_A\mathcal{M}^H$ that is finitely generated as an A-module.
Note that ${}_A\mathcal{M}^H$ satisfies condition (†) if A is semisimple; it follows from the Fundamental Theorem for Hopf modules ([11, 1.9.4]) that ${}_H\mathcal{M}^H$ satisfies condition (†).

PROPOSITION 4.3. *Assume that ${}_A\mathcal{M}^H$ satisfies condition (†), and that one of the two following conditions holds:*

1. *$(-)^{\mathrm{co}H} : \mathcal{M}^H \to {}_k\mathcal{M}$ is exact;*
2. *A and H are commutative, and $(-)^{\mathrm{co}H} : {}_A\mathcal{M}^H \to {}_B\mathcal{M}$ is exact.*

If $M \in {}_A\mathcal{M}^H$ is finitely generated as an A-module, then it is a projective object in ${}_A\mathcal{M}^H$

Proof. 1) The functor ${}_A\mathrm{Hom}^H(M, -)$ is exact since it is the composition of the exact functors ${}_A\mathrm{Hom}(M, -)$ and $(-)^{\mathrm{co}H}$.
2) It follows from condition (†) and the fact that H and A are commutative that

$$ {}_A\mathrm{Hom}(M, -) : \ {}_A\mathcal{M}^H \to {}_A\mathcal{M}^H $$

is exact. Now $(-)^{\mathrm{co}H} : {}_A\mathcal{M}^H \to {}_B\mathcal{M}$ and the restriction of scalars functor ${}_B\mathcal{M} \to {}_k\mathcal{M}$ are both exact, and ${}_A\mathrm{Hom}^H(M, -)$ is again exact, being the composition of three exact functors. □

COROLLARY 4.4. *Let A and H be as in Proposition 4.3, and assume moreover that A is left noetherian. Then every $M \in {}_A\mathcal{M}^H$ which is finitely generated as an A-module is the direct sum of a family of simple subobjects that are finitely generated as A-modules, and consequently M is a semisimple object in ${}_A\mathcal{M}^H$.*

Proof. Let N be a subobject of M in ${}_A\mathcal{M}^H$. Then N and M/N are finitely generated A-modules, since A is left noetherian. It follows from Proposition 4.3 that N and M/N are projective objects in ${}_A\mathcal{M}^H$, and the exact sequence

$$ 0 \to N \to M \to M/N \to 0 $$

splits in ${}_A\mathcal{M}^H$. □

Let V be a right H-subcomodule of $M \in {}_A\mathcal{M}^H$. It is clear that AV is a subobject of M in ${}_A\mathcal{M}^H$.

COROLLARY 4.5. *Let A and H be as in Corollary 4.4. Then ${}_A\mathcal{M}^H$ is a semisimple category.*

Proof. It is well-known (see e.g. [11, 5.1.1]) that every $m \in M$ is contained in a finite-dimensional H-subcomodule V_m of M. AV_m is then finitely generated as an A-module, and, by Corollary 4.4, the direct sum in ${}_A\mathcal{M}^H$ of a family of simple subobjects of AV_m (and of M), all finitely generated as A-modules. In particular, every $m \in M$ is contained in a simple subobject of M, and this implies that M is the sum of a family of simple subobjects. Since the intersection of two simple subobjects is trivial, it follows that this sum is a direct sum. □

COROLLARY 4.6. *Let A be left noetherian semisimple and H cosemisimple. Then every object M of ${}_A\mathcal{M}^H$ is a direct sum in ${}_A\mathcal{M}^H$ of a family of simple subobjects of M finitely generated as A-modules. Hence M is a semisimple object in ${}_A\mathcal{M}^H$ and ${}_A\mathcal{M}^H$ is a semisimple category.*

REFERENCES

[1] E. Abe, "Hopf Algebras", Cambridge University Press, Cambridge, 1977.

[2] S. Caenepeel, "Brauer groups, Hopf algebras and Galois theory", *K-Monographs Math.* **4**, Kluwer Academic Publishers, Dordrecht, 1998.

[3] S. Caenepeel, G. Militaru, and Shenglin Zhu, "Frobenius and separable functors for generalized module categories and nonlinear equations", *Lecture Notes in Math.* **1787**, Springer Verlag, Berlin, 2002.

[4] S. Dăscălescu, C. Năstăsescu and Ş. Raianu, "Hopf algebras: an Introduction", *Monographs Textbooks in Pure Appl. Math.* **235**, Marcel Dekker, New York, 2001.

[5] S. Donkin, On projective modules for algebraic groups, *J. London Math. Soc.* **54** (1996), 75–88.

[6] I. Doraiswamy, Projectivity of modules over rings with suitable group action, *Comm. Algebra* **10** (1982), 787–795.

[7] J. J. García, A. Del Río, On flatness and projectivity of a ring as a module over a fixed subring, *Math. Scandin.* **76** (1995), 179–192.

[8] T. Guédénon, Projectivity and flatness of a module over the subring of invariants, *Comm. Algebra* **29** (2001), 4357–4376.

[9] J. C. Jantzen, "Representations of algebraic groups", *Pure Appl. Math.* **131**, Academic Press, Boston, 1987.

[10] A. R. Magid, Picard group of rings of invariants, *J. Pure Appl. Algebra* **17** (1980), 305–311.

[11] S. Montgomery, "Hopf algebras and their actions on rings", American Mathematical Society, Providence, 1993.

[12] D. Ştefan, F. Van Oystaeyen, The Wedderburn-Malcev Theorem for comodule algebras, *Comm. Algebra* *27* (1999), 3569–3581.

[13] M. E. Sweedler, "Hopf algebras", Benjamin, New York, 1969.

A BRIEF INTRODUCTION TO COALGEBRA REPRESENTATION THEORY

William Chin
DePaul University
Chicago, Illinois 60614
USA
wchin@condor.depaul.edu

Abstract

In this survey, we review fundamental properties of coalgebras and their representation theory. Following J.A. Green we present the block theory of coalgebras using indecomposable injectives comodules. Using the cohom and cotensor functors we state Takeuchi-Morita equivalence and use it to sketch the proof of existence of "basic" coalgebras, due to the author and S. Montgomery. This leads to a discussion of theory of path coalgebras, quivers and representations. Results concerning almost split sequences are discussed. Some quantum and algebraic group examples are given.

1 Introduction

This survey article is aimed at algebraists who are not necessarily specialists in coalgebras and Hopf algebras. As coalgebras are the unions of their finite dimensional subcoalgebras, their representation theory can be viewed as a generalization of the theory of finite dimensional algebras. We will see that many fundamental results extend to coalgebras.

We begin by reviewing some of the most basic definitions and properties of coalgebras and their representations, with the nonspecialist in mind. Some of this material is covered in standard texts [Abe, Mo, Sw], though perhaps in different ways. We mainly follow the treatment in [Gr], with updated terminology. We discuss local finiteness, simple comodules, the coradical filtration and coradically graded coalgebras, and pointed coalgebras in section 2. We proceed in section 3 is to see how the structure theory for finite dimensional algebras extends to coalgebras, with injectives comodules playing a role closely analogous to the role of projectives in module theory. We see that block theory extends to coalgebras, and then discuss the Ext-quivers of coalgebras, and path coalgebras of arbitrary quivers. In a final subsection we describe a special case of the Brauer correspondence for modular coalgebras.

We continue in section 4 by discussing the cohom and cotensor functors, the adjoint pair that dualize hom and tensor for module categories. These functors yield a category equivalence theory for comodules due to Takeuchi [Tak], now known as *Morita-Takeuchi equivalence.* This in turn allows the construction of an equivalent *basic coalgebra* in section 5, which is pointed over an algebraically closed field. The construction of a basic coalgebra first appeared in [Sim], and reappeared [CMo] where it was studied further. It follows that an arbitrary coalgebra is equivalent to a suitably large subcoalgebra of the path coalgebra of its quiver (5.1). In the hereditary case, we get the entire path coalgebra (5.2). Representations of path coalgebras can be regarded as quiver representations that are locally nilpotent (5.3)

Examples drawn from quantum and algebraic group theory are given in section 6 . Quantum and algebraic groups provide an example of a setting where coalgebras and comodules are pertinent. When the base field is infinite, comodules correspond to *rational* representations of the algebraic group. Recent work [CKQ] addresses the transpose and the existence of almost split sequences for comodules. We discuss this work in section 7, and present a special case, which allows for the construction of almost split sequences in the category of finite-dimensional comodules. This result enables the construction of the Auslander-Reiten quiver.

We conclude with a remark from [Tak] characterizing comodule categories among abelian k-categories. The reader may find hypotheses required an abelian k-category to be a comodule category to be surprisingly mild.

Conventions:

k a fixed base field

$\otimes = \otimes_k$
$C = (C, \Delta, \varepsilon)$ a coalgebra over k
$M = (M, \rho)$ a right C-comodule
${}^B\mathcal{M}^C$ the category of B, C -bicomodules
$\mathcal{M}^C$ the category of right C -comodules
$^* = Hom_k(\_, k)$
$Hom^C(M, M') = Hom_{\mathcal{M}^C}(M, M')$

2 Coalgebras and Comodules

Definition A *coalgebra* is a vector space C with a comultiplication $\Delta : C \otimes C \to C$ and a counit map $\epsilon : C \to k$ satisfying

(a) coassociativity $(id \otimes \Delta)\Delta = (\Delta \otimes id)\Delta$
(b) counitary property $(id \otimes \varepsilon)\Delta = (\varepsilon \otimes id)\Delta = id_C$.

Thus a coalgebra is obtained by dualizing the associative multiplication map $A \otimes A \to A$ and unit map $k \to A$. So a finite dimensional coalgebra is the linear dual of a finite dimensional algebra (and vice-versa). While this duality might lend an air of redundancy to coalgebra theory, there are properties that are satisfied by infinite dimensional coalgebras which are denied infinite dimensional algebras.

Comodules are similarly defined by dualizing the definition of module.

Definition A *right C-comodule* for a coalgebra C is a vector space M with a comodule structure map $\rho = \rho_M : M \to M \otimes C$ satisfying

(a) $(id \otimes \Delta)\rho = (\rho \otimes id)\rho$
(b) $(id \otimes \varepsilon)\rho = id_M$.

We shall assume that comodules are on the right unless we say otherwise.

If M happens to be a subspace of C, then (with $\rho = \Delta$),we have that M is a subcomodule if $\rho(M) \subset M \otimes C$; here M is said to be a *right coideal* of C.

By further analogy, we say that a subspace $D \subset C$ is a *subcoalgebra* if $\Delta(D) \subset D \otimes D$. If $I \subset \ker \varepsilon$ satisfies $\Delta(I) \subset I \otimes C + C \otimes I$, we say that I is a *coideal.*

A linear map between comodules $f : M \to N$ is *a comodule homomorphism* if $(f \otimes id)\rho_M = \rho_N f$. Let $Hom^C(M, N)$ denote the space of comodule morphisms for $M, N \in \mathcal{M}^C$.

The fundamental homomorphism theorems hold as one would guess. The category $\mathcal{M}^C$ of right comodules is an abelian category.

Coalgebras generalize finite dimensional algebras because of the following fact, sometimes known as the

2.1 "Fundamental Theorem of Coalgebras"

Proposition 1 *Every coalgebra is the sum of its finite-dimensional subcoalgebras.*

We shall prove this fact soon, after surveying some notation and results. We follow [Gr] here.

Let X and Y be vector spaces with bases $\{x_j\}$ and $\{y_k\}$ respectively. Let $u \in X \otimes Y$ and express

$$u = \sum u_{ik} \otimes y_k = \sum x_j \otimes v_{ji}.$$

where $u_{ik} \in X$ and $v_{ji} \in Y$ are uniquely determined. Define

$$L(u) = span\{u_{ik}\} \text{ and } R(u) = span\{v_{ji}\}$$

Note that $L(u)$ and $R(u)$ are finite dimensional. The definitions of L and R extend in the obvious way to subsets of $X \otimes Y$ by taking sums. Obviously $L(U)$ and $R(U)$ are finite dimensional if U is.

Now let M be a right C-comodule. It is easy to see that for any subset $U \subset M$, $L(\rho(U))$ is the subcomodule of M generated by U (i.e. the intersection of subcomodules containing U). In particular, if U is a right comodule, then $L(\rho(U)) = U$. These facts follow from the counitary property (b) for comodules.

Now let M be a comodule with basis $\{m_i\}$. Write $\rho(m_i) = \sum_t m_t \otimes c_{ti}$, for some uniquely determined $c_{ti} \in C$, and observe that coassociativity implies that

$$\rho(c_{ij}) = \sum_t c_{it} \otimes c_{tj}.$$

Let $R(\rho(M))$ be denoted by $cf(M)$, which is known as the *coefficient space* of M. It is the subspace spanned by the c_{ti} as just defined and it is evident from the equation just displayed that

1. $cf(M)$ is a subcoalgebra of C .
2. M is a $cf(M)$-comodule that is finite dimensional if M is finite dimensional.

Now for the proof of the Proposition. Let $c \in C$ and let $M = L(\Delta(c))$ be the right subcomodule generated by c. The counitary property $id_C = (\varepsilon \otimes id_C)\Delta$ implies that $c \in cf(M)$. Thus every element of C is contained in a finite dimensional subcoalgebra of C.

We shall denote the subcoalgebra of generated by c by (c).

Remarks: It can be shown that $(c) = R(\Delta(L(\Delta(c)))$, the coefficient space of the right subcomodule generated by c. Also (c) can be expressed as $C^* \rightharpoonup c \leftharpoonup C^*$, using the left and right "hit" actions of C^* (see [Mo]). The coefficient space is a notion dual to the annihilator of module; in fact $cf(M)^{\perp}$ is the annihilator of M as a left C^*-module.

2.2 Simple Comodules

A comodule is said to be *simple* if it has no proper nontrivial subcomodules. A coalgebra is said to be *simple* if it has no proper nontrivial subcoalgebras. By results above, these simple objects are finite dimensional.

Let S be a simple comodule, and let $D = cf(S)$. Then by dualizing Artinian ring theory, it is not hard to see that:

- D is a simple subcoalgebra of C and
- D is isomorphic as a comodule to the direct sum of $\dim_{End(S)}(S)$ copies of S.

2.3 The coradical filtration

Let $C_0 = corad(C)$ denote the sum of the simple subcoalgebras of C. It is also the socle (=sum of simple subcomodules) of C as a comodule, on either side.

Let C_n be defined inductively by

$$C_n = \Delta^{-1}(C_{n-1} \otimes C + C \otimes C_0)$$

This is known as the *coradical filtration* of C. It is equal to the socle series of C as a comodule. More generally we can describe the socle series of M by $M_0 = \rho^{-1}(M \otimes C_0)$ and

$$M_n = \rho^{-1}(M_{n-1} \otimes C + M \otimes C_0)$$

Here M_0 equals the direct sum of the simple subcomodules of M (as our terminology suggests).

The next Lemma is a useful property of the degree one term.

Lemma 2 *(see [Mo1], 5.3.1) Let $f : C \to D$ be coalgebra map. Then f is monic if and only if its restriction to C_1 is monic.*

A coalgebra that is an $\mathbb{N}$-graded vector space $C = \oplus C(n)$ is said to be ($\mathbb{N}$-) graded if $\Delta C(n) \subset \sum_i C(i) \otimes C(n-i)$ for all $n \in \mathbb{N}$. Basic results concerning graded coalgebras can be found in [NT]. A graded coalgebra is said to be *coradically graded* [CMu] if $C_0 = C(0)$ and $C_1 = C(0) \oplus C(1)$. The coradical filtration can then be expressed in terms of the grading. Coradically graded coalgebras are a special case of strictly graded coalgebras [Sw], and have recently been generalized in [AS].

Proposition 3 (CMu) *If C is a coradically graded coalgebra, then $C_n = \oplus_{i \leq n} C(i)$ for all $n \in \mathbb{N}$.*

The next result is used in [CMu] to find the coradical filtration for quantized enveloping algebras (see Example 6d below). The statement here is a corrected version of [CMu, 2.3] where the hypotheses originally only required the coalgebras C, D to be bialgebras.

Proposition 4 (CMu2) *Let C and D be pointed Hopf algebras with the same coradical A. Then $H = C \otimes_A D$ is coradically graded, where $H(m) = \sum_{i=0}^{m} H(i) \otimes H(m-i)$.*

2.4 Pointed coalgebras and skew primitives

C is said to be *pointed* if every simple subcoalgebra is of dimension one.

Define the *group-like* elements of C to be

$$G(C) = \{g \in C | \Delta(g) = g \otimes g\}.$$

The kg, $g \in G(C)$ are precisely the one dimensional subcoalgebras, so the span of the $G(C)$ is C_0 if and only if C is pointed. If C is cocommutative ($\Delta = twist \circ \Delta$) and k is algebraically closed, then a coalgebraic version of the Nullstellensatz says that C is pointed. A related fact says that the coordinate Hopf algebra of an affine algebraic group is pointed if and only if it is solvable. Other examples include (quantized) enveloping algebras. See example (d) at the end of this article. $G(C)$ is a group in case C is a Hopf algebra, and thus C_0 is a group algebra.

Assume that C is pointed. We describe how the skew primitive and group-like elements make up the first two terms of the coradical filtration.

We have $C_0 = kG(C)$. Let $g, h \in G(C)$ and set

$$P_{g,h} = \{c \in C | \Delta(c) = g \otimes c + c \otimes h\}$$

The elements of $P_{g,h}$ are called (g, h)-*skew primitives*. Choose a vector space complement $P'_{g,h}$ for $k(g - h)$ in $P_{g,h}$ (the "nontrivial" ones).

The Taft-Wilson theorem (see [Mo1], 5.4.1)states that

$$C_1 = C_0 \oplus \sum_{g,h \in G(C)} P'_{g,h}.$$

3 Structure theory

3.1 Injectives

Let X be any k-space. We make $X \otimes C$ into a C-comodule via the map $id_X \otimes \Delta$. If M is a comodule, we write $(M) \otimes C$ to denote the comodule with (M) being the underlying vector space of M (whose comodule structure is ignored). This "free" comodule is just the direct sum of $\dim M$ copies of C.

It is known that the category of comodules is a locally finite abelian category (see section 8 below, if interested), and thus has enough injectives. Let's make this more concrete.

Theorem 5 *(a) C is an injective comodule*
(b) $\mathcal{M}^C$ has enough injectives
(c) direct sums and direct summands of injective comodules are injective.

Proof. We prove (a) and (b). Let $f' \in Hom^C(M,C)$ and define $f \in Hom_k(M,k)$ by setting $f = \varepsilon \circ f'$. We can recover f' from f by seeing that $f' = (f \otimes id_C)\rho$. This yields a natural isomorphism between the functors $Hom^C(-,C)$ and $Hom_k(-,k) : \mathcal{M}^C \rightsquigarrow Mod(k)$. Since the latter is exact, so too is the former. This shows that C is injective. Similarly any direct sum of copies of C is injective.

Next we show that every comodule embeds in an injective comodule. To see this consider the map $\rho : M \to (M) \otimes C$. It is straightforward to check that this is an embedding of comodules. For instance the counitary property immediately implies that ρ is a monomorphism. This proves (b).

Remarks: Part (c) of the theorem above might seem surprising since the statement is false for modules in general. Generally, the direct product of comodules does not have a comodule structure.

The category $\mathcal{M}^C$ does not necessarily have enough projectives. If $C = span\{x_i | i = 0,1,2,\cdots\}$ is the divided power coalgebra, with

$$\Delta(x_n) = \sum_{i=0}^{n} x_i \otimes x_{n-i}.$$

Then $\mathcal{M}^C$ has no projectives (so it doesn't have enough). Let $C_n = span\{x_i | i = 0,1,2,\cdots,n\}$. It is not hard to see that a projective comodule would have $C^* = \varprojlim C_n^*$ as a homomorphic image as a C^*-module. But C^* is a not rational as a C^*-module.

Injective Hulls exist in the "usual" sense: Every comodule M is contained in a maximal essential extension, which is minimal with respect to being injective and containing M. This comodule is denoted by $I(M)$. As usual, $I(M) \cong I(M_0)$.The proofs are similar to the module case.

A key Lemma in constructing injective hulls is

Lemma 6 *Let $I \in \mathcal{M}^C$ be injective. If $e_0 = e_0^2 \in End^C(I_0)$, then there exists $e = e^2 \in End^C(I)$ extending e_0.*

The proof involves inductively constructing idempotents $e_n \in End^C(I_n)$, so that e_{n+1} extends e_n for $n = 0, 1, 2, \ldots$ This is done by a generalization of Brauer's famous idempotent lifting procedure. The point here is that the sequence $\{e_n\}$ specifies an idempotent endomorphism of I.

3.2 Indecomposable Injectives

Let $\mathcal{G}$ be a full set of simple comodules in $\mathcal{M}^C$. For each $\mathbf{g} \in \mathcal{G}$, let $m(\mathbf{g})$ denote the multiplicity of $\mathbf{g}$ in C. The coefficient space of $\mathbf{g}$ is a cosemisimple coalgebra and as a right comodule is isomorphic to the direct sum of $m(\mathbf{g}) = \dim_{End(\mathbf{g})} \mathbf{g}$ copies of $\mathbf{g}$. It is known [Gr] that

Theorem 7 *The $I(\mathbf{g})$ form a full set of indecomposable injectives in $\mathcal{M}^C$. As right C-comodules,*

$$C \cong \oplus_{\mathbf{g}\in\mathcal{G}} I(\mathbf{g})^{m(\mathbf{g})}.$$

This generalizes the structure theory for finite-dimensional algebras where injective indecomposables replace projective indecomposables. We see next that block theory generalizes as well.

3.3 Blocks and Quivers

Define the (*Ext-*) *quiver* of C to be the directed graph $Q(C)$ with vertices $\mathcal{G}$ and $dim_k Ext^1(\mathbf{h}, \mathbf{g})$ arrows (possibly infinite) from $\mathbf{g}$ to $\mathbf{h}$. Notation as in 3.2 above. The blocks of C are the vertex sets of components of the graph (ignoring directionality) $Q(C)$. In other words, the blocks are the equivalence classes of the equivalence relation on $\mathcal{G}$ generated by arrows. This equivalence relation is the same as the one in [Gr].

Let $bl(\mathbf{g})$ denote the block containing $\mathbf{g}$, so that $\mathcal{G} = \dot{\cup}_{\mathbf{g}\in T} bl(\mathbf{g})$ where T is a set of representatives of blocks of C. In a manner dual to the decomposition theory for finite dimensional algebras, the blocks determine the coalgebra decomposition of C.

Theorem 8 *([Gr]) Let* $C(\mathbf{g}) = \sum_{h\in bl(g)} cf(I(\mathbf{h}))$ *Then* $C(\mathbf{g}) = C(\mathbf{h})$ *for all* $\mathbf{h} \in bl(\mathbf{g})$. *Also*
(a) $C(\mathbf{g}) \cong \bigoplus_{h\in bl(\mathbf{g})} I(\mathbf{h})^{m(\mathbf{h})}$ *as right comodules, and*
(b) $C = \oplus_{\mathbf{g}\in T} C(\mathbf{g})$

As a consequence, $C(\mathbf{g})$ is the largest indecomposable subcoalgebra containing $\mathbf{g}$. Also C is indecomposable as a coalgebra if and only if its quiver is (topologically, ignoring directionality) connected.

When C is a pointed Hopf algebra, $G = G(C)$ is a group, $C(k1_G)$ is a Hopf subalgebra and $P := bl(k1_G)$ is a normal subgroup of G. Furthermore, C is isomorphic as an algebra to a crossed product $C(k1_G)\#_t G/P$, see [Mo2].

3.4 Quivers for Pointed Coalgebras

Assume C is pointed. Then the simple subcoalgebras are in bijection with the grouplikes $G = G(C)$. Furthermore, there are $dim_k\ P'_{g,h}$ arrows (see 2.4) from g to h, for all $g, h \in G$. This can be seen by considering the extension

$$0 \to kg \to kg + kd \to kh \to 0$$

corresponding to the (g, h)-skew primitive d. The ordinary $(1,1)$- primitives for a bialgebra thus correspond to loops.

The enveloping algebra $U(\mathbf{g})$ of a Lie algebra $\mathbf{g}$ is a cocommutative pointed Hopf algebra with quiver consisting of a single vertex and $dim\ \mathbf{g}$ loops.

We shall discuss more examples near the end of this article in section 7.

3.5 Modular Theory

A tool in the computation of the structure of indecomposable injectives for some coalgebras is Green's [Gr1] coalgebraic generalization of the *Brauer correspondence*. Let us summarize this in a nice special case (that is relevant to example 6c below). Let C be a coalgebra defined over a field $\mathbf{K}$, which is the quotient field of R, a discrete valuation ring with maximal ideal m. Let $k = R/m$ be the residue field. Assume C contains an R-lattice C_0 with $C = C_0 \otimes_R K$, and write $\overline{C} = C_0 \otimes_R k$.

Now assume that

- C is cosemisimple with $\{V_j \mid j \in \mathcal{J}\}$ a complete set of simple right comodules.
- The simple C-comodules V_j are absolutely simple (i.e., remain simple upon any field extension of $\mathbf{K}$).

Let $\overline{V}_j$ be a specialization of the simple comodule to k (which generally is no longer simple).

Let S_i, $i \in \mathcal{I}$ denote a complete set of simple comodules for C_k and let I_i be the injective hull of S_i. Let d_{ji} denote the multiplicity of S_i in $\overline{V}_j$ and let c_{ki} denote the multiplicity of S_i in I_k $(i, k \in I)$. The matrices $\mathbf{d} = (d_{ji})$ and $\mathbf{c} = (c_{ki})$ are known respectively as the *decomposition* and *Cartan* matrices. These matrices are infinite in general.

Assume further that

- $End(V_j) = \mathbf{K}$ and $End(S_i) = k$ for all i, j.

J. A. Green's result says (^{t} =transpose):

Theorem 9 $\mathbf{c} = \mathbf{d}^t \cdot \mathbf{d}$

4 Morita-Takeuchi Equivalence

4.1 Cotensor and Cohom

Define the cotensor product $\square_C$ as follows. Let $M \in \mathcal{M}^C$, and $N = (N, \lambda) \in {}^C\mathcal{M}$ and define

$$M \square_C N = \{\alpha \in M \otimes N | (\rho \otimes id_N)(\alpha) = (id_M \otimes \lambda)(\alpha)\},$$

which also can be expressed as the appropriate coequalizer.

Lemma 10 *With* $\square = \square_C$ *and a subcoalgebra* $D \subset C$,
$M \square C_n \cong M_n$ *for all* n.
$M \square C \cong M$
$M \square D \cong \rho^{-1}(M \otimes D)$.

The last statement is proved by seeing that the comodule embedding $\rho : M \to M \otimes D$ has image isomorphic $M \square D$. The first two statements are special cases.

Takeuchi solves the problem of determining when there is a left adjoint to $-\square_C N$, $N \in {}^C\mathcal{M}$, see below. The left adjoint functor gives rise to coendomorphism coalgebras that extend (and dualize) the endomorphism rings of finite dimensional modules.

Definitions Let $M, N \in \mathcal{M}^C$ and write $\{N_i\}$ be the directed system of finite dimensional subcomodules of N, ordered by inclusion. Define

$$cohom^C(M,N) = \varinjlim Hom(N_i, N)^*.$$

We write $coend^C(M)$ for $co\hom^C(M,M)$
We say that $M \in \mathcal{M}^C$ is:

- *finitely cogenerated* if M embeds in a finite direct sum of copies of C.
- *quasi-finite* if $Hom(F,M)$ is finite dimensional for finite dimensional $F \in \mathcal{M}^C$.
- *a cogenerator* (for $\mathcal{M}^C$) if C embeds in a direct sum of copies of M.

We note that quasi-finite implies finitely cogenerated.

Proof. It suffices to show that $Hom^C(F,C)$ is finite dimensional for finite dimensional $F \in \mathcal{M}^C$. But we showed in the proof of Theorem 3 that $Hom^C(-,C)$ is naturally isomorphic to $()^* = Hom_k(-,k)$. The assertion follows immediately.

Also, one can observe that M is quasi-finite if and only if $Hom^C(g,M)$ is finite dimensional for simple $g \in \mathcal{M}^C$, which holds if and only if every simple has finite multiplicity in M_0. It is now easy to produce examples of quasi-finite, non-finitely cogenerated comodules.

Adjunction Property: If $X \in {}^B\mathcal{M}^C$ is quasi-finite in M^C, then $-\Box_B X : \mathcal{M}^B \to \mathcal{M}^C$ has left adjoint $cohom^C(X,-) : \mathcal{M}^C \to \mathcal{M}^B$ [Tak, 1.9]. The cohom functor is characterized by the adjunction property.

The following characterizes equivalent comodule categories.

Theorem 11 *([Tak]) Let B, C be coalgebras and let $E \in {}^B\mathcal{M}^C$. The following are equivalent:*
(a) $-\Box_B E : \mathcal{M}^B \to \mathcal{M}^C$ is an equivalence of categories.
(b) E is a quasi-finite injective cogenerator in $\mathcal{M}^C$ and $B \cong coend^C(E)$.

In the case that (a) and (b) are true say that B and C are "(Morita-Takeuchi) equivalent" coalgebras and write $B \sim C$. The inverse functor can be expressed as

$$-\Box_C co\hom(E,C) \approx co\hom(E,-).$$

It is shown [Tak] on the other hand that if $\mathcal{M}^B$ and $\mathcal{M}^C$ are equivalent categories, then the bicomodule E as in the statement of the theorem exists.

Here is a lemma that is useful in reducing things to finite dimensional subobjects.

Lemma 12 *Let D be a subcoalgebra of C and $E \in \mathcal{M}^C$; set $F = \rho_E^{-1}(E \otimes D)$ $(\cong E\square_C D)$. Then*
(a) E injective implies F injective in $\mathcal{M}^D$.
(b) E finitely cogenerated and D finite dimensional implies F finite dimensional.

A key observation in proving this lemma ([CMo]) is that $Hom^D(Y, E) = Hom^D(Y, F)$ for all $Y \in M^D$.

Remark We can see how $coend(M)$ is the direct limit of coalgebras and so is a coalgebra.. (In [Tak] this is deduced from a universal property of *cohom* instead.) Write E as the directed union of finite dimensional subcomodules E_i as before. Put

$$C_i = cf(E_i)$$
$$F_i = \rho_E^{-1}(E \otimes C_i)$$

as in the lemma above. Then as above,

$$Hom^C(F_i, E) = Hom^C(F_i, F_i);$$

so:

$$\begin{aligned} cohom(E, E) &= \varinjlim Hom(E_i, E)^* \\ &= \varinjlim Hom(F_i, E)^* \\ &= \varinjlim End(F_i)^* \end{aligned}$$

where the second equality holds because of the cofinality of $\{F_i\}$. Thus $cohom(E, E)$ is a direct limit of finite dimensional coalgebras.

It follows similarly that $E \in {}^{coend^C(E)}\mathcal{M}^C$.

5 Basic Coalgebras

We can write $C \in \mathcal{M}^C$ as a direct sum of indecomposable injectives with multiplicities. Let E denote the direct sum of the indecomposables injectives where each indecomposable occurs with multiplicity one. Clearly E is injective, called a "basic" injective for C. Define

$$B = B_C = coend^C(E).$$

By Theorem 9, B is Morita-Takeuchi equivalent to C. The coalgebra B is *basic* in the sense that

Theorem 13 *([Sim], see also [CMo]) The simple subcoalgebras of B are duals of finite dimensional division algebras.*

As in the remark above, $B = \varinjlim End(F_i)^*$. Using the fact that F_i is injective in M^{C_i}, and in fact *basic*, the theorem is proved by reducing to the finite-dimensional case, where the result is known.

Corollary 14 *([CMo]) If k is algebraically closed, then B is pointed.*

Remark 1 *In [CG], it is shown that two coalgebras are equivalent if and only if their basic coalgebras are isomorphic. Moreover it is shown that the basic coalgebra of C is $B = e \rightharpoonup C \leftharpoonup e$ for an $e \in C^*$ which acts on the left and right by the usual hits. The category equivalence sends a right C-comodule to the right B-comodule $e \rightharpoonup M$. It can be seen that Morita-Takeuchi equivalence can be expressed with idempotents, just as is the case with artinian algebras. The theory of Morita-Takeuchi equivalence can be thus simplified.*

5.1 Path Coalgebras

Let Q be a quiver (not necessarily finite) with vertex set Q_0 and arrow set Q_1. The *path coalgebra kQ* of Q is defined to be the span of all paths in Q with coalgebra structure

$$\Delta(p) = \sum_{p=p_2p_1} p_2 \otimes p_1$$
$$\varepsilon(p) = \delta_{|p|,0}$$

where p_2p_1 is the concatenation $a_t a_{t-1} ... a_{s+1} a_s ... a_1$ of the paths $p_2 = a_t a_{t-1} ... a_{s+1}$ and $p_1 = a_s ... a_1$ $(a_i \in Q_0)$. Here $|p| = t$ denotes the length of p and the starting vertex of a_{i+1} is the end of a_i.

Thus *vertices* are group-like elements, and if a is an arrow $g \leftarrow h$, with $g, h \in Q_0$, then a is a $(g,h)-$ skew primitive. It is apparent that kQ is pointed with coradical $(kQ)_0 = kQ_0$ and the degree one term of the coradical filtration is $(kQ)_1 = kQ_0 \oplus kQ_1$. More generally, the coradical filtration of kQ is given by

$$(kQ)_n = span\{p||p| \leq n\}.$$

Alternatively, kQ can be defined as the cotensor coalgebra [Ni] associated to the kQ_0, kQ_0-bicomodule kQ_1.

If kQ is finite dimensional, it is the dual of the usual path algebra (or its opposite). Note that if there are infinitely many vertices, the path algebra lacks a identity element; the path coalgebra always has a counit.

Now let C be an arbitrary coalgebra, let B be the associated basic coalgebra and assume k is algebraically closed. Then B is pointed and $Q(C) = Q(B)$. From the universal property of cotensor coalgebras [Ni], it follows that there exists a coalgebra map $B \to kQ(C)$ which is a bijection on the degree one subcoalgebras. By lemma 2, this map is an embedding. Thus we obtain a coalgebraic version (with no finiteness restrictions) of a fundamental result of Gabriel for finite dimensional algebras:

Corollary 15 *([CMo]) Every coalgebra C over an algebraically closed field is Morita-Takeuchi equivalent to a subcoalgebra of $kQ(C)$ containing $kQ(C)_1$*

5.2 Hereditary Coalgebras

A coalgebra C is said to be *hereditary* [NTZ] if homomorphic images of injective comodules are injective. Pointed hereditary coalgebras are exactly path coalgebras of the form kQ for some quiver Q [Ch]. With the aid of the preceding result we also have

Theorem 16 (Ch) *Every hereditary coalgebra C over an algebraically closed field is Morita-Takeuchi equivalent to the path coalgebra $kQ(C)$ of its quiver $Q(C)$.*

5.3 Representations of Path Coalgebras

We define a representation of a quiver Q as usual by assigning a vector space V_g to each vertex g and a linear map $f_a : V_g \to V_h$ corresponding to each arrow $a : g \to h$. By composing f's we obtain maps f_p for all nonempty paths p. The category of representations is denoted by $rep(Q)$. It is a standard fact that $rep(Q)$ is equivalent to the category of modules over the path algebra of Q (at least when Q is finite).

Let C denote the path coalgebra kQ, and let M be a C-comodule. To M we associate an object of $rep(Q(C)) = rep(Q)$ as follows. Let I be the coideal spanned by paths of length at least 1, and define $\pi : C \to C/I \cong C_0$ as the projection. Set $\rho_0 = (id_M \otimes \pi) \circ \rho$. Define

$$V_g = \{v \in M | \rho_0(m) = m \otimes g\}$$

and define $f_p : V_g \to V_h$ for paths p (from g to h) by

$$\rho(v) = \sum f_p(m) \otimes p.$$

where the sum is over paths p starting at g. Since the paths are linearly independent, the f_p are well-defined and we get a representation of Q. Note that the sum has only finitely many nonzero terms. Checking details, we find that $\mathcal{M}^{kQ}$ corresponds to the subcategory of "locally nilpotent" representations.

Proposition 17 *(see [CKQ]) For any quiver Q, $\mathcal{M}^{kQ}$ is equivalent to the full subcategory of $rep(Q)$ consisting of representations (V, f) such that for all $g \in Q_0$ and $v \in V_g$, $f_p(v) = 0$ for all but finitely many paths p (starting at g).*

Let Q be the quiver with a single vertex g and a single loop a. Let $V_g = k$ and $f_a = \lambda\cdot$, $\lambda \in k$. This representation does not correspond to a kQ-comodule unless $\lambda = 0$ (though of course it is a module over the path algebra $k[a]$).

We would like two mention the articles [Sim2, NS] that contain interesting results concerning certain types of path coalgebras.

6 Examples

a. Let C be the coordinate Hopf algebra of an affine connected simply connected semisimple algebraic group G or the quantum variant where q is not a root of 1 (see [CP]). If k is of characteristic zero, then C is cosemisimple. So the blocks are singletons.

b. Let C be the coordinate algebra of $SL(2)$. If k is of characteristic zero, then C is cosemisimple. In positive characteristic, the blocks are infinite and infinite in number. The quiver structure is somewhat complicated, and is given by "p-reflections" see [Cl].

c. Let $C = k_\zeta[SL(2)]$ be the q-analog of the coordinate algebra of $SL(2)$ (see [CP]), where q is specialized to a root of unity of odd order l. Assume k is of characteristic zero. The quiver structure is given as follows: For each nonnegative integer r, there is a unique simple module $L(r)$ of highest weight r (closely analogous to the simple highest weight modules for nonquantum groups). These comodules exhaust the simple comodules.

Write

$$r = r_1 l + r_0$$

where $0 \leq r_0 < l$. Define an "l-reflection" $\tau : \mathbf{Z} \to \mathbf{Z}$ by

$$\tau(r) = (r_1 - 1)l + l - r_0 - 2$$

if $r_0 \neq l-1$, and $\tau(r) = r$ if $r_0 = l-1$. Put $\sigma = \tau^{-1}$ (perhaps after checking that τ is a bijection).

Theorem 18 *([CK]) Set $I(r) = I(L(r))$ for all integers $r \geq 0$.*
(a) If $r_0 = l-1$, then $I(r) = L(r)$.
(b) If $r < l-1$, then $I(r)$ has socle series with factors

$$L(r), L(\sigma(r)), L(r).$$

(c) If $r > l$ (and $r_0 \neq l-1$), then $I(r)$ has socle series with factors

$$L(r), L(\sigma(r)) \oplus L(\tau(r)), L(r).$$

This determines the quiver as having vertices labelled by nonnegative integers and with arrows $r \leftrightarrows s$ in case $r_0 \neq l-1$ if and only if $s = \tau(r)$. In case $r_0 = l-1$, the block of r is a singleton (equivalently $L(r)$ is injective). Thus the nontrivial block containing $L(r)$, $r < l-1$ has quiver

$$r \leftrightarrows \tau(r) \leftrightarrows \tau^2(r) \leftrightarrows \cdots$$

The injective indecomposable comodules are all finite dimensional, in contrast to the injectives for the ordinary (nonquantum) modular coordinate coalgebra. Also, the result shows that the coradical filtration is of length 2 (i.e. $C = C_2$). It follows that C is *semiperfect* (see [Lin]) in the sense that every finite dimensional comodule has a projective cover.

These results are obtained using the modular result at the end of section 3. We sketch here a short proof of (b) and (c) above. The simple comodules for $C = k_\zeta[SL(2)]$ are known to be highest weight modules, say with weight $r \in \mathbb{N}$, which can be specialized to C. The resulting $C-$comodules are the simple ones $L(r)$ in case $r_1 = l-1$ or $r < l$; otherwise they have composition series $L(\rho(r)) \leq L(r)$ [CK]. Fixing $r < l-1$, we can list the simples of the nontrivial block containing $L(r)$ as the comodules $S_i = L(\tau^i(r))$, $i \in \mathbb{N}$. Thus the decomposition matrix $\mathbf{d} = (d_{ij})$ for this block is given by $d_{ii} = 1 = d_{i+1,i}$, and 0's elsewhere. By the Brauer correspondence in section 3, the Cartan matrix $\mathbf{c} = (c_{ij})$ is given by $c_{ii} = 2$ and $c_{i+1,i} = c_{i,i+1} = 1$, and 0's elsewhere. Now the conclusions (b) and (c) follow once we know the there are no self-extensions of simples. This fact follows by following a standard argument in the nonquantum modular theory, e.g. [Ja, 2.14].

In [CK], we also determine the indecomposable injectives for the quantum analogs of coordinate coalgebras of 2×2 matrices and general linear group at odd roots of unity. The situation is more complicated for these coalgebras.

d. Let C be the quantized enveloping algebra $U_q(\mathbf{g})$, associated to a finite dimensional complex simple Lie algebra $\mathbf{g}$ of rank n, defined over a field k of characteristic zero. See [Lu1] and e.g., [Mo], [CP] for definitions.

Assume q is specialized to an element of k which is not a root of unity (of odd order and prime to 3 if $\mathbf{g}$ is of type G_2). Then C is generated as an algebra by group-likes $K_i^{\pm 1}$ and the $(K_i, 1)-$ skew primitives E_i and K_iF_i, $i = 1, ..., n$. C is pointed with $G(C)$ being a free abelian group of rank n. It turns out that the only skew primitives are the "obvious" ones, spanned by

gE_i, gF_i, and the "trivial" ones $g - h$, $(g, h \in G(C))$. (The nonsimple root vectors are not skew-primitive.) Thus the vertex set is $G(C)$, and arrows are

$$gK_i \rightrightarrows g,$$

for any $g \in G(C)$, $1 \leq i \leq n$.

This result was obtained in [CMu] when q is transcendental over the rationals. Further progress was made (working with quantized coordinate algebras) in [Mus]. E. Müller [Mü] solved the problem of determining the coradical filtration more generally, including versions for specializations to roots of unity. His methods rely on Lusztig's construction of quantized enveloping algebras [Lu]. A recent generalization appears in [AS].

7 Almost Split Sequences

In [CKQ], we investigated the existence of almost split sequences for comodules given a fixed right or left-hand term. To construct the Auslander-Reiten quiver, one needs to be able to iterate the construction in some subcategory. We present a situation where this can be done in the category of finite-dimensional comodules.

A coalgebra is defined to be *right semiperfect* [Lin] if every simple *left* C-comodule has an finite-dimensional injective hull (equivalently, every finite-dimensional right comodule has a projective cover). In the context of group schemes, (left and right) semiperfect coalgebras were called *virtually linearly reductive* in [D1,D2].

Theorem 19 (CKQ) *Let C be a right semiperfect coalgebra such that* $soc(I(S)/S)$ *is of finite length for all simple right C-comodules. Then the category of finite-dimensional right C-comodules has almost split sequences.*

In the context of group schemes, (left and right) semiperfect coalgebras were called *virtually linearly reductive* in [D1,D2]. The theorem applies to right semiperfect coalgebras whose Ext-quivers have only finitely many arrows ending at each vertex. A special case of this are coalgebras whose Ext-quivers have only finitely many arrows ending at each vertex and only finitely many paths starting at each vertex. Hence, finite-dimensional comodules over subcoalgebras of path coalgebras of such quivers have almost split sequences.

The theorem applies to $k_\zeta[SL(2)]$, which is semiperfect, as in 6c above, (though not to the path coalgebra of its quiver). More generally, let $k_\zeta[G]$ be a quantized coordinate algebra at the root of unity ζ as in [APW]. Then by [loc. cit., section 9] (see also [AD] for a different proof), $k_\zeta[G]$ is a semiperfect coalgebra. It should be noted that a Hopf algebra is semiperfect as a coalgebra if and only if it is right semiperfect.

8 Abelian Categories

We append a remark from [Tak].

Let $\mathcal{A}$ be an abelian k-category that

- has exact directed colimits and has a set of objects of finite length which generate $\mathcal{A}$ (i.e. $\mathcal{A}$ is "locally finite"), and such that
- $\mathcal{A}(S, S)$ is finite dimensional for all simple objects S.

Takeuchi [Tak] says such categories $\mathcal{A}$ are of "of finite type". He shows that

Theorem 20 *An abelian* k*-category* $\mathcal{A}$ *is equivalent to a comodule category* $\mathcal{M}^C$ *for some coalgebra* C *if and only if* $\mathcal{A}$ *is of finite type.*

The proof mirrors the construction of the basic coalgebra. Let E be a an injective cogenerator, which is the direct sum of injective objects, each occurring with multiplicity one. Then E is isomorphic to the direct limit of its finite length subobjects E_i. Letting $C = \varinjlim \mathcal{A}(E_i, E)^*$ works.

References

[Abe] H. H. Andersen, Polo, Patrick and K. Wen, Representations of quantum algebras. Invent. Math. 104 (1991), no. 1, 1–59.

[APW] E. Abe, Hopf Algebras, Cambridge Univ. Press, 1980.

[AD] N. Andruskiewitsch and S. Dascalescu, Cofrobenius Hopf algebras and the coradical filtration, Math Z., to appear.

[AS] N. Andruskiewitsch and H-J. Schneider, A characterization of quantum groups, preprint.

[Ch] W. Chin, Hereditary and path coalgebras. Comm. Algebra 30 (2002), no. 4, 1829–1831.

[CK] W. Chin and L. Krop, Injective Comodules for 2×2 quantum matrices, Comm. Alg (2000) no. 4, 2043–2057.

[CKQ] W. Chin, M. Kleiner and D. Quinn, Almost split sequences for comodules, J. Alg. 249 (2002) no. 1, 1-19.

[CMo] W. Chin and S. Montgomery, Basic Coalgebras, *in* Modular Interfaces (Riverside, CA, 1995) AMS/IP Studies in Advanced Math. vol.4, Providence RI, (1997) 41-47.

[CMu] W. Chin and I. Musson, The coradical filtration for quantized enveloping algebras, J. London Math Soc.,vol. 53(2) 50-62. Corrigenda: "The coradical filtration for quantized enveloping algebras" J. London Math. Soc. (2) 61 (2000), no. 1, 319–320.

[Cl] E. Cline, Ext^1 for SL_2,Comm. Alg. 7(1), 107-111.

[CP] V. Chari and A. Pressley, a Guide to Quantum Groups, Cambridge University Press, 1994.

[CG] J. Cuadra, J. Gómez-Torrecillas, Idempotents and Morita-Takeuchi theory. Comm. Algebra 30 (2002), no. 5, 2405–2426.

[D1] S. Donkin, On the existence of Auslander-Reiten sequences of group representations I, Alg. Rep. Theory, 1,(1998) 97-127.

[D2] S. Donkin, On the existence of Auslander-Reiten sequences of group representations II, Alg. Rep. Theory, 1,(1998) 215-253.

[Gr] J.A. Green, Locally finite representations, Journal of Alg. 41 (1976), 137-171.

[Ja] Jantzen, J.C., Representations of Algebraic Groups, Academic Press 1987.

[Lin] B. I.-P. Lin, Semiperfect coalgebras, J. Alg. 49 (1977), no. 2, 357-373.

[Lu] G. Lusztig, an Introduction to Quantum Groups, *in* Prog. Math., Vol. 110, Birkhauser, Basel, 1993.

[Lu1] G. Lusztig, Modular representations and quantum groups, in "Classical Groups and Related Topics, pp. 55-77, Contemp. Math., Vol. 82, AMS, Providence, RI, 1989.

[Mo1] S. Montgomery, Hopf Algebras and their Actions on Rings, AMS, CBMS NO. 82, 1993.

[Mo2] S. Montgomery, Indecomposable coalgebras, simple comodules, and pointed Hopf algebras, Proc. A.M.S. (123), 2343-2351.

[Mus] I.M. Musson, Link between cofinite prime ideals in multiparameter quantized enveloping algebras, Israel J. Math 100(1997), 285-308.

[Mü] E. Müller, Some topics on Frobenius-Lusztig kernels I, J. Alg. 206 (1998), 624-644.

[Ni] W. Nichols, Bialgebras of type I, Comm. Alg. 6(15) (1978), 1521-1552.

[NS] S. Nowak, D. Simson, Locally Dynkin quivers and hereditary coalgebras whose left comodules are direct sums of finite dimensional comodules. Comm. Algebra 30 (2002), no. 1, 455–476.

[NT] C. Nastasescu, B. Torrecillas, Graded coalgebras. Tsukuba J. Math. 17 (1993), no. 2, 461–479.

[NTZ] C. Nastasescu, B. Torrecillas, Y. H. Zhang, Hereditary coalgebras. Comm. Algebra 24 (1996), no. 4, 1521–1528.

[Sim] D. Simson, On the structure of pure semisimple Grothendieck categories, Cahiers de Topologie et Geom. Diff. 33 (1982) 397-406.

[Sim2] Coalgebras, comodules, pseudocompact algebras and tame comodule type, Colloq. Math., 90(2001) no. 1, 101-150.

[Sw] Sweedler, M., Hopf Algebras, Benjamin, New York 1969.

[Tak] M. Takeuchi, Morita Theorems for categories of comodules, J. Fac. Sci. Univ. Tokyo 24 (1977), 629-644.

Some examples of integrals for bialgebras

S. Dăscălescu *
Kuwait University, Faculty of Science
Dept. Mathematics, PO BOX 5969
Safat 13060, Kuwait
e-mail: sdascal@mcs.sci.kuniv.edu.kw

Abstract

For any integer $n \geq 2$ we construct several bialgebras H with zero space of left integrals and n-dimensional space of right integrals. For any such bialgebra H we compute the injective envelopes of the simple left and right H-comodules, showing in particular that H is left and right semiperfect. We prove that H is right co-Frobenius, but not left co-Frobenius.

0 Introduction and preliminaries

The concepts of left and right integral for a bialgebra were introduced by Larson and Sweedler in [6], [12]. For finite dimensional Hopf algebras, the spaces $\int_l$ of left integrals and $\int_r$ of right integrals have dimension 1. Sullivan proved in [11] the uniqueness of the integrals, i.e. $dim \int_l \leq 1$ and $dim \int_r \leq 1$ in any Hopf algebra. A better understanding of Hopf algebras with non-zero integrals was possible after Lin studied in [7] several finiteness properties of coalgebras. A coalgebra C is called left (right) semiperfect if the category of left (right) C-comodules has enough projectives. This is equivalent to the fact that the injective envelope of any right (left) simple C-comodule has finite dimension.

A coalgebra C is called left (right) co-Frobenius if there exists an injective morphism $\phi : C \to C^*$ of left (right) C^*-modules. The properties of being semiperfect or co-Frobenius are not left-right symmetric.

*On leave from University of Bucharest, Facultatea de Matematică. Research supported by Grant SM 10/01 of the Research Administration of Kuwait University.

A left (right) co-Frobenius coalgebra is left (right) semiperfect, but the converse is not true. However, for a Hopf algebra H we have that H is left semiperfect $\Leftrightarrow$ H is right semiperfect $\Leftrightarrow$ H is left co-Frobenius $\Leftrightarrow$ H is right co-Frobenius $\Leftrightarrow$ $\int_l \neq 0 \Leftrightarrow \int_r \neq 0$. Thus the existence of the integrals can be proved by any of the above equivalent coalgebra properties for H. These coalgebra properties have been further investigated in [4], where it was also emphasized the simple but useful fact that $\int_l$ is just $Hom_{H^-}(H,k)$, the space of all left H-comodule morphisms from H to the trivial H-comodule k.

A new shorter proof for the uniqueness of the integrals was given in [10] by using homological methods. One more proof was given in [9]. A different approach was presented in [2], where it was proved that for a coalgebra C that is right co-Frobenius and left semiperfect we have that $dim\ Hom_{C^-}(C,M) \leq dim\ M$ for any finite dimensional left C-comodule M. In particular this proves the uniqueness of left integrals for bialgebras that are right co-Frobenius and left semiperfect.

In this paper we construct a family of infinite dimensional bialgebras $H(m,n,\lambda)$, where m,n are positive integers, $n \geq 2$ and λ is a primitive mn-th root of unity. We start with the monoid algebra of an infinite monoid generated by one element, then adjoin a skew-primitive element by an Ore extension and then factor a bi-ideal. We show that the space of left integrals of $H(m,n,\lambda)$ is zero, while the space of right integrals of $H(m,n,\lambda)$ has dimension n. We explain this phenomenon by the fact that $H(m,n,\lambda)$ is left and right semiperfect (this is proved by an explicit computation of the injective envelopes of all simple left and right comodules), $H(m,n,\lambda)$ is right co-Frobenius, but it is not left co-Frobenius. This shows that for bialgebras that are right co-Frobenius and left semiperfect, the uniqueness holds for left integrals, but not necessarily for right integrals. Our construction is similar to the one in [1], where a large class of Hopf algebras with non-zero integrals was constructed.

We work over a field k which contains enough roots of unity. For the basic notions and properties connected to coalgebras, comodules, bialgebras and Hopf algebras we refer to [3] and [8].

1 The construction of some bialgebras

We work over a field k. Let M be the infinite multiplicative monoid generated by one element g, i.e. $M = \{1, g, g^2, \ldots\}$ (note that M is isomorphic to the additive monoid of non-negative integers). Then the monoid algebra kM has a natural bialgebra structure if we make all elements of M to be grouplike elements, this is $\Delta(g^i) = g^i \otimes g^i$ and $\varepsilon(g^i) = 1$ for any $i \geq 0$. Let m and n be two positive integers, $n \geq 2$, and let λ be a primitive mn-th root of 1 in k. Then we have an algebra automorphism ϕ of kM defined by $\phi(g^i) = \lambda^i g^i$ for any i. Then we can form the Ore extension $(kM)[X,\phi]$, which is like the polynomial ring over kM in one indeterminate X with the multiplication modified according to $Xg = \phi(g)gX = \lambda gX$. By using the universal property for Ore extensions (see for example [1, Lemma 1.1]) we can extend the bialgebra structure of kM to $(kM)[X,\phi]$ by setting

$$\Delta(X) = g^m \otimes X + X \otimes 1, \quad \varepsilon(X) = 0$$

Then since $(X \otimes 1)(g^m \otimes X) = \lambda^m (g^m \otimes X)(X \otimes 1)$, and λ^m is a primitive n-th root of 1, the quantum binomial formula and the fact that for a primitive n-th root q of 1 and any $0 < i < n$, all quantum binomial coefficients $\binom{n}{i}_q$ are zero (see [5, Section IV.2] or [3, Lemma 5.6.2]), show that $\Delta(X^n) = g^{mn} \otimes X^n + X^n \otimes 1$. Hence the ideal generated by X^n is also a coideal of $(kM)[X,\phi]$, and we can consider the factor bialgebra $H(m,n,\lambda) = (kM)[X,\phi]/(X^n)$. The bialgebra $H(m,n,\lambda)$ can be presented by generators g, x subject to

$$xg = \lambda gx, \quad x^n = 0$$

$$\Delta(g) = g \otimes g, \quad \Delta(x) = g^m \otimes x + x \otimes 1$$

$$\varepsilon(g) = 1, \quad \varepsilon(x) = 0$$

For the rest of the paper we will write $H = H(m,n,\lambda)$. H has a basis consisting of all elements of the form $g^i x^j$, with $i \geq 0$ and $0 \leq j \leq n-1$. By using the quantum binomial formula we see that

$$\Delta(g^i x^j) = \sum_{0 \leq r \leq j} \binom{j}{r}_{\lambda^m} g^{i+m(j-r)} x^r \otimes g^i x^{j-r}$$

Let us denote $e_{i,j} = (1/(j)_{\lambda^m}!) g^i x^j$ for any i, j. Note that since λ^m is primitive of order n, then for any $j < n$ we have that $(j)_{\lambda^m}! \neq 0$. We also denote by $E_{i,j} \in H^*$ the map such that $E_{i,j}(e_{u,v}) = \delta_{i,u}\delta_{j,v}$ for any u, v (we use Kronecker's delta). The comultiplication formula above becomes

$$\Delta(e_{i,j}) = \sum_{0 \leq r \leq j} e_{i+m(j-r),r} \otimes e_{i,j-r} \tag{1}$$

Proposition 1.1 *Let $H_0 \subseteq H_1 \subseteq \ldots$ be the coradical filtration of H. Then for any $0 \leq j < n$ we have that H_j is the space spanned by all $e_{i,r}$ with $r \leq j$. In particular $H_{n-1} = H$ and the coradical of H is $H_0 = kM$.*

Proof: The comultiplication formula (1) shows by a standard inductive argument that

$$\wedge^j(kM) = < e_{i,r} \mid i \geq 0, r \leq j-1 \}$$

where by $\wedge$ we denote the usual wedge for subspaces of coalgebras. In particular $\wedge^{n-1}(kM) = H$, which implies by [3, Exercise 3.1.12] that $H_0 \subseteq kM$. Since obviously $kM \subseteq H_0$, we obtain that $H_0 = kM$, and the rest follows immediately from here. ∎

Remark 1.2 *Let C be the coalgebra defined in [7]. C has a basis $\{ g_i, d_i \mid i \in \mathbf{Z}, i \geq 0 \}$ and has the comultiplication and counit defined by*

$$\begin{aligned} \Delta(g_i) &= g_i \otimes g_i \\ \Delta(d_i) &= g_i \otimes d_i + d_i \otimes g_{i+1} \\ \varepsilon(g_i) &= 1 \\ \varepsilon(d_i) &= 0 \end{aligned}$$

for any $i \geq 0$.

We see that C is isomorphic as a coalgebra to $H(1, 2, -1)^{cop}$.

Proposition 1.3 *$H(m, n, \lambda) \simeq H(m', n', \lambda')$ as bialgebras if and only if $m = m', n = n'$ and $\lambda = \lambda'$. The group of bialgebra automorphisms of $H(m, n, \lambda)$ is isomorphic to the multiplicative group k^*.*

Proof: If $f : H(m, n, \lambda) \to H(m', n', \lambda')$ is a bialgebra isomorphism, then it induces a bialgebra isomorphism between the coradicals. Since the only automorphism of the monoid M is the identity map, we see that $f(g) = g$. By looking at the skew-primitive elements in $H(m, n, \lambda)$ and $H(m', n', \lambda')$, we see that $g^m = g^{m'}$, so $m = m'$, and $f(x) = \alpha x' + \beta(g^m - 1)$ for some $\alpha, \beta \in k$. Here we denoted by x' the skew-primitive generator in $H(m', n', \lambda')$. Apply f to $xg = \lambda gx$ and find that

$$\alpha\lambda' gx' + \beta g(g^m - 1) = \alpha\lambda gx' + \lambda\beta g(g^m - 1)$$

which shows that $\lambda = \lambda'$ and $\beta = 0$. This also proves the second part. ∎

2 Finiteness properties of the bialgebra H

For any $i \geq 0$ define the subspace

$$V_i = < e_{i,j} \mid 0 \leq j \leq n-1 >$$

Proposition 2.1 *V_i is the injective envelope $E_{H-}(kg^i)$ of the left H-comodule kg^i.*

Proof: Equation (1) shows that V_i is a left H-coideal of H. Moreover, $H = \oplus_{i\geq 0} V_i$, a direct sum of left H-comodules, showing that each V_i is an injective left H-comodule. Clearly $kg^i = ke_{i,0} \subseteq V_i$, so we have that $kg^i \subseteq E_{H-}(kg^i) \subseteq V_i$. If $E_{H-}(kg^i) \neq V_i$, then $V_i = E_{H-}(kg^i) \oplus X$ for some non-zero comodule X, and since X must contain a simple subcomodule, we find that

$$s(V_i) = s(E_{H-}(kg^i)) \oplus s(X) \neq s(E_{H-}(kg^i)) = kg^i$$

where by $s(M)$ we denote the socle of a comodule (the sum of all simple subcomodules). This is a contradiction since $s(H) = H_0$, so the only simple subcomodule of V_i is kg^i. ∎

Corollary 2.2 *H is a right semiperfect coalgebra.*

We note that the above result also follows directly from the fact that H is right co-Frobenius, which we will prove later, but we wanted to see what the injective envelopes of the simple left comodules are.

Proposition 2.3 *The injective envelopes $E_{-H}(kg^i)$ of the right H-comodules kg^i can be described as follows.*

- *$E_{-H}(kg^i) = kg^i =< e_{i,0} >$ for any $0 \leq i < m$.*
- *$E_{-H}(kg^i) =< e_{i,0}, e_{i-m,1} >$ for any $m \leq i < 2m$.*
- *In general for any $0 \leq p \leq n-2$ and any $pm \leq i < (p+1)m$ we have*

$$E_{-H}(kg^i) =< e_{i,0}, e_{i-m,1}, \dots, e_{i-pm,p} >$$

Thus $E_{-H}(kg^i)$ has dimension $[\frac{i}{m}] + 1$ (where $[a]$ denotes the greatest integer in a).

- *$E_{-H}(kg^i) =< e_{i,0}, e_{i-m,1}, \dots, e_{i-(n-1)m,n-1} >$ for $i \geq (n-1)m$.*

Proof: Using again the equation (1) we see that for any $0 \leq p \leq n-2$ and any $pm \leq i < (p+1)m$, the space $< e_{i,0}, e_{i-m,1}, \dots, e_{i-pm,p} >$ is a right coideal of H, and also that for any $i \geq (n-1)m$ the space $< e_{i,0}, e_{i-m,1}, \dots, e_{i-(n-1)m,n-1} >$ is also a right coideal of H. Moreover, H is the direct sum of all these subspaces, showing that each such subspace is an injective right H-comodule. To see that these subspaces are indeed the injective envelopes of the simple right subcomodules kg^i one proceeds as in the the proof for the left hand side case from Proposition 2.1. ∎

Corollary 2.4 *H is a left semiperfect coalgebra.*

Remark 2.5 *It is interesting to note that H has m (non-isomorphic) simple subcomodules that are injective, thus with injective envelope of dimension 1, m simple subcomodules with injective envelope of dimension 2, and so on, up to m simple subcomodules with injective envelope of dimension $n-1$, and finally there are infinitely many simple subcomodules with injective envelope of dimension n.*

Proposition 2.6 *H is right co-Frobenius as an algebra, but it is not left co-Frobenius.*

Proof: Let us denote by $\rightarrow$ and $\leftarrow$ the usual left and right actions of H^* on H, i.e. $h^* \rightarrow h = \sum h^*(h_2)h_1$, $h \leftarrow h^* = \sum h^*(h_1)h_2$ for any $h \in H, h^* \in H^*$. Then for any $i, u \geq 0$, $0 \leq j, v < n$ we have

$$e_{i,j} \leftarrow E_{u,v} = \sum_{0 \leq r \leq j} E_{u,v}(e_{i+m(j-r),r})e_{i,j-r}$$

This shows that $e_{i,j} \leftarrow E_{u,v} = e_{i,j-v}$ in the case where $v \leq j$ and $u = i + m(j-v)$, and $e_{i,j} \leftarrow E_{u,v} = 0$ in any other case.

On the other hand

$$E_{u,v} \rightarrow e_{i,j} = \sum_{0 \leq r \leq j} E_{u,v}(e_{i,j-r})e_{i+m(j-r),r}$$

shows that $E_{u,v} \rightarrow e_{i,j} = e_{i+mv,j-v}$ if $u = i$ and $v \leq j$, and $E_{u,v} \rightarrow e_{i,j} = 0$ in any other case.

Finally

$$(E_{p,q}E_{u,v})(e_{i,j}) = \sum_{0 \leq r \leq j} E_{p,q}(e_{i+m(j-r),r})E_{u,v}(e_{i,j-r})$$

shows that $E_{p,q}E_{u,v} = E_{u,q+v}$ if $u + mv = p$ and $q + v < n$, and $E_{p,q}E_{u,v} = 0$ in any other case.

Now define $\phi : H \rightarrow H^*$ by $\phi(e_{i,j}) = E_{i+mj,n-1-j}$ for any i, j. We prove that ϕ is right H^*-linear, i.e. $\phi(e_{i,j} \leftarrow E_{u,v}) = \phi(e_{i,j})E_{u,v}$ for any i, j, u, v. Fix some i, j. If $v \leq j$ and $u = i + m(j-v)$, then

$$\phi(e_{i,j} \leftarrow E_{u,v}) = \phi(e_{i,j-v}) = E_{i+m(j-v),n-1-j+v}$$

On the other hand, since $n-1-j+v < n$ and $i+m(j-v)+mv = i+mj$, we have

$$\begin{aligned} \phi(e_{i,j})E_{u,v} &= E_{i+mj,n-1-j}E_{i+m(j-v),v} \\ &= E_{i+m(j-v),n-1-j+v} \end{aligned}$$

and the desired relation is satisfied.

If either $v > j$ or $u \neq i + m(j - v)$, then $e_{i,j} \leftarrow E_{u,v} = 0$. Since either $n - 1 - j + v \geq n$ or $i + m(j - v) + mv \neq i + mj$, we see that $\phi(e_{i,j})E_{u,v} = 0$, so again $\phi(e_{i,j} \leftarrow E_{u,v}) = \phi(e_{i,j})E_{u,v}$.

We clearly have that ϕ is injective since $i + mj = i' + mj'$ and $n - 1 - j = n - 1 - j'$ imply $i = i'$ and $j = j'$. We have thus proved that H is right co-Frobenius.

We show now that H is not left co-Frobenius. We see that if $\phi : H \to H^*$ is a morphism of left H^*-modules, then for any i, j we have that

$$\begin{aligned}\phi(1)(e_{i,j}) &= \phi(1)(e_{i,j} \leftarrow E_{i+mj,0}) \\ &= (E_{i+mj,0}\phi(1))(e_{i,j}) \\ &= \phi(E_{i+mj,0} \rightarrow 1)(e_{i,j}) \\ &= \phi(E_{i+mj,0} \rightarrow e_{0,0})(e_{i,j}) \\ &= 0\end{aligned}$$

Therefore $\phi(1) = 0$ and ϕ can not be injective. This means that H is not left co-Frobenius. ∎

3 Integrals of H

Recall that a left (respectively right) integral for the bialgebra H is an element $T \in H^*$ satisfying $h^*T = h^*(1)T$ (respectively $Th^* = h^*(1)T$) for any $h^* \in H^*$. If we apply both sides of the equation to elements of H, this relation is equivalent to $\sum T(c_2)c_1 = T(c)1$ (respectively $\sum T(c_1)c_2 = T(c)1$) for any $c \in H$. The aim of this section is to prove the following.

Proposition 3.1 *The space of left integrals of H is zero, and the space of right integrals of H is $< E_{0,0}, E_{0,1}, \dots, E_{0,n-1} >$, and thus has dimension n.*

Proof: Let us first look for the left integrals T. Then for $c = g^i$ with $i > 0$ we obtain $T(g^i)g^i = T(g^i)1$, so $T(g^i) = T(e_{i,0}) = 0$. Then for $c = e_{i,1}$, $i \geq 0$, we obtain

$$T(e_{i,1})e_{i+m,0} + T(e_{i,0})e_{i,1} = T(e_{i,1})1$$

For $i = 0$ this implies that $T(e_{0,1}) = T(e_{0,0}) = 0$. For $i > 0$, it implies that $T(e_{i,1}) = 0$. Thus $T(e_{i,j}) = 0$ for any $j \leq 1$. We continue by induction, showing that for any $j \leq n - 1$ we have that $T(e_{i,r}) = 0$ for any $r \leq j$. Indeed, if the claim is true for $j' < j$, then apply the relation $\sum T(c_2)c_1 = T(c)1$ for $c = e_{i,j}$ (any i), and find that

$$T(e_{i,j})e_{i+mj,0} + T(e_{i,j-1})e_{i+m(j-1),1} + \dots + T(e_{i,0})e_{i,j} = T(e_{i,j})1$$

Since all but the first terms in the left hand side are zero by the induction hypothesis, we obtain $T(e_{i,j})e_{i+mj,0} = T(e_{i,j})1$, implying $T(e_{i,j}) = 0$. Therefore the space of left integrals is zero.

Let us take now a right integral T of H. Use the relation $\sum T(c_1)c_2 = T(c)1$ for $c = e_{i,0} = g^i$, with $i > 0$, and find that $T(e_{i,0}) = 0$. We prove by induction on $j \leq n-1$ that $T(e_{i,j}) = 0$ for any $i > 0$. Indeed, if we assume that this is true for $j' < j$, apply $\sum T(c_1)c_2 = T(c)1$ for $c = e_{i,j}$, where $i > 0$. By using equation (1), this gives

$$\sum_{0 \leq r \leq j} T(e_{i+m(j-r),r})e_{i,j-r} = T(e_{i,j})1 \tag{2}$$

and by the induction hypothesis it becomes $T(e_{i,j})e_{i,0} = T(e_{i,j})1$, showing that $T(e_{i,j}) = 0$ for any $i > 0$.

Now if $i = 0$, (2) is equivalent to $T(e_{0,j})e_{0,0} = T(e_{0,j})1$, which holds true. We have thus proved that the space of right integrals of H is $< E_{0,0}, E_{0,1}, \ldots, E_{0,n-1} >$. ∎

References

[1] M. Beattie, S. Dăscălescu, and L. Grünenfelder, Constructing pointed Hopf algebras by Ore extensions, J. Algebra **225** (2000), 743-770.

[2] S. Dăscălescu, C. Năstăsescu, B. Torrecillas, Co-Frobenius Hopf algebras: integrals, Doi-Koppinen modules and injective objects, J. Algebra **220** (1999), 542-560.

[3] S. Dăscălescu, C. Năstăsescu, Ş. Raianu, Hopf algebras: An introduction, Pure and Applied Math. **235** (2000), Marcel Dekker.

[4] Y. Doi, Homological coalgebra, J. Math. Soc. Japan **33** (1981), 31-50.

[5] C. Kassel, Quantum Groups, *Graduate Texts in Mathematics* **155**, Springer Verlag, Berlin, 1995.

[6] R. G. Larson, M. E. Sweedler, An associative orthogonal form for Hopf algebras, Amer. J. Math. **91** (1969), 75-93.

[7] B. J. Lin, Semiperfect coalgebras, J. Algebra **49** (1977), 357-373.

[8] S. Montgomery, Hopf algebras and their actions on rings, CBMS, Vol. 82, AMS, 1993.

[9] Ş. Raianu, An easy proof for the uniqueness of integrals, in Hopf algebras and quantum groups, Proc. of the Brussels Conference

1998, Lect. Notes Pure Appl. Math. 209, 237-240 (2000), Marcel Dekker.

[10] D. Ştefan, The uniqueness of integrals: a homological approach, Comm. Algebra **23** (1995), 1657-1662.

[11] J. B. Sullivan, The uniqueness of integrals for Hopf algebras and some existence theorems of integrals for commutative Hopf algebras, J. Algebra **19** (1971), 426-440.

[12] M. E. Sweedler, integrals for Hopf algebras, Ann. of Math. **89** (1969), 323-335.

BI-FROBENIUS ALGEBRAS AND GROUP-LIKE ALGEBRAS

YUKIO DOI
OKAYAMA UNIVERSITY, FACULTY OF EDUCATION,
OKAYAMA 700-8530, JAPAN

ABSTRACT. Bi-Frobenius algebras (or bF algebras) were recently introduced by the author and Takeuchi, as a generalization of finite dimensional Hopf algebras. In this paper we introduce and study a notion of a group-like algebra, which is an interesting example of BF algebras. The concept generalizes Bose-Mesner algebras of (non-commutative) association schemes.

INTRODUCTION

Bi-Frobenius algebras, or briefly bF algebras, were introduced by the author and Takeuchi [DT]. They generalize finite dimensional Hopf algebras. In [DT] and [D] some basic structure theory of bF algebras was presented. In this paper we introduce and study group-like algebras. A group-like algebra is a finite dimensional algebra over a field k with a distinguished k-basis $\mathbf{B}$ containing 1 which has certain specified properties (given in 2.2 below). The concept generalizes a Bose-Mesner algebra of an association scheme. It becomes a bF algebra in a natural way. Thus we can use the theory of bF algebras to study group-like algebras. For example, the sum of elements in $\mathbf{B}$ is a right (and left) integral.

In Section 1, after reviewing the definition and some basic properties of bF algebras, we develop some general theory. In particular we show that –under certain assumptions– the irreducible characters are orthogonal. In Section 2 we define the concept of a group-like algebra. Section 3 consists of a list of examples of group-like algebras. We determine the structure of group-like algebras of dimension 2 or 3. In section 4 we consider a subset $N \subset \mathbf{B}$ having certain specified properties. Then the k-span kN becomes a group-like algebra. In the case of $|N| = |\mathbf{B}| - 1$ we determine the structure of the ring extension $kN \subset A$.

We work over a field k; unadorned Hom and $\otimes$ are assumed to be taken over k. Let H be an algebra and coalgebra with comultiplication Δ and counit ε. For $h \in H$ we represent $\Delta(h) \in H \otimes H$ by $\Delta(h) = \sum h_{(1)} \otimes h_{(2)}$. The dual algebra $H^* = \text{Hom}\,(H, k)$ has a two-sided H-module structure

$$(h \rightharpoonup f)(x) = f(xh), \quad (f \leftharpoonup h)(x) = f(hx)$$

for all $h,\ x \in H$ and $f \in H^*$. And H has a two-sided H^*-module structure

$$f \rightharpoonup h = \sum h_{(1)} f(h_{(2)}), \quad h \leftharpoonup f = \sum f(h_{(1)}) h_{(2)}$$

for all $f \in H^*$ and $h \in H$.

1. Bi-Frobenius algebras: Preliminaries

Definition 1.1. Let H be a finite dimensional algebra and coalgebra over a field k, and $\phi \in H^*$, $t \in H$. Define $S : H \to H$ by

$$S(h) = t \leftharpoonup (h \rightharpoonup \phi) = \sum \phi(t_{(1)}h)t_{(2)}, \quad h \in H.$$

Then (H, ϕ, t, S) is called a *bi-Frobenius algebra* (or *bF algebra*) if
(BF1) $\varepsilon(hh') = \varepsilon(h)\varepsilon(h')$, $\forall h, h' \in H$ and $\varepsilon(1) = 1$,
(BF2) $\Delta(1) = 1 \otimes 1$,
(BF3) $\{\phi \leftharpoonup h | h \in H\} = H^*$,
(BF4) $\{t \leftharpoonup f | f \in H^*\} = H$,
(BF5) $S(hh') = S(h')S(h)$,
(BF6) $\Delta(S(h)) = \sum S(h_{(2)}) \otimes S(h_{(1)})$.
S is called the *bF antipode* of H.

We summarize some basic properties of bF algebras from [DT] and [D].

Fact 1.2. Let (H, ϕ, t, S) be a bF algebra. Then
(1) S is a bijection, in particular $S(1) = 1$ and $\varepsilon(S(h)) = \varepsilon(h)$, by (BF5,6).
(2) [D, Lemma 1.2] Conversely the bijectivity of S implies the conditions (BF3,4). That is, (BF5,6) + H the bijectivity of S I $\Rightarrow$ (BF3,4).
(3) $\sum \overline{S}(t_{(2)})\phi(t_{(1)}h) = h = \sum \phi(h\overline{S}(t_{(2)}))t_{(1)}$, $\forall h \in H$, here $\overline{S}$ denotes the composition inverse of S.
(4) $\sum h\overline{S}(t_{(2)}) \otimes t_{(1)} = \sum \overline{S}(t_{(2)}) \otimes t_{(1)}h$, $\forall h \in H$ and hence the element

$$v := \sum \overline{S}(t_{(2)})t_{(1)}$$

(called the *volume* of H) is in the center $Z(H)$. Consequently if v is invertible, then H is a separable algebra (and hence a semisimple algebra), since $\sum v^{-1}\overline{S}(t_{(2)}) \otimes t_{(1)}$ is a separable idempotent. A good reference about Frobenius algebras and Hopf algebras is [Sch].
(5) $\sum \phi(xy_{(1)})S(y_{(2)}) = \sum \phi(x_{(1)}y)x_{(2)}$, $\forall x, y \in H$.
(6) t is a right integral in H, i.e., $th = t\varepsilon(h)$, $\forall h \in H$. The space of right (left) integrals in H is one-dimensional.
(7) ϕ is a right integral in H^*, i.e., $\sum \phi(h_{(1)})h_{(2)} = \phi(h)1$, $\forall h \in H$. The space of right (left) integrals in H^* is one-dimensional.
(8) $\phi(t) = \phi(\overline{S}(t)) = 1$.

Theorem 1.3. *Let (H, ϕ, t, S) be a bF algebra such that the volume v is invertible. Then*
(1) $\varepsilon(t) \neq 0$, $t = S(t) = \overline{S}(t)$, and $v = \sum t_{(1)}S(t_{(2)})$
(2) Define $\mu : H \to H$ by

$$\mu(h) = \sum v^{-1}\overline{S}(t_{(2)})ht_{(1)}, \quad h \in H.$$

We have $\mu(H) = Z(H)$.
(3) If $S^2 = id$, then

$$\mu(xy) = \mu(yx) \ (x, y \in H), \quad H = Z(H) \oplus [H, H].$$

Proof. (1) We have

$$\varepsilon(v) = \sum \varepsilon(\overline{S}(t_{(2)})\varepsilon(t_{(1)}) = \sum \varepsilon(t_{(2)})\varepsilon(t_{(1)}) = \varepsilon(t).$$

This implies $\varepsilon(t) \neq 0$, since v is invertible. Let $\alpha : H \to k$ be the right modular function of H, i.e., $ht = \alpha(h)t, \forall h$. Then

$$t(ht) = \alpha(h)t^2 = \alpha(h)\varepsilon(t)t, \ (th)t = \varepsilon(h)t^2 = \varepsilon(h)\varepsilon(t)t.$$

It follows from $\varepsilon(t) \neq 0$ that $\alpha = \varepsilon$. Thus t is also a left integral. So $\overline{S}(t)$ is a right integral and $\overline{S}(t) = ct$ for some $c \in k$. Now $c = c\phi(t) = \phi(ct) = \phi(\overline{S}(t)) = 1$, by Fact 1.2(8). Thus $t = S(t) = \overline{S}(t)$. Then

$$\sum t_{(1)} \otimes t_{(2)} = \sum S(t_{(2)}) \otimes S(t_{(1)}).$$

Hence

$$(v =) \ \sum \overline{S}(t_{(2)})t_{(1)} = \sum t_{(1)}S(t_{(2)}).$$

(2) Fact 1.2(4) gives $\mu(H) \subset H$. If $h \in Z(H)$, then

$$\mu(h) = \sum v^{-1}\overline{S}(t_{(2)})t_{(1)}h = v^{-1}vh = h.$$

(3) Since $S(t) = t$ and $\overline{S} = S$ we have $\sum S(t_{(2)}) \otimes t_{(1)} = \sum t_{(1)} \otimes S(t_{(2)})$, and hence $\sum yt_{(1)} \otimes S(t_{(2)}) = \sum t_{(1)} \otimes S(t_{(2)})y$, by Fact 1.2(4).
Thus for any $x \in H$,

$$\sum S(t_{(2)})xyt_{(1)} = \sum S(t_{(2)})yxt_{(1)}.$$

Hence $\mu(xy) = \mu(yx)$ and so $[H, H] \subset Ker\ \mu$.

Conversely if $\mu(h) = \sum v^{-1}t_{(1)}h\overline{S}(t_{(2)}) = 0$, then
$\sum[v^{-1}h\overline{S}(t_{(2)}),\ t_{(1)}] = \sum v^{-1}h\overline{S}(t_{(2)})t_{(1)} - \sum v^{-1}t_{(1)}h\overline{S}(t_{(2)}) = h - 0 = h$.
Hence $Ker\ \mu = [H, H]$. □

We remark in above (1) that $\varepsilon(t) \neq 0$ does not imply the invertibility of v in general (see in Remark 3.5 below).

Definition 1.4 (Hadamard products). Let (H, ϕ, t, S) be a bF algebra. For $x, y \in H$, define

$$x \star y = \sum \phi(y_{(1)}\overline{S}(x))y_{(2)} = \sum \phi(y\overline{S}(x_{(2)}))x_{(1)}.$$

$x \star y$ is characterized by the equality

$$\phi \leftharpoonup (x \star y) = (\phi \leftharpoonup x)(\phi \leftharpoonup y),$$

for, $(\phi \leftharpoonup x)(\phi \leftharpoonup y)(z) = \sum \phi(xz_{(1)})\phi(yz_{(2)}) = \phi(y(\sum \phi(xz_{(1)})z_{(2)})) =^{1.2(5)}$ $\phi(y(\sum \phi(x_{(1)}z)\overline{S}(x_{(2)}))) = \sum \phi(y\overline{S}(x_{(2)}))\phi(x_{(1)}z)$. Hence $(H, \star)$ is an (associative) algebra with identity t, and the following property satisfies (see [DT]):

$$\varepsilon(S(\mathcal{N}(x)) \star y) = \phi(xy), \ \forall x, y \in H$$

where $\mathcal{N}$ denotes the Nakayama automorphism for ϕ.

Assume that the base field k is algebraically closed and (H, ϕ, t, S) be a bF algebra such that v is invertible and $S^2 = id$. We show that the analog of orthogonality of characters is true. We first define a bilinear form on H^* as follows:

$$(f|g) = \sum f(v^{-1}S(t_{(2)}))g(t_{(1)}), \quad f,\ g \in H^*.$$

Note that

$$(\phi \leftharpoonup x \mid \phi \leftharpoonup y) = \sum \phi(xv^{-1}S(t_{(2)}))\phi(yt_{(1)})$$
$$= \phi(xv^{-1}S(\sum \phi(yt_{(1)})t_{(2)})) = \phi(xv^{-1}S^2(y)) = \phi(xv^{-1}y).$$

So if we write $\phi_v = \phi \leftharpoonup v$, we have

(*) $(\phi_v \leftharpoonup x \mid \phi_v \leftharpoonup y) = \phi_v(xy)$.

Theorem 1.5. *Let $\{\chi_1, \chi_2, \dots, \chi_l\}$ be the complete set of irreducible characters of H. Then $(\chi_i|\chi_j) = \delta_{ij}$.*

Proof. Let

$$H = e_1H \oplus e_2H \oplus \cdots \oplus e_lH$$

be a decomposition of H into simple components, where the $\{e_i\}$ are primitive central idempotents. We write

$$\rho_{e_iH} : H \to End(e_iH)$$

for the representation of H on e_iH. Thus $\rho_{e_iH}(h)(x) = xh$ for $x \in e_iH$ and $h \in H$. Let $\chi_{e_iH} : H \to k$ denote the character of ρ, that is, $\chi_{e_iH} = tr \circ \rho_{e_iH}$. We claim that $\chi_{e_iH} = \phi_v \leftharpoonup e_i$.

For any $x \in e_iH, x = e_ix = \sum e_i\overline{S}(t_{(2)})\phi(t_{(1)}x)$. Thus $xh = \sum e_i\overline{S}(t_{(2)})h\phi(t_{(1)}x)$. Now using the isomorphism $End(e_iH) \simeq (e_iH)^* \otimes e_iH$, we see that $\rho_{e_iH}(h)$ corresponds to $\sum \phi \leftharpoonup t_{(1)} \otimes e_i\overline{S}(t_{(2)})h$. Thus,

$$\chi_{e_iH}(h) = \sum \phi(t_{(1)}e_i\overline{S}(t_{(2)})h) = \phi(ve_ih) = (\phi_v \leftharpoonup e_i)(h),$$

where here we have used that e_i is central and $v = \sum t_{(1)}\overline{S}(t_{(2)})$ by Theorem 1.3(1).

Now we may write $e_iH = V_i \oplus V_i \oplus \cdots \oplus V_i$ (n_i-times) where V_i is a minimal right ideal in e_iH with $dim\ V_i = n_i$. We may assume that χ_i is precisely the character of the representation of H on V_i. Then $\chi_{e_iH} = n_i\chi_i$. It follows that $n_i \neq 0$ in k, since $n_i \sum \chi_i(v^{-1}\overline{S}(t_{(2)}))t_{(1)} = \sum(\phi_v \leftharpoonup e_i)(v^{-1}\overline{S}(t_{(2)}))t_{(1)} = \phi(e_i\overline{S}(t_{(2)}))t_{(1)} =^{Fact1.2(3)} e_i \neq 0$.

Now using (*) we have

$$(\chi_i|\chi_j) = \frac{1}{n_in_j}(\chi_{e_iH}|\chi_{e_jH}) = \frac{1}{n_in_j}(\phi_v \leftharpoonup e_i \mid \phi_v \leftharpoonup e_j)$$
$$= \frac{1}{n_in_j}\phi_v(e_ie_j) = \frac{1}{n_in_j}\phi_v(e_i)\delta_{ij} = \frac{1}{n_in_j}\chi_{e_iH}(1)\delta_{ij} = \delta_{ij}.$$

□

2. Group-like algebras

We begin with a theorem that motivates the notion of group-like algebra.

Theorem 2.1. *Let A be a finite dimensional algebra over a field k and $\mathbf{B} = \{b_0, b_1, \dots, b_d\}$ a k-basis of A with identity element $1_A = b_0$. Let $\varepsilon : A \to k$ be an algebra map and $S : A \to A$ an anti-algebra automorphism such that*

(i) $\varepsilon \circ S = \varepsilon$,

(ii) For all i, $\varepsilon(b_i) \neq 0$,

(iii) $S(b_i) \in \mathbf{B}$ *(then i^* is defined by $b_{i^*} = S(b_i)$).*
Define $\phi \in A^* = Hom(A, k)$ *and* $t \in A$ *by*

$$\phi(b_i) = \delta_{i0} \text{ (Kronecker delta) and } t := b_0 + b_1 + \ldots + b_d$$

and regard A as a coalgebra via $\Delta(b_i) = \frac{1}{\varepsilon(b_i)}\, b_i \otimes b_i$. *Then:*
(A, ϕ, t, S) is a bF algebra if and only if the following (iv) satisfies
(iv) For all i, j, $p_{ij}^0 = \delta_{ij^*}\varepsilon(b_i)$,
here p_{ij}^k denotes the structure constant for $\mathbf{B}$, i.e., $b_i b_j = \sum_{k=0}^d p_{ij}^k b_k$.
In this case, we have that $S^2 = id$.

Proof. Note that S is an anti-algebra and anti-coalgebra automorphism by (i).
Now $S(x) = \sum \phi(t_{(1)}x)t_{(2)},\ \forall x \in A$
$\Leftrightarrow\ b_{j^*} = \sum \phi(t_{(1)}b_j)t_{(2)},\ \forall j$
$\Leftrightarrow\ b_{j^*} = \sum_{i=0}^d \frac{1}{\varepsilon(b_i)}\ \phi(b_i b_j) b_i,\ \forall j$
$\Leftrightarrow\ b_{j^*} = \sum_{i=0}^d \frac{1}{\varepsilon(b_i)}\ p_{ij}^0 b_i$
$\Leftrightarrow\ p_{ij}^0 = \delta_{ij^*}\varepsilon(b_i)$.

Thus the theorem follows from Fact 1.2(2). Finally the fact $S^2 = id$ follows from [DT, 3.5(b)]. But we give here a direct proof (due to referee): As $\phi(b_{i^*}) = 0 = \phi(b_i)$ for $i \neq 0$ and $\phi(b_{0^*}) = 1 = \phi(b_0)$, it follows that $\phi = \phi \circ S = \phi \circ \overline{S}$. Then

$$\begin{aligned} S(x) &= \sum \phi(t_{(2)}x)t_{(1)} \quad (\Delta \text{ is cocommutative}) \\ &= \sum (\phi \circ \overline{S})(t_{(2)}x)t_{(1)} \\ &= \sum \phi(\overline{S}(x)\overline{S}(t_{(2)}))t_{(1)} \\ &= \overline{S}(x) \quad (\text{by Fact } 1.2(3)) \end{aligned}$$

□

This result suggests the following definition.

Definition 2.2. A *group-like algebra* (or *generalized group algebra*) is a 4-tuple $(A, \varepsilon, \mathbf{B}, S)$, where A is a finite dimensional algebra over a field k, $\varepsilon : A \to k$ an algebra map, $\mathbf{B} = \{b_0 = 1, b_1, \ldots, b_d\}$ a k-basis of A, $S : \mathbf{B} \to \mathbf{B}, b_i \mapsto b_{i^*}$ an involution (i.e. $i^{**} = i$) satisfying the following conditions:
(G1) $\varepsilon(b_{i^*}) = \varepsilon(b_i) \neq 0,\ \forall i$,
(G2) $p_{ij}^k = p_{j^*i^*}^{k^*},\ \forall i, j, k$, (here p_{ij}^k denotes the structure constant for $\mathbf{B}$)
(G3) $p_{ij}^0 = \delta_{ij^*}\varepsilon(b_i),\ \forall i, j$.
We say that H is *symmetric* if $S = id$. In this case, it is clearly a commutative algebra.

Let A be the Bose-Mesner algebra of an association scheme ([BI],[Z]). Then A becomes a group-like algebra over $\mathbb{C}$ in a natural way, here S is the matrix transpose. Thus group-like algebras can be regarded as a non-commutative analogue of Kawada's character algebras ([BI]).

Remark 2.3. Let $(A, \varepsilon, \mathbf{B}, S)$ be a group-like algebra.
(a) From (G2) it follows that S induces an anti-multiplicative morphism of A. Hence $0^* = 0$ and, by (G1), $\varepsilon \circ S = \varepsilon$. It follows from Theorem 2.1 that A endowed with the coproduct that makes all the elements $\{b_i/\varepsilon(b_i) : i = 1, \cdots, d\}$ group-like, is a

bF algebra. Moreover A is a symmetric algebra, since $\phi(b_i b_j) = p_{ij}^0 = p_{ji}^0 = \phi(b_j b_i)$.
(b) By Fact 1.2(3) we have for $x \in H$,

$$x = \sum_{i=0}^{d} \lambda_i b_i \;\Rightarrow\; \lambda_i = \frac{\phi(x b_{i^*})}{\varepsilon(b_i)}.$$

(c) Since $t := 1 + b_1 + b_2 + \cdots + b_d$ is a right integral, $tb_i = \varepsilon(b_i)t$ and so
(G4) $\quad p_{0i}^k + p_{1i}^k + \cdots + p_{di}^k = \varepsilon(b_i)$.
Moreover t is a left integral by $S(t) = t$. Thus
(G4') $\quad p_{i0}^k + p_{i1}^k + \cdots + p_{id}^k = \varepsilon(b_i)$.
The volume of A is given by

$$v = 1 + b_{1^*} b_1/\varepsilon(b_1) + b_{2^*} b_2/\varepsilon(b_2) + \cdots + b_{d^*} b_d/\varepsilon(b_d).$$

(d) The Hadamard product is as follows:

$$b_j \star b_i = \frac{\phi(b_i b_{j^*}) b_i}{\varepsilon(b_i)} = \frac{p_{ij^*}^0 b_i}{\varepsilon(b_i)} = \delta_{ij} b_i.$$

In the next section we will need one more relationship between the structure constants of a group-like algebra.

Lemma 2.4. *We have*
(G5) $\quad p_{ij}^k \varepsilon(b_k) = p_{kj^*}^i \varepsilon(b_i)$

Proof. By (G3) we have

$$\phi((b_i b_j) b_k) = \phi(p_{ij}^0 b_k + p_{ij}^1 b_1 b_k + \cdots + p_{ij}^d b_d b_k)$$
$$= p_{ij}^0 \delta_{k0} + p_{ij}^1 \delta_{1k^*} \varepsilon(b_1) + \cdots + p_{ij}^d \delta_{dk^*} \varepsilon(b_d) = p_{ij}^{k^*} \varepsilon(b_{k^*}).$$

And similarly $\phi(b_i(b_j b_k)) = \phi(p_{jk}^0 b_i + p_{jk}^1 b_i b_1 + \cdots + p_{jk}^d b_i b_d) = p_{jk}^{i^*} \varepsilon(b_i)$. Hence $p_{ij}^{k^*} \varepsilon(b_{k^*}) = p_{jk}^{i^*} \varepsilon(b_i) = p_{k^* j^*}^i \varepsilon(b_i)$ by (G2). Replacing k^* to k we get the lemma. □

Definition 2.5 (Skew group ring). Let $(A, \varepsilon, \mathbf{B} = \{b_0 = 1, b_1, \ldots, b_d\}, S)$ be a group-like algebra and G a finite group. Given a group homomorphism $\varphi : G \to Aut(A)$. We write $g \triangleright x$ for $\varphi(g)(x)$, $(g \in G,\ x \in A)$. We assume that for all $g \in G$ and $b_i \in \mathbf{B}$,
(1) $g \triangleright b_i \in \mathbf{B}$
(2) $\varepsilon(g \triangleright b_i) = \varepsilon(b_i)$
(3) $S(g \triangleright b_i) = g \triangleright S(b_i)$.

Then the usual skew group ring $A * G$ is a group-like algebra:
As a vector space, $A * G := A \otimes kG$, and if we write $x \sharp y$ for the element $x \otimes y$, then its multiplication is defined by

$$(b_i \sharp g)(b_j \sharp h) = b_i (g \triangleright b_j) \sharp gh, \quad g,\ h \in G.$$

One see from Theorem 2.1 that $A * G$ is a group-like algebra with basis $\{b_i \sharp g;\ i = 0, 1, \ldots, d,\ g \in G\}$, where

$$\varepsilon(b_i \sharp g) = \varepsilon_A(b_i), \quad S(b_i \sharp g) = (g^{-1} \triangleright S_A(b_i)) \sharp g^{-1}.$$

Indeed, for $y \in G$,

$$\sum_{i,g} \tfrac{1}{\varepsilon(b_i)} (\phi_A \otimes \phi_G)((b_i \sharp g)(b_j \sharp y))\, b_i \sharp g = \sum_{i,g} \tfrac{1}{\varepsilon(b_i)} \phi_A(b_i(g \triangleright b_j)) \phi_G(gy)(b_i \sharp g)$$
$$= \tfrac{1}{\varepsilon(b_i)} p_{ik}^0 b_i \sharp y^{-1} \ (\textit{say } b_k = g \triangleright b_j) = (g^{-1} \triangleright S(b_j)) \sharp y^{-1} = S(b_j \sharp y).$$

3. Examples of group-like algebras

Example 3.1 (2-dimensional case). Let q be any non-zero element in k. Denote by $A_q(2)$ the algebra with k-basis $\{1, b\}$ and with multiplication

$A_q(2)$	1	b
1	1	b
b	b	$q+(q-1)b$

Then $A_q(2)$ is a group-like algebra with $S = id$ and $\varepsilon(b) = q$. Any 2-dimensional group-like algebra is of this form by (G4) and (G4').

The volume is $v = 2 + \frac{q-1}{q}b$. It is invertible if and only if $\varepsilon(t) = 1 + q \neq 0$. The inverse is given by $v^{-1} = \frac{(q^2+1)+(1-q)b}{(q+1)^2}$.

Example 3.2 (3-dimensional non-symmetric case). Let q be any non-zero element in k. Assume the characteristic of k, char(k), is not 2. Denote by $A_q(3)$ the algebra with k-basis $\{1, b_1, b_2\}$ and with multiplication

$A_q(3)$	1	b_1	b_2
1	1	b_1	b_2
b_1	b_1	$\frac{q-1}{2}b_1 + \frac{q+1}{2}b_2$	$q + \frac{q-1}{2}(b_1 + b_2)$
b_2	b_2	$q + \frac{q-1}{2}(b_1 + b_2)$	$\frac{q+1}{2}b_1 + \frac{q-1}{2}b_2$

Then $A_q(3)$ is a group-like algebra with $S(b_1) = b_2$ and $q := \varepsilon(b_1) = \varepsilon(b_2) \neq 0$. The volume is $v = 3 + \frac{q-1}{q}(b_1 + b_2)$. Ii is invertible if and only if $\varepsilon(t) = 2q + 1 \neq 0$. The inverse is given by

$$v^{-1} = \frac{2q^2 + 1 + (1-q)(b_1 + b_2)}{(2q+1)^2}.$$

Theorem 3.3. *If char(k) $\neq 2$, then any 3-dimensional non-symmetric group-like algebra is of the above form. If char(k) $= 2$, the multiplication table is as follows:*

	1	b_1	b_2
1	1	b_1	b_2
b_1	b_1	$\alpha b_1 + (1+\alpha)b_2$	$1 + \alpha(b_1 + b_2)$
b_2	b_2	$1 + \alpha(b_1 + b_2)$	$(1+\alpha)b_1 + \alpha b_2$

where $\varepsilon(b_1) = \varepsilon(b_2) = 1$.

Proof. We may assume $S(b_1) = b_2$ and $q := \varepsilon(b_1) = \varepsilon(b_2) \neq 0$ by (G1). By (G3),

$$p_{11}^0 = p_{22}^0 = 0,\ p_{12}^0 = p_{21}^0 = q$$

And by (G2),

$$p_{11}^1 = p_{22}^2,\ p_{12}^1 = p_{12}^2,\ p_{21}^1 = p_{21}^2,\ p_{22}^1 = p_{11}^2$$

Also (G5) gives

$$p_{12}^2 = p_{21}^1,\ p_{11}^1 = p_{12}^1.$$

Thus we have

	1	b_1	b_2
1	1	b_1	b_2
b_1	b_1	$\alpha b_1 + \beta b_2$	$q + \alpha(b_1 + b_2)$
b_2	b_2	$q + \alpha(b_1 + b_2)$	$\beta b_1 + \alpha b_2$

Now using (G4),

$$q = 1 + 2\alpha = \alpha + \beta.$$

So if $char(k) \neq 2$, then $\alpha = \frac{q-1}{2}$, $\beta = \frac{q+1}{2}$. If $char(k) = 2$, then $q = 1,\ \beta = 1 + \alpha$.

Finally we check the associativity of multiplication. We have :

$$(b_1b_1)b_2 = \frac{q(q-1)}{2} + \frac{q^2+1}{2}b_1 + \frac{q^2-q}{2}b_2 = b_1(b_1b_2)$$

and

$$(b_1b_2)b_2 = \frac{q(q-1)}{2} + \frac{q^2-q}{2}b_1 + \frac{q^2+1}{2}b_2 = b_1(b_2b_2).$$

The other cases are obvious from commutativity. Hence the multiplication is associative. □

Theorem 3.4 (3-dimensional symmetric case). *Let β, p, q be elements in k with $p \neq 0$, $q \neq 0$. Denote by $A^{\beta}_{p,q}(3)$ the algebra with k-basis $\{1, b_1, b_2\}$ and with multiplication*

$A^{\beta}_{p,q}(3)$	1	b_1	b_2
1	1	b_1	b_2
b_1	b_1	$p + (p-1-\beta-q)b_1 + \beta p b_2$	$\beta q b_1 + (p - \beta p)b_2$
b_2	b_2	$\beta q b_1 + (p - \beta p)b_2$	$q + (q - \beta q)b_1 + (q - 1 - p + \beta p)b_2$

where $\varepsilon(b_1) = p$ and $\varepsilon(b_2) = q$ and $\beta \in k$.

Proof. Since $S = id$, the multiplication is commutative. Let $p := p^0_{11} = \varepsilon(b_1)$ and $q := p^0_{22} = \varepsilon(b_2)$. By (G5), $p^2_{11}q = p^1_{21}p$. So if we let $\beta := p^2_{11}/p = p^1_{21}/q$, then

$$p^2_{11} = \beta p, \quad p^1_{21} = \beta q.$$

It follows from (G4) and (G4') that

$$p^1_{11} = p - 1 - \beta q,\ p^2_{12} = p - \beta p,\ p^1_{22} = q - \beta q,\ p^2_{22} = q - 1 - p + \beta p.$$

Now it is easy to check that $(b_1b_1)b_2 = b_1(b_1b_2)$ and $(b_1b_2)b_2 = b_1(b_2b_2)$ hold. So the multiplication is associative from commutativity. □

The volume is

$$v = 1 + \frac{b_1b_1}{p} + \frac{b_2b_2}{q} = 3 + \frac{2p - 1 - \beta(p+q)}{p}b_1 + \frac{q - p - 1 + \beta(p+q)}{q}b_2.$$

It is invertible if and only if

$$\frac{(p+q+1)^2}{pq} \cdot \{\beta^2(p+q)^2 + 2\beta(q - p - pq - p^2) + (p+1)^2\} \neq 0.$$

For example, if $k = \mathbb{R}$ (the real field) and $p, q > 0$, then v is always invertible, since $D/4 = (q - p - pq - p^2)^2 - (p+q)^2(p+1)^2 = -4pq(1 + p + q) < 0$.

Let (H, ϕ, t, S) be a bF algebra. A map $\sigma : H \to H$ is called a *Hopf antipode* if it is a convolution-inverse of identity, i.e.,

$$\sum \sigma(h_{(1)})h_{(2)} = \varepsilon(h)1 = \sum h_{(1)}\sigma(h_{(2)}),\ h \in H.$$

BF algebras need not have Hopf antipodes:

Remark 3.5. The multiplication table of $A^0_{p,q}(3)$ is as follows:

$A^0_{p,q}(3)$	1	b_1	b_2
1	1	b_1	b_2
b_1	b_1	$p+(p-1)b_1$	pb_2
b_2	b_2	pb_2	$q+qb_1+(q-p-1)b_2$

It is easy to see that the element b_2 is not invertible. It follows from $\Delta(b_2) = \frac{1}{q}b_2 \otimes b_2$ that $A^0_{p,q}(3)$ does not have a Hopf antipode.

The volume is $v = 3 + \frac{2p-1}{p}b_1 + \frac{q-p-1}{q}b_2$. It is invertible if and only if $\varepsilon(t) = 1+p+q \neq 0$ and $1+p \neq 0$. In this case, it is easy to verify that

$$e_0 = \frac{1+b_1+b_2}{1+p+q}, \quad e_1 = \frac{p-b_1}{1+p}, \quad e_2 = 1-e_0-e_1 = \frac{q+qb_1-(1+p)b_2}{(1+p)(1+p+q)}$$

are orthogonal primitive idempotents whose sum is 1.

Theorem 3.6. *Any 4-dimensional group-like algebra is commutative as an algebra.*

Proof. Let $(H, \varepsilon, \{1, b_1, b_2, b_3\}, S)$ be a (4-dimensional) group-like algebra. We may assume that $S(b_1) = b_3$ and $S(b_2) = b_2$. Let $p := p^0_{13} = \varepsilon(b_1) = \varepsilon(b_3) \neq 0$ and $q := \varepsilon(b_2) \neq 0$. Using (G2) and (G5),

$$p^3_{33} = p^1_{11} \overset{(G5)}{=} p^1_{13} \overset{(G5)}{=} p^3_{13} = p^1_{31}.$$

Let $\alpha := p^3_{33}$ and $\beta := p^3_{11} = p^1_{33}$. Then, by (G4), (G4'),

$$p^1_{12} = p^1_{21} = p^3_{23} = p^3_{32} = p - 1 - 2\alpha,$$
$$p^3_{12} = p^3_{21} = p^1_{23} = p^1_{32} = p - \alpha - \beta,$$
$$p^1_{22} = p^3_{22} = q - 2p + 1 + 3\alpha + \beta.$$

Applying ε to $b_1^2 = \alpha b_1 + p^2_{11}b_2 + \beta b_3$ and $b_3^2 = \beta b_1 + p^2_{33}b_2 + \alpha b_3$, we have $p^2 = (\alpha+\beta)p + p^2_{11}q$. Hence $p^2_{11} = p^2_{33} = \frac{p}{q}(p-\alpha-\beta)$. Similarly $p^2_{13} = p^2_{31} = \frac{p}{q}(p-1-2\alpha)$. Thus we obtain the following commutative table:

	1	b_1	b_2	b_3
1	1	b_1	b_2	b_3
b_1	b_1	$\alpha b_1 + lb_2 + \beta b_3$	$fb_1 + nb_2 + gb_3$	$p + \alpha b_1 + mb_2 + \alpha b_3$
b_2	b_2	$fb_1 + nb_2 + gb_3$	$q + hb_1 + p^2_{22}b_2 + hb_3$	$gb_1 + nb_2 + fb_3$
b_3	b_3	$p + \alpha b_1 + mb_2 + \alpha b_3$	$gb_1 + nb_2 + fb_3$	$\beta b_1 + lb_2 + \alpha b_3$

where $f = p - 1 - 2\alpha$, $g = p - \alpha - \beta$, $h = (q - 2p + 1 + 3\alpha + \beta)$,
$l = \frac{p}{q}(p - \alpha - \beta)$, $m = \frac{p}{q}(p - 1 - 2\alpha)$, $n = \frac{p}{q}(q - 2p + 1 + 3\alpha + \beta)$,
$p^2_{22} = q - 1 - \frac{2p}{q}(q - 2p + 1 + 3\alpha + \beta)$. □

We can show in a similar way that any 5-dimensional group-like algebra with $S = (14)(23)$ is commutative. It remains open for the case of $S = (14)$.

The next is an example of a group-like algebra which is not commutative.

Example 3.7. Let $\mathrm{char}(k) \neq 2$ and $q \neq 0 \in k$.

!	1	b_1	b_2	b_3	b_4	b_5
1	1	b_1	b_2	b_3	b_4	b_5
b_1	b_1	$q^-b_1 + q^+b_2$	$q + q^-(b_1 + b_2)$	b_5	$qb_3 + q^-(b_4 + b_5)$	$q^+b_4 + q^-b_5$
b_2	b_2	$q + q^-(b_1 + b_2)$	$q^+b_1 + q^-b_2$	b_4	$q^-b_4 + q^+b_5$	$qb_3 + q^-(b_4 + b_5)$
b_3	b_3	b_4	b_5	1	b_1	b_2
b_4	b_4	$q^-b_4 + q^+b_5$	$qb_3 + q^-(b_4 + b_5)$	b_2	$q + q^-(b_1 + b_2)$	$q^+b_1 + q^-b_2$
b_5	b_5	$qb_3 + q^-(b_4 + b_5)$	$q^+b_4 + q^-b_5$	b_1	$q^-b_1 + q^+b_2$	$q + q^-(b_1 + b_2)$

where $S(b_1) = b_2,\ S(b_i) = b_i\ (i = 3, 4, 5)$ and

$$\varepsilon(b_j) = q\ (j = 1, 2, 4, 5),\ q^- = \frac{q-1}{2},\ q^+ = \frac{q+1}{2},\ \varepsilon(b_3) = 1.$$

For the associativity of multiplication, see in 3.8 below. Note that $q = 1$ gives kS_3 (the group algebra of symmetric group of degree 3), via $b_1 = (123),\ b_2 = (132),\ b_3 = (23),\ b_4 = (13),\ b_5 = (12)$. We have

$$\varepsilon(t) = 4q + 2, \quad v = 6 + \frac{2q-2}{q}(b_1 + b_2).$$

If $4q + 2 \neq 0$ (i.e., $q \neq -\frac{1}{2}$), then v is invertible and the inverse is given by

$$v^{-1} = \frac{1}{2(2q+1)^2}\{(2q^2+1) - (q-1)(b_1 + b_2)\}.$$

A full set of non-isomorphic irreducible representations is as follows (due to Wakui): 1-dimensional: ε and ε' where

$$\varepsilon'(1) = 1,\ \varepsilon'(b_1) = \varepsilon'(b_2) = q,\ \varepsilon'(b_3) = -1,\ \varepsilon'(b_4) = \varepsilon'(b_5) = -q.$$

2-dimensional:

$$\rho(1) = \begin{pmatrix} 1 & 0 \\ 0 & 1 \end{pmatrix},\ \rho(b_1) = \frac{1}{2}\begin{pmatrix} -1 & -1 \\ 1+2q & -1 \end{pmatrix},\ \rho(b_2) = \frac{1}{2}\begin{pmatrix} -1 & 1 \\ -(1+2q) & -1 \end{pmatrix},$$

$$\rho(b_3) = \begin{pmatrix} 1 & 0 \\ 0 & -1 \end{pmatrix},\ \rho(b_4) = \frac{1}{2}\begin{pmatrix} -1 & -1 \\ -(1+2q) & 1 \end{pmatrix},\ \rho(b_5) = \frac{1}{2}\begin{pmatrix} -1 & 1 \\ 1+2q & 1 \end{pmatrix}.$$

We next determine the center $Z(H)$, using μ in Theorem 1.3. Since $\mu(b_1) = \mu(b_2) = \frac{1}{2}(b_1+b_2)$, $\mu(b_3) = \frac{1}{2q+1}(b_3+b_4+b_5)$ and $\mu(b_4) = \mu(b_5) = \frac{q}{2q+1}(b_3+b_4+b_5)$, $Z(H)$ has a k-basis $\{c_0 = 1,\ c_1 = b_1 + b_2,\ c_3 = b_3 + b_4 + b_5\}$. Moreover we have $c_1^2 = 2q + (2q-1)c_1,\ c_1c_2 = 2qc_2,\ c_2^2 = (2q+1)(1+c_1)$ and so $Z(H) = A^0_{2q,2q+1}(3)$.

Remark 3.8. Put $A = A_q(3)$ with basis $\{c_0 = 1, c_1, c_2\}$. Let $C_2 = \{1, g\}$ be the cyclic group of order 2. Define a C_2-action on A by

$$g \triangleright 1 = 1,\ g \triangleright c_1 = c_2,\ g \triangleright c_2 = c_1.$$

We then construct the skew group ring $A_q(3) * C_2$, which coincides with the one as in Example 3.7, via

$$1 = 1 \sharp 1,\ b_1 = c_1 \sharp 1,\ b_2 = c_2 \sharp 1,\ b_3 = 1 \sharp g,\ b_4 = c_2 \sharp g,\ b_5 = c_1 \sharp g.$$

Example 3.9 (Tensor product). Let $(A, \varepsilon, \{b_0 = 1, b_1, \dots, b_d\}, S)$ and $(A', \varepsilon', \{c_0 = 1, c_1, \dots, c_h\}, S')$ be group-like algebras. Then $A \otimes A'$ is also a group-like algebra, via $\varepsilon \otimes \varepsilon',\ \{b_i \otimes c_j\}_{i,j},\ S \otimes S'$. For example, the following is the multiplication table of $A_p(2) \otimes A_q(2)$.

	1	b_1	b_2	b_3
1	1	b_1	b_2	b_3
b_1	b_1	$p + (p-1)b_1$	b_3	$pb_2 + (p-1)b_3$
b_2	b_2	b_3	$q + (q-1)b_2$	$qb_1 + (q-1)b_3$
b_3	b_3	$pb_2 + (p-1)b_3$	$qb_1 + (q-1)b_3$	$pq + (p-1)qb_1 + p(q-1)b_2 + (p-1)(q-1)b_3$

Example 3.10 (Wreath product). In the above situation, we assume that $\varepsilon(t_A) \neq 0$ where $t_A = b_0 + b_1 + \cdot + b_d$. Then the linear span of

$$1 \otimes 1,\ b_1 \otimes 1,\ \cdots,\ b_d \otimes 1,\ t_A \otimes c_1,\ \cdots,\ t_A \otimes c_h$$

forms a subalgebra of $A \otimes A'$ and is a group-like algebra with respect to the above basis. This is called the *wreath product* of A and A', and denoted by $A \wr A'$. We have that $\dim A \wr A' = \dim A + \dim A' - 1$. (Thanks to Akihide Hanaki for suggesting this definition.)

4. Closed subsets of bases

Definition 4.1. Let $(A, \varepsilon, \mathbf{B} = \{b_0 = 1, b_1, \ldots, b_d\}, S)$ be a group-like algebra. A subset $N \subset \mathbf{B}$ is *closed* if
(i) $1 \in N$,
(ii) $b_i,\ b_j \in N,\ b_k \notin N \ \Rightarrow\ p_{ij}^k = 0$, i.e., $b_i b_j \in kN$,
(iii) $b_i \in N \ \Rightarrow\ b_{i^*} \in N$.

Then $(kN, \varepsilon|_{kN}, N, S|_{kN})$ is clearly a group-like algebra. We say that A –with basis $\mathbf{B}$– is *primitive* if $\mathbf{B}$ has no closed subsets other than $\{1\}$ and $\mathbf{B}$ itself. For example, $A_q(2)$ is primitive. $A_q(3)$ is primitive if and only if $q \neq -1$. $A_{p,q}^{\beta}$ is primitive if and only if $\beta = 0$. Remark that the condition (iii) cannot be removed, for example, $b_{1^*} = b_2 \notin \{1,\ b_1\}$ in $A_{-1}(3)$.

Remark 4.2. Let $(A, \varepsilon, \mathbf{B} = \{b_0 = 1, b_1, \ldots, b_d\}, S)$ be a group-like algebra. Regard A as a coalgebra via $\Delta(b_i) = \frac{1}{\varepsilon(b_i)} b_i \otimes b_i$. Since $\{b_i/\varepsilon(b_i)\}$ are group-like elements we know that any subcoalgebra of A is of the form kN for some subset N of $\mathbf{B}$ (see [Sw, p16, Exercise 1]). Hence we get that $K \subset A$ is a subalgebra and a subcoalgebra with $S(K) \subset K$ if and only if K is of the form kN for some closed subset N.

Lemma 4.3. *Let $(A, \varepsilon, \mathbf{B} = \{b_0 = 1, b_1, \ldots, b_d\}, S)$ be a group-like algebra and N a closed subset of $\mathbf{B}$. Let $N' := \mathbf{B} \setminus N$ and $b_i \in N,\ b_k \in N'$. Then $b_k b_i$ and $b_i b_k$ belong to kN'.*

Proof. We show that $\phi(b_k b_i b_{j^*}) = 0$ for $b_j \in N$, then by Remarks 2.3(b) we get $b_k b_i \in kN'$. Since $b_i b_{j^*} \in kN$ it suffices to show that $\phi(b_k x) = 0, \forall x \in kN$. We may take $x = b_i \in N$, and

$$\phi(b_k b_i) = p_{ki}^0 = \delta_{ki^*} \varepsilon(b_k) = 0.$$

Similarly we have $b_i b_k \in kN'$. □

Theorem 4.4 ("+1"embedding). *(1) Let $(A, \varepsilon, \{b_0 = 1, b_1, \ldots, b_{d+1}\}, S)$ be a $d+2$-dimensional group-like algebra such that $N = \{b_0, b_1, \ldots, b_d\}$ is a closed subset. Then we have*
(*) $\quad b_i b_{d+1} = b_{d+1} b_i = \varepsilon(b_i) b_{d+1}\ (0 \leq i \leq d), \quad b_{d+1} b_{d+1} = qt + (q - \varepsilon(t)) b_{d+1}$
where $t = 1 + b_1 + \cdots + b_d$ and $q = \varepsilon(b_{d+1})\%$
(2) Let $(A, \varepsilon, \{b_0 = 1, b_1, \ldots, b_d\}, S)$ be any $d+1$-dimensional group-like algebra. Let $A' := A \oplus kb_{d+1}$ and $q \neq 0 \in k$. Define the multiplication for b_{d+1} as (*) above. Then A' is a group-like algebra.

Proof. (1) follows from Lemma 4.3 and (G4).
(2) We must show that the multiplication is associative. We check the equality $(b_{d+1}b_i)b_j = b_{d+1}(b_ib_j)$, $0 \le i,\ j \le d$.

$$\begin{aligned} b_{d+1}(b_ib_j) &= b_{d+1}(p_{ij}^0 1 + p_{ij}^1 b_1 + \cdots + p_{ij}^d b_d) \\ &= (p_{ij}^0 + p_{ij}^1 \varepsilon(b_1) + \cdots + p_{ij}^d \varepsilon(b_d))b_{d+1} \\ &= \varepsilon(p_{ij}^0 + p_{ij}^1 b_1 + \cdots + p_{ij}^d b_d)b_{d+1} \\ &= \varepsilon(b_ib_j)b_{d+1} = \varepsilon(b_i)\varepsilon(b_j)b_{d+1} \\ &= (b_{d+1}b_i)b_j \end{aligned}$$

The other cases are easy to check. □

This algebra will be denoted by $A[b_{d+1};q]$. We have $A_{p,q}^0(3) = A_p(2)[b_2;q]$. One can easily prove that if $\varepsilon(t_A) \neq 0$, then

$$A[b_{d+1};q] = A \wr A_{q'}(2), \text{ where } q' = q/\varepsilon(t_A).$$

"+2"embeddings are quite complicated and there are several types.

Example 4.5. Let $(A, \varepsilon, \mathbf{B} = \{b_0 = 1, b_1, \dots, b_d\}, S)$ be a group-like algebra such that $\varepsilon(t) \neq 0$. Define $A^{+2} := A \oplus kb_{d+1} \oplus kb_{d+2}$ by:
(non-symmetric embedding)

$$b_ib_j = b_jb_i = \varepsilon(b_i)b_j \ (1 \le i \le d,\ d+1 \le j \le d+2)$$

	b_{d+1}	b_{d+2}
b_{d+1}	$\varepsilon(t)b_{d+2}$	$\varepsilon(t)t$
b_{d+2}	$\varepsilon(t)t$	$\varepsilon(t)b_{d+1}$

$$\varepsilon(b_{d+1}) = \varepsilon(b_{d+2}) = \varepsilon(t),\ S(b_{d+1}) = b_{d+2}.$$

Then A^{+2} is a group-like algebra.

Assume that $\mathrm{char}(k) \neq 2$. Define $A^{+2'} := A \oplus kb_{d+1} \oplus kb_{d+2}$ by
(symmetric extension)

$$b_ib_j = b_jb_i = \varepsilon(b_i)b_j \ (1 \le i \le d,\ d+1 \le j \le d+2)$$

	b_{d+1}	b_{d+2}
b_{d+1}	$\varepsilon(t)(2t + b_{d+2})$	$\varepsilon(t)(b_{d+1} + b_{d+2})$
b_{d+2}	$\varepsilon(t)(b_{d+1} + b_{d+2})$	$\varepsilon(t)(2t + b_{d+1})$

$$\varepsilon(b_{d+1}) = \varepsilon(b_{d+2}) = 2\varepsilon(t),\ S(b_{d+1}) = b_{d+1},\ S(b_{d+2}) = b_{d+2}.$$

Then $A^{+2'}$ is also a group-like algebra.

Using certain specified closed set N we can construct another "+2"embeddings:

Theorem 4.6. *Let $(A, \varepsilon, \mathbf{B} = \{b_0 = 1, b_1, \dots, b_d\}, S)$ be a group-like algebra and N a closed subset of $\mathbf{B}$ such that $\varepsilon(t) = 2\varepsilon(t_N)$ and $N'N' \subset kN$. Define $A' := A \oplus kb_{d+1} \oplus kb_{d+2}$ with multiplication:*

$$b_{d+1}b_i = b_ib_{d+1} = \varepsilon(b_i)b_{d+1},\ b_{d+1}b_k = b_kb_{d+1} = \varepsilon(b_k)b_{d+2}$$

$$b_{d+2}b_i = b_ib_{d+2} = \varepsilon(b_i)b_{d+2},\ b_{d+2}b_k = b_kb_{d+2} = \varepsilon(b_k)b_{d+1}$$

for all $b_i \in N$ and $b_k \in N' = \mathbf{B} \setminus N$ and
(non-symmetric type)

	b_{d+1}	b_{d+2}
b_{d+1}	$qt_{N'} + \frac{q-\varepsilon(t_N)}{2}(b_{d+1}+b_{d+2})$	$qt_N + \frac{q-\varepsilon(t_N)}{2}(b_{d+1}+b_{d+2})$
b_{d+2}	$qt_N + \frac{q-\varepsilon(t_N)}{2}(b_{d+1}+b_{d+2})$	$qt_{N'} + \frac{q-\varepsilon(t_N)}{2}(b_{d+1}+b_{d+2})$

(symmetric type)

	b_{d+1}	b_{d+2}
b_{d+1}	$qt_N + \frac{q-\varepsilon(t_N)}{2}(b_{d+1}+b_{d+2})$	$qt_{N'} + \frac{q-\varepsilon(t_N)}{2}(b_{d+1}+b_{d+2})$
b_{d+2}	$qt_{N'} + \frac{q-\varepsilon(t_N)}{2}(b_{d+1}+b_{d+2})$	$qt_N + \frac{q-\varepsilon(t_N)}{2}(b_{d+1}+b_{d+2})$

where q is any non-zero element in k, $t_N = \sum_{b_i \in N} b_i$, and $t_{N'} = t - t_N$.

Proof. (sketch) To check the associativity, we use the following facts:
(a) $b_i \in N \Rightarrow t_N b_i = t_N\varepsilon(b_i) = b_i t_N,\ t_{N'} b_i = t_{N'}\varepsilon(b_i) = b_i t_{N'}$,
(b) $b_k \in N' \Rightarrow t_N b_k = t_{N'}\varepsilon(b_k) = b_k t_N,\ t_{N'} b_k = t_N\varepsilon(b_k) = b_k t_{N'}$.

(a) follows from the fact that t and t_N are integrals. To prove (b), note

$$t_N b_k + t_{N'} b_k = t b_k = t\varepsilon(b_k) = t_N\varepsilon(b_k) + t_{N'}\varepsilon(b_k).$$

But $t_N b_k \in kN'$ by Lemma 4.3, and $t_{N'} b_k \in kN$ by assumption. So (b) follows by comparing the both sides of the above equation. □

The following is an example of a non-symmetric "+2"embedding, where $A = k1 \oplus kb_1 \oplus kb_2$ and $N = \{1, b_1\}$ with $q = 3 = \varepsilon(t_N)$.

	1	b_1	b_2	b_3	b_4
1	1	b_1	b_2	b_3	b_4
b_1	b_1	$2+b_1$	$2b_2$	$2b_3$	$2b_4$
b_2	b_2	$2b_2$	$3+3b_1$	$3b_4$	$3b_3$
b_3	b_3	$2b_3$	$3b_4$	$3b_2$	$3+3b_1$
b_4	b_4	$2b_4$	$3b_3$	$3+3b_1$	$3b_2$

Acknowledgments: The author wishes to thank the referee for many valuable comments. The author also would like to thank Professor Akihide Hanaki for many helpful conversations.

References

[BI] E. Bannai and T. Ito, H Algebraic Combinatorics I: Association Schemes I, Benjamin-Cumming, Menlo Park CA, 1984.

[D] Y. Doi, *Substructures of bi-Frobenius algebras*, J. Algebra 256 (2002) 568-582.

[DT] Y. Doi and M. Takeuchi, *BiFrobenius algebras*, Contemporary Mathematics 267 (2000), 67-97.

[Sch] H.-J. Schneider, *Lectures on Hopf algebras*, Trabajos Mat. 31/95, IMAF, Universidad Nac. Cordoba, Argentina, 1995. Available at www.mate.uncor.edu/andrus.

[Sw] M. E. Sweedler, *Hopf algebras*, Benjamin, New York, 1969.

[Z] P.-H. Zieschang, *An algebraic approach to association schemes*, Lecture Notes in Mathematics 1628, Springer-Verlag, 1996.

Bialgebras and Realizations

R. L. Grossman
R. G. Larson
University of Illinois at Chicago
Chicago, Illinois, U.S.A.

In this paper we consider when a linear functional on a bialgebra is realized by the action of the bialgebra on a finite object. This depends on whether the action of the bialgebra on the functional is finite. We consider two specific cases: the Myhill–Nerode Theorem, which gives a condition for a language to be accepted by a finite automaton, and Fliess' Theorem, which gives a condition for for the input/output maps of a control system to be realized by action on a finite dimensional state space.

If H is a bialgebra, then the linear dual H^* is an algebra with

$$pq(h) = \sum_{(h)} p(h_{(1)})q(h_{(2)})$$

for $h \in H$, p, $q \in H^*$. The algebra H^* is a left and right module algebra over H via

$$\begin{aligned} (h \rightharpoonup p)(k) &= p(kh) \\ (p \leftharpoonup h)(k) &= p(hk). \end{aligned}$$

If A is a left or right H-module algebra, we say that H measures A to itself. If the bialgebra H measures the algebra A to itself, then the elements of $P(H)$ act as derivations of A.

The Myhill–Nerode Theorem says that a language is accepted by a finite automaton if a certain equivalence relation defined on the set of all words has only finitely many equivalence classes. In this situation the bialgebra H is the semigroup algebra of the semigroup of all words in the alphabet, and the functional $p \in H^*$ is the characteristic function of the language. We show that the language is accepted by a finite automaton if and only if $H \rightharpoonup p$ is finite dimensional. Our treatment of the Myhill–Nerode Theorem is adapted from [4].

Fliess' Theorem says that if H is the universal enveloping algebra of a free Lie algebra and $p \in H^*$, then p is differentially produced by a finite rank augmented algebra R if and only if $P(H) \rightharpoonup p$ is finite dimensional. Our presentation of Fliess' Theorem is adapted from [3].

Our approach of studying realizations using bialgebras also has been used for the study of hybrid systems [3] and the study of data mining [5].

1 Myhill–Nerode Theorem

The Myhill–Nerode Theorem is traditionally stated as in Theorem 3.1 of [6]: If Σ is a finite alphabet, and $L \subseteq \Sigma^*$ is a language, define an equivalence relation on Σ^* by $w \sim w'$ if and only if if and only if $wz \in L$ exactly when $w'z \in L$ for all $z \in \Sigma^*$. The Myhill–Nerode Theorem states that L is accepted by a finite automaton if and only if the equivalence relation $\sim$ partitions Σ^* into finitely many equivalence classes.

Let M be a finite automaton accepting the language $L \subseteq \Sigma^*$. That is, M has a finite set of states S, an initial state $s_0 \in S$, and a set $F \subseteq S$ of accepting states. The word $w \in \Sigma^*$ is *accepted* by M if and only if $s_0 \cdot w \in F$.

Σ^* is a semigroup with operation concatenation and with identity the empty string ϵ. The bialgebra version of the Myhill–Nerode Theorem is as follows.

Theorem 1.1 (Myhill–Nerode) *Let G be a semigroup with unit, let k be a field, and let $H = kG$. Let $p \in H^*$. Then the following are equivalent:*

1) $\dim(H \rightharpoonup p)$ *is finite and there exists a non zero polynomial* $Q(X) \in k[X]$ *such that* $Q(p) = 0$;

2) $\dim(H \rightharpoonup p \leftharpoonup H)$ *is finite and there exists a non zero polynomial* $Q(X) \in k[X]$ *such that* $Q(p) = 0$;

3) $\dim(p \leftharpoonup H)$ *is finite and there exists a non zero polynomial* $Q(X) \in k[X]$ *such that* $Q(p) = 0$;

4) *there exists a finite dimensional commutative left H-module algebra R with augmentation α and $f \in R$ such that $p(h) = \alpha(h \cdot f)$ for all $h \in H$;*

5) *there exists an augmented commutative left H-module algebra $R \subseteq H^*$ which is isomorphic to the algebra of k-valued functions on some finite set S and $f \in R$ such that $p(h) = \alpha(h \rightharpoonup f)$ for all $h \in H$.*

PROOF. It is immediate that (2) implies (1).

We prove that (3) implies (2). Let $I = \{ h \in H \mid (p \leftharpoonup H) \leftharpoonup h = 0 \}$. Then I is a two sided ideal in H. Since $\dim(p \leftharpoonup H)$ is finite, and H/I is isomorphic to a subalgebra of $\mathrm{End}_k(p \leftharpoonup H)$, it follows that $\dim(H/I)$ is finite. Since I is a two sided ideal and $p(I) = p \leftharpoonup I(1) = 0$, it follows that $H \rightharpoonup p \leftharpoonup H \subseteq I^\perp \cong (H/I)^*$ is finite dimensional.

We prove that (4) implies (3). We first show that $p \leftharpoonup H$ is finite dimensional. Let $r_1, \ldots, r_n$ be a basis for R. Then there exist $x_1, \ldots, x_n \in H^*$ such that

$$h \cdot f = \sum_{i=1}^{n} x_i(h) r_i, \qquad \text{for all } h \in H.$$

Now

$$\begin{aligned} (p \leftharpoonup l)(h) &= p(lh) \\ &= \alpha(lh \cdot f) \\ &= \alpha(l \cdot h \cdot f) \\ &= \sum_{i=1}^{n} x_i(h) \alpha(l \cdot r_i), \end{aligned}$$

for all $h, l \in H$. Therefore $p \leftharpoonup H \subseteq \sum k x_i$ is finite dimensional.

We next show that there exists a polynomial $Q(X)$ such that $Q(p) = 0$. Since R is finite dimensional and $f \in R$, there exists a polynomial $Q(X)$ such that $Q(f) = 0$. Let $w \in \Sigma^*$. Then $r \mapsto w \cdot r$ is an algebra endomorphism. Therefore $Q(w \cdot f) = w \cdot Q(f) = 0$. Since $p(w) = \alpha(w \cdot f)$ and $\alpha : R \rightarrow k$ is an algebra homomorphism, it follows that

$$Q(p(w)) = Q(\alpha(w \cdot f)) = \alpha(Q(w \cdot f)) = 0.$$

Since H^* can be identified with the algebra of k-valued functions on Σ^*, it follows that $Q(p) = 0$.

It is immediate that (5) implies (4).

We finally prove that (1) implies (5). Since $q \mapsto (w \rightharpoonup q)$ is an algebra endomorphism of H^*, and since p satisfies $Q(p) = 0$, it follows that $Q(w \rightharpoonup p) = 0$. Therefore $H \rightharpoonup p \subseteq H^*$ is finite dimensional and spanned by algebraic elements, so generates a finite dimensional commutative algebra $R \subseteq H^*$ which is a left H-module algebra. Since H^* is the algebra of functions on Σ^* it contains no non zero nilpotent elements. Therefore R is semisimple. Therefore R is a direct sum of finitely many field extensions of k. Since H^* is the direct product of copies of k, all of these field extensions must be k. Therefore R is isomorphic to the set of functions from S, the set of maximal ideals of R, to k. Let $f = p \in R$. Then it immediate that $p(h) = \alpha(f \rightharpoonup h)$ for all $h \in H$, where $\alpha(q) = q(1)$ for all $q \in H^*$. This completes the proof of the theorem.

We now discuss the connection between Theorem 1.1 and the traditional form of the Myhill–Nerode Theorem. In the traditional case, the function p is the characteristic function of the language L being considered, and takes only the values 0 and 1 on elements of Σ^*, and so always satisfies the polynomial $Q(X) = X^2 - X$. We will use this fact freely in the following discussion.

Condition (5) of Theorem 1.1 is equivalent to the assertion that the language L is accepted by a finite automaton. The set S of maximal ideals of R is the set of states of the automaton. Σ^* acts on S as follows: if $w \in \Sigma^*$, since $r \mapsto w{\cdot}r$ is an algebra homomorphism, it induces a map $S \rightarrow S$ on the set of maximal ideals of R. The augmentation $\alpha : R \rightarrow k$ gives a maximal ideal which is the initial state. The function $f \in R$ is the characteristic function of some subset of S which is the set of accepting states.

We now consider Condition (1) of Theorem 1.1. Define the equivalence relation $\sim$ on Σ^* by $w \sim w'$ if and only if $q(w) = q(w')$ for all $q \in H \rightharpoonup p$. In other words, $w \sim w'$ if and only if $p(wz) = p(w'z)$ for all $z \in \Sigma^*$, if and only if $wz \in L$ exactly when $w'z \in L$ for all $z \in \Sigma^*$. The subalgebra of H^* generated by $H \rightharpoonup p$, which is finite dimensional if and only if $H \rightharpoonup p$ is finite dimensional, is the algebra of all functions on the equivalence classes of this equivalence relation. Therefore Condition (1) of Theorem 1.1 is equivalent to the assertion that this equivalence relation has finite index.

The Myhill–Nerode Theorem is a realization theorem, in that it describes when a formal language is realized as the language recognized by a finite automaton.

2 Fliess' Theorem

In this section we prove a realization theorem for input-output maps of control systems. We prove the algebraic parts of the classical results of Fliess [1], [2]

We use the following definition. If H is a bialgebra, we say that $p \in H^*$ is *differentially produced by the algebra R with the augmentation* ϵ if

1. there is right H-module algebra structure on R;

2. there exists $f \in R$ satisfying $p(h) = \epsilon(f \cdot h)$.

If H is the universal enveloping algebra of a Lie algebra, we will characterize those $p \in H^*$ which are differentially produced by finite rank algebras.

Let H denote the free associative algebra in the symbols $E_1,\ldots,E_M$ over the field k.

If the coalgebra structure on H is defined by

$$\Delta(E_i) = 1 \otimes E_i + E_i \otimes 1$$

then H is the universal enveloping algebra of the free Lie algebra on $E_1,\ldots,E_M$. H^* is isomorphic to a formal power series algebra in infinitely many variables.

Differentially produced functionals arise naturally when studying control systems with inputs and outputs. For example, let R denote the field of rational functions in the variables $x_1, \ldots, x_N$ with coefficients from the field k, and let $E_1, \ldots, E_M$ denote M derivations of R. The control system

$$\begin{aligned} \dot{z}(t) &= \sum_{i=1}^{M} u_i(t) E_i(z(t)), \\ z(0) &= z^0 \in k^N \end{aligned} \tag{1}$$

together with an observation function $f \in R$

$$f : k^N \longrightarrow k \tag{2}$$

naturally specifies an input-output map, which is defined by sending the input functions

$$t \to u_1(t), \ \ldots, \ t \to u_M(t)$$

to the output function

$$t \to f(z(t)).$$

The properties of the input-output map are captured by the formal series $\sum_\mu c_\mu \mu$, where $c_\mu = E_{\mu_k} \cdots E_{\mu_1} f(z(0))$.

This series is often called the generating series, while the data consisting of a control system with inputs, together with an observation, are called a state space realization of the input-output map.

If $p \in H^*$ is the formal series associated with such an input-output map, then it is differentially produced. Conversely, we can ask which formal series $p \in H^*$ have the property that there is a control system and an observation function which realizes it as above; that is, which formal series p are differentially produced?

We say that $p \in H^*$ has *finite Lie rank* if $\dim P(H) \rightharpoonup p$ is finite.

Theorem 2.1 (Fliess) *Let H be a primitively generated bialgebra over a field of characteristic* 0. *Let $p \in H^*$. Then the following are equivalent:*

1) *p has finite Lie rank;*

2) *p is differentially produced by some augmented k-algebra R for which* $\dim (\operatorname{Ker} \epsilon)/(\operatorname{Ker} \epsilon)^2$ *is finite;*

3) *p is differentially produced by a subalgebra of H^* which is isomorphic to $k[[x_1, \ldots, x_N]]$, the algebra of formal power series in N variables.*

PROOF. We first prove that (1) implies (3). Given a fixed $p \in H^*$, we define three basic objects:

$$\begin{aligned} L &= \{ h \in P(H) \mid h \rightharpoonup p = 0 \} \\ J &= HL \\ J^\perp &= \{ q \in H^* \mid q(j) = 0 \text{ for all } j \in J \}. \end{aligned}$$

Since $L \subseteq P(H)$, it follows that J is a coideal, that is, $\Delta(J) \subseteq J \otimes H + H \otimes J$. Therefore $J^\perp \cong (H/J)^*$ is a subalgebra of H^*. We will show that

$J^{\perp}$ is isomorphic to a formal power series algebra, and will construct derivations of this ring which will be used to realize the input-output map defined by p.

Lemma 2.2 *If* $\dim P(H) \rightharpoonup p = N$, *then* $J^{\perp}$ *is a subalgebra of* H^* *satisfying*

$$J^{\perp} \cong k[[x_1, \dots, x_N]].$$

PROOF. The sub Lie algebra L has finite codimension N. Choose a basis $\{e_1, e_2, \dots\}$ of $P(H)$ such that $\{e_{N+1}, e_{N+2}, \dots\}$ is a basis of L. Note that if $\bar{e}_i$ is the image of e_i under the quotient map $P(H) \to P(H)/L$, then $\{\bar{e}_1, \dots, \bar{e}_N\}$ is a basis for $P(H)/L$.

By the Poincaré–Birkhoff–Witt Theorem, H has a basis of the form

$$\{ e_{i_1}^{\alpha_{i_1}} \cdots e_{i_k}^{\alpha_{i_k}} \mid i_1 < \cdots < i_k \text{ and } 0 < \alpha_{i_r} \}.$$

Since L is a sub Lie algebra of $P(H)$, and the basis $\{e_i\}$ of $P(H)$ has been chosen so that $e_i \in L$ for $i > N$, it follows that the operation of putting monomials in standard form which is used in the proof of the Poincaré–Birkhoff–Witt Theorem will map elements of $J = HL$ to linear combinations of monomials of the form

$$e^{\alpha} = e_{i_1}^{\alpha_{i_1}} \cdots e_{i_k}^{\alpha_{i_k}}$$

with at least one $i_r > N$. Therefore J has a basis of such monomials. It follows that

$$\{ \bar{e}_1^{\alpha_1} \cdots \bar{e}_N^{\alpha_N} \mid \alpha_1, \dots, \alpha_N \geq 0 \}$$

is a basis for H/J. It now follows that the elements of the form

$$x_{\alpha} = \frac{x^{\alpha}}{\alpha!} = \frac{x_{i_1}^{\alpha_{i_1}} \cdots x_{i_k}^{\alpha_{i_k}}}{\alpha_{i_1}! \cdots \alpha_{i_k}!}$$

with all $1 \leq i_r \leq N$ are in $J^{\perp} \subseteq H^*$. Note that these elements satisfy

$$x_{\alpha}(e^{\beta}) = \begin{cases} 1 & \text{if } \alpha = \beta \\ 0 & \text{otherwise.} \end{cases}$$

Indeed, $J^{\perp}$ consists precisely of the completion in the finite topology of the span of such elements. In other words,

$$J^{\perp} \cong k[[x_1, \dots, x_N]],$$

completing the proof.

We will use the following notation and facts from the proof of Lemma 2.2: Suppose that $\{e_1, \ldots, e_N, \ldots\}$ is a basis for $P(H)$ such that $\{e_{N+1}, \ldots\}$ is a basis for L. Let $\{e^\alpha\}$ be the corresponding Poincaré–Birkhoff–Witt basis. Denote $J^\perp$ by R. Then $R \cong k[[x_1, \ldots, x_N]]$, and $x_1^{\alpha_1} \cdots x_N^{\alpha_N}/\alpha_1! \cdots \alpha_N!$ equals the element of the dual (topological) basis of H^* to the Poincaré–Birkhoff–Witt basis $\{e^\alpha\}$ of H, corresponding to the basis element $e_1^{\alpha_1} \cdots e_N^{\alpha_N}$.

We now collect some properties of the ring of formal power series R which will be necessary for the proof of the theorem.

Lemma 2.3 *Assume $p \in H^*$ has finite Lie rank, and let $R \subseteq H^*$, $e_\alpha \in H$, and $x^\alpha \in R$ be as above. Define*

$$f = \sum_{\alpha=(\alpha_1,\ldots,\alpha_N)} c_\alpha x^\alpha \in R,$$

where $c_\alpha = \dfrac{p(e^\alpha)}{\alpha!}$. Then

1. *H measures R to itself via $\leftharpoonup$;*

2. *$p(h) = \epsilon(f \leftharpoonup h)$ for all $h \in H$.*

PROOF. We begin with the proof of (1). Since H measures H^* to itself and $R \subseteq H^*$, we need show only that $R \leftharpoonup H \subseteq R$. Take $r \in R$, $h \in H$ and $j \in J$. We have $(r \leftharpoonup h)(j) = r(hj)$. Since J is a left ideal, $hj \in J$, so $r(hj) = 0$, so $r \leftharpoonup h \in J^\perp = R$.

We now prove (2). Let $e^\alpha = e_{i_1}^{\alpha_{i_1}} \cdots e_{i_k}^{\alpha_{i_k}}$ be a Poincaré–Birkhoff–Witt basis element of H. Since $e^\alpha \in J$ unless $\{i_1, \ldots, i_k\} \subseteq \{1, \ldots, N\}$, $p(e^\alpha) = 0$ unless $\{i_1, \ldots, i_k\} \subseteq \{1, \ldots, N\}$. Also $\epsilon(f \leftharpoonup e^\alpha) = f \leftharpoonup e^\alpha(1) = f(e^\alpha) = 0$ unless $\{i_1, \ldots, i_k\} \subseteq \{1, \ldots, N\}$. Now suppose $\{i_1, \ldots, i_k\} \subseteq \{1, \ldots, N\}$. We have in this case that $p(e^\alpha) = \alpha! c_\alpha = f(e^\alpha) = f \leftharpoonup e^\alpha(1) = \epsilon(f \leftharpoonup e^\alpha)$. Since $\{e^\alpha\}$ is a basis for H, this completes the proof of the lemma.

Corollary 2.4 *Under the assumptions of Lemma 2.3, $f = p$.*

Lemmas 2.2 and 2.3 yield that (1) implies (3) in Theorem 2.1. It is immediate that (3) implies (2).

We now complete the proof of Theorem 2.1 by proving that (2) implies (1).

Let $x_1, \ldots, x_N$ be chosen so that $\{\bar{x}_1, \ldots, \bar{x}_N\}$ is a basis for $(\operatorname{Ker}\epsilon)/(\operatorname{Ker}\epsilon)^2$. If $f \in R$ and $h \in H$, then

$$f \cdot h = q_0(h)1 + \sum_{i=1}^{N} q_i(h)x_i + g(h),$$

where $q_i \in H^*$ and $g(h) \in (\operatorname{Ker}\epsilon)^2$. Let $l \in P(H)$. Since H measures R to itself and $\Delta(l) = 1 \otimes l + l \otimes 1$, the map $f \mapsto f \cdot l$ is a derivation of R. Now let $f \in R$ be the element such that

$$p(h) = \epsilon(f \cdot h).$$

Then

$$\begin{aligned} f \cdot hl &= (f \cdot h) \cdot l \\ &= q_0(h)1 \cdot l + \sum_{i=1}^{N} q_i(h)x_i \cdot l + g(h) \cdot l. \end{aligned}$$

Since the map $f \mapsto f \cdot l$ is a derivation, $1 \cdot l = 0$; since $g(h) \in (\operatorname{Ker}\epsilon)^2$, $g(h) \cdot l \in \operatorname{Ker}\epsilon$. It follows that

$$\begin{aligned} l \rightharpoonup p(h) &= p(hl) \\ &= \epsilon(f \cdot hl) \\ &= \sum_{i=1}^{N} q_i(h)\epsilon(x_i \cdot l). \end{aligned}$$

Therefore $P(H) \rightharpoonup p \subseteq \sum_{i=1}^{N} kq_i$, so p has finite Lie rank. This completes the proof of Theorem 2.1

References

[1] M. Fliess, Réalisation locale des systèmes non linéaires, algèbres de Lie filtrées transitives et séries génératrices non commutatives, *Invent. Math.*, **71** (1983) 521–537.

[2] M. Fliess, Nonlinear realization theory and abstract transitive Lie algebras, *Bull. Amer. Math. Soc.*, (NS) **2** (1980), 444–446.

[3] R. L. Grossman and R. G. Larson, The realization of input-output maps using bialgebras, *Forum Mathematicum* **4** (1992) 109–121.

[4] R. L. Grossman and R. G. Larson, An algebraic approach to hybrid systems, *Theoretical Computer Science*, **138** (1995), 101–112.

[5] R. L. Grossman and R. G. Larson, *An Algebraic Approach to Data Mining: Some Examples*, Proceedings of the 2002 IEEE International Conference on Data Mining (ICDM 2002), December 2002, Maebashi City, Japan.

[6] J. E. Hopcroft and J. D. Ullman, *Formal Languages and their Relation to Automata,* Addison-Wesley, Reading, 1969.

[7] M. E. Sweedler, *Hopf algebras,* W. A. Benjamin, New York, 1969.

Relatively Free Coalgebras

Mikhail Kochetov

Algebra and Logic Research Unit
University of Saskatchewan
Saskatoon, SK, Canada S7N 5E6
kochetov@math.usask.ca

Abstract

We study varieties of coassociative coalgebras and the relatively free coalgebras associated to them. For any variety defined by multilinear identities, we give an explicit construction of its free coalgebras in terms of suitable completed tensor algebras, extending such representation given by R. Block and P. Leroux for free coassociative coalgebras.

0 Introduction and Notation

The notion of (co)free coalgebra is a natural dualization of the notion of free algebra. Free coassociative coalgebras were introduced by M. Sweedler in [9], where he also proved their existence. Then R. Block and P. Leroux in [5] found a more explicit construction of free coassociative coalgebras using completed tensor algebras. Their approach was modified by G. Griffing in [6] to obtain a construction of a free non-coassociative coalgebra. In general, a free coalgebra of any variety can be naturally imbedded into a free non-coassociative coalgebra, but this does not help us understand its structure unless the image of this imbedding is explicitly described. We will restrict our attention to coassociative coalgebras, although we will occasionally mention analogous results for the non-coassociative case. In Section 1 we will discuss possible approaches to the definition of a polynomial identity for coalgebras and the notion of a variety of coalgebras. In Section 2 we will present the construction of R. Block and P. Leroux and see a connection between identities and free coalgebras. Section 3 is devoted to extending this construction, which allows us to describe explicitly the free coalgebra of any multilinear variety inside the free coassociative coalgebra.

Throughout $\mathbb{k}$ will denote the ground field. All vector spaces, algebras, coalgebras, tensor products, etc. will be considered over $\mathbb{k}$. Unless stated otherwise, algebras will be assumed associative and unital and coalgebras coassociative and counital. For a vector space V, V^* and $T(V)$ will denote its dual vector space and tensor algebra, respectively. For $v \in V$, $f \in V^*$, we use the symmetric notation $\langle f, v\rangle$ for $f(v)$. The dual space V^* has a natural topology defined by the fundamental system of neighbourhoods of 0 of the form

$$U_{v_1,\ldots,v_m} = \{f \in V^* | \langle f, v_k\rangle = 0, \ \forall k = 1, \ldots, m\},$$

where $v_1, \ldots, v_m \in V$, $m \in \mathbb{N}$.

Given an algebra A with multiplication $m : A \otimes A \to A$ and unit $u : \mathbf{k} \to A$, the largest subspace of A^* whose image under m^* lies in $A^* \otimes A^*$ (viewed as a subspace of $(A \otimes A)^*$) is the so called *finite dual*:

$$A^\circ = \{f \in A^* \mid f(I) = 0 \text{ for some ideal } I \triangleleft A \text{ with } \dim A/I < \infty\}.$$

A° is a coalgebra with comultiplication $\Delta = m^*$ and counit $\varepsilon = u^*$ (restricted to A°). The algebra A is called *residually finite-dimensional* if its ideals of finite codimension intersect to 0.

We will use the so-called *sigma notation* for coalgebras. Let C be any coalgebra with comultiplication $\Delta : C \to C \otimes C$. For any $c \in C$, we write:

$$\Delta c = \sum c_{(1)} \otimes c_{(2)}.$$

In sigma notation, coassociativity means that

$$\sum c_{(1)} \otimes c_{(2)_{(1)}} \otimes c_{(2)_{(2)}} = \sum c_{(1)_{(1)}} \otimes c_{(1)_{(2)}} \otimes c_{(2)},$$

so we simply write $\sum c_{(1)} \otimes c_{(2)} \otimes c_{(3)} = \Delta_3 c$. Iterating this procedure gives, for any $n \geq 2$,

$$\Delta_n c = \sum c_{(1)} \otimes \ldots \otimes c_{(n)},$$

where $\Delta_2 = \Delta$. We will also use the convention that $\Delta_1 = id_C$ and $\Delta_0 = \varepsilon$.

1 Identities and Varieties of Coalgebras

It is natural to define the notion of a polynomial identity for coalgebras using their duality with algebras.

Definition 1.1. Let C be a coalgebra over a field $\mathbf{k}$, $F(X_1, \ldots, X_n)$ a polynomial in n noncommuting variables with coefficients in $\mathbf{k}$. We say that $F = 0$ (or just F) is an identity for the coalgebra C, if it is an identity for the dual algebra C^*.

Using duality, we immediately observe that if a coalgebra C satisfies the identity $F = 0$, then any subcoalgebra and any factorcoalgebra of C satisfies this identity. If each coalgebra of some family satisfies $F = 0$, then so does the direct sum of this family.

If F is a multilinear polynomial, then it is possible to give an intrinsic definition when a coalgebra satisfies $F = 0$, i.e. a definition that would not involve the dual algebra.

A multilinear polynomial of degree n has the form:

$$F(X_1, \ldots, X_n) = \sum_{\pi \in S_n} \lambda_\pi X_{\pi(1)} \ldots X_{\pi(n)},$$

where S_n is the group of permutations and $\lambda_\pi \in \mathbf{k}$. It will be convenient to identify F with the element $\sum_{\pi \in S_n} \lambda_\pi \pi$ of the group algebra $\mathbf{k}S_n$.

For any vector space V, there are natural right and left actions of S_n on $V^{\otimes n}$:

$$\begin{aligned} (v_1 \otimes \ldots \otimes v_n) \cdot \pi &= v_{\pi(1)} \otimes \ldots \otimes v_{\pi(n)}, \\ \pi \cdot (v_1 \otimes \ldots \otimes v_n) &= v_{\pi^{-1}(1)} \otimes \ldots \otimes v_{\pi^{-1}(n)}. \end{aligned}$$

Then the fact that an algebra A satisfies a multilinear identity $F = 0$ can be written as follows:

$$m_n(A^{\otimes n} \cdot F) = 0,$$

where $m_n : A^{\otimes n} \to A$ is the multiplication of A. The following dual definition for coalgebras is due to Yu. Bahturin.

Definition 1.2. Let C be a coalgebra, $F = \sum_{\pi \in S_n} \lambda_\pi X_{\pi(1)} \ldots X_{\pi(n)}$ a multilinear polynomial. We say that C satisfies the identity $F = 0$ if

$$F \cdot (\Delta_n C) = 0,$$

where $\Delta_n : C \to C^{\otimes n}$ is the comultiplication of C.

So a multilinear identity of degree n can be viewed as a sort of symmetry condition on the tensors $\Delta_n c$, for all $c \in C$. The proof of the following proposition is straightforward (see [7]).

Proposition 1.3. *For multilinear identities, Definitions 1.1 and 1.2 are equivalent.* ∎

Using the sigma notation, the fact that a coalgebra C satisfies the multilinear identity $F = 0$ as in Definition 1.2 can be written as follows:

$$\sum_{\pi \in S_n} \lambda_\pi \sum x_{(\pi^{-1}(1))} \otimes \ldots \otimes x_{(\pi^{-1}(n))} = 0, \qquad \forall x \in C.$$

For example, cocommutativity can be expressed like this:

$$\sum x_{(1)} \otimes x_{(2)} - \sum x_{(2)} \otimes x_{(1)} = 0.$$

By Definition 1.1, any finite-dimensional coalgebra C satisfies the standard identity

$$\sum_{\pi \in S_n} \operatorname{sgn}(\pi) \sum x_{(\pi^{-1}(1))} \otimes \ldots \otimes x_{(\pi^{-1}(n))} = 0,$$

for any $n > \dim C$.

The following observation [7, Proposition 1.2] provides a way of constructing infinite-dimensional PI-coalgebras, i.e. coalgebras with a nontrivial identity.

Proposition 1.4. *If an algebra A satisfies the identity $F = 0$, then so does the coalgebra A°. The converse holds if A is residually finite-dimensional.* ∎

As shown in [2, Theorem 4.2], the first statement of this proposition still holds for non-associative algebras if the definition of A° is properly extended to this case. Namely, A° is the largest "good subspace" of A^*, where a subspace $V \subset A^*$ is called "good" if $m^*(V) \subset V \otimes V$.

Now we turn our attention to varieties of coalgebras. By analogy with algebras, it is natural to give the following definition. Let $\mathcal{F}$ denote the free (associative and unital) algebra with countably many generators, so $\mathcal{F} = T(V)$, where $V = \langle X_1, X_2, \ldots \rangle$. Then any polynomial identity, no matter in how many variables, can be viewed as an element of $\mathcal{F}$.

Definition 1.5. Let S be a subset of $\mathcal{F}$. The *variety of coalgebras* defined by S is the class $\mathrm{cVar}(S)$ of all coalgebras that satisfy each identity from S.

As is well-known, varieties of algebras are in one-to-one correspondence with T-ideals in $\mathcal{F}$ as follows. If $\mathfrak{A} = \mathrm{Var}(S)$ is the variety defined by some $S \subset \mathcal{F}$, then the set $\mathcal{I}(\mathfrak{A})$ of all identities satisfied by each algebra from $\mathfrak{A}$

is the T-ideal generated by S. Therefore, if $J \subset \mathcal{F}$ is a T-ideal, then for the variety of algebras $\mathfrak{A} = \mathrm{Var}(J)$ we have $\mathcal{I}(\mathfrak{A}) = J$. Conversely, if $\mathfrak{A}$ is a variety, then $\mathfrak{A} = \mathrm{Var}(\mathcal{I}(\mathfrak{A}))$.

This one-to-one correspondence with T-ideals also holds for varieties of coalgebras, at least over an infinite field, as shown by the following proposition [7, Corollary 1.4].

Proposition 1.6. *Suppose the ground field* $\mathbf{k}$ *is infinite. Let* $\mathfrak{C}$ *be the variety of coalgebras defined by a set of identities* S. *Then the* T*-ideal* $\mathcal{I}(\mathfrak{C})$ *of identities of* $\mathfrak{C}$ *is generated by* S *as a* T*-ideal. In other words, the consequences of the system of identities* S *are the same for coalgebras as they are for algebras.*

Proof. First of all, since the base field $\mathbf{k}$ is infinite, any T-ideal J is graded. It follows that the algebra $\mathcal{F}/J$ is residually finite-dimensional. Indeed, for any $F(X_1, \ldots, X_n) \notin J$, we need to find an ideal of finite codimension $J' \supset J$ such that $F \notin J'$. Since J is graded, we can set J' equal to the ideal generated by J, $X_{n+1}, X_{n+2}, \ldots$, and by all monomials in $X_1, \ldots, X_n$ of degree $d+1$, where d is the maximum degree of monomials occuring in F.

Now let J be an arbitrary T-ideal containing our set S. By Proposition 1.4, the coalgebra $D = (\mathcal{F}/J)^{\circ}$ satisfies the same identities as the algebra $\mathcal{F}/J$, so $\mathcal{I}(D) = J$. Since $S \subset J$, D is in the variety $\mathfrak{C}$, hence $J = \mathcal{I}(D) \supset \mathcal{I}(\mathfrak{C})$. Therefore, $\mathcal{I}(\mathfrak{C})$ is the smallest T-ideal containing S. ■

Varieties of algebras are characterized by the following classical theorem of Birkhoff [3].

Theorem 1.7. *Let* $\mathfrak{A}$ *be a nonempty class of algebras. Then* $\mathfrak{A}$ *is a variety iff* $\mathfrak{A}$ *is closed under isomorphisms, subalgebras, factoralgebras, and direct products.* ■

It turns out that the analog of Theorem 1.7 does not hold for coalgebras. Clearly, any variety of coalgebras is closed under isomorphisms, subcoalgebras, factorcoalgebras, and direct sums. However, not every such class is a variety.

Example 1.8. The class *Grp* of all coalgebras spanned by group-like elements is closed under the four operations just listed, but it is not a variety.

Proof. If C is spanned by group-like elements, then any subcoalgebra of C is just a span of a subset of these elements. Also any homomorphic image of C is spanned by the images of these elements, which are either group-like or zero. Clearly, the class *Grp* is also closed under isomorphisms and direct sums.

Grp is not a variety, because it is properly contained in the variety *Cocomm* of all cocommutative coalgebras (with counit), which does not have any proper subvarieties other than $\{0\}$. ■

Example 1.9. The class *Pnt* of all pointed coalgebras is closed under the four operations listed above, but it is not a variety.

Proof. Recall that a coalgebra is called pointed if all its simple subcoalgebras are one-dimensional or, equivalently, its coradical is spanned by group-like elements. Thus *Pnt* is closed under isomorphisms and subcoalgebras. It is also known that a homomorphic image of a pointed coalgebra is pointed (see e.g. [8, Corollary 5.3.5]). Finally, [8, Lemma 5.6.2(1)] says that if $C = \sum_i C_i$, where $C_i \subset C$ are subcoalgebras, then any simple subcoalgebra of C lies in one of the C_i, hence a sum of pointed coalgebras is pointed.

The class *Pnt* does not satisfy any nontrivial identity. This can be seen, for instance, from the fact that a Taft's algebra of dimension n^2 [10] is pointed and does not satisfy any identity of degree $< 2n$ [7, Proposition 2.1]. Hence *Pnt* is a proper subclass of the variety *Coalg* of all (coassociative) coalgebras, which is not contained in any proper subvariety. ■

Definition 1.10. We will use the term *pseudo-variety* for any nonempty class of coalgebras closed under isomorphisms, subcoalgebras, factorcoalgebras, and direct sums.

Thus *Grp* and *Pnt* are pseudo-varieties which are not varieties. As we have seen, varieties of coalgebras are in one-to-one correspondence with T-ideals in the free algebra $\mathcal{F}$ with countably many generators $X_1, X_2, \ldots$. In Section 3 we will establish a one-to-one correspondence between pseudo-varieties and T-subcoalgebras in the free coalgebra "with countably many cogenerators".

To conclude this section, we give a characterization of varieties of coalgebras among pseudo-varieties [7, Proposition 1.7].

Theorem 1.11. *A nonempty class of coalgebras $\mathfrak{C}$ is a variety iff it is closed under isomorphisms, subcoalgebras, factorcoalgebras, and direct sums, and in addition, for any coalgebra C from $\mathfrak{C}$ and any subalgebra $A \subset C^*$, A° belongs to $\mathfrak{C}$.* ■

2 Free Coassociative Coalgebra

A polynomial identity $F(X_1, \ldots, X_n)$ can be considered as an element of the free (associative) algebra with n generators, i.e. the tensor algebra $T(V)$, where $V = \langle X_1, \ldots, X_n \rangle$.

In order to find a link between identities of coalgebras and free coalgebras, we first define the latter. Free (coassociative) coalgebras (which more precisely should be called "cofree coalgebras") were introduced by M. Sweedler in [9]. They are defined by the following universal property, which is dual to the universal property of tensor algebras.

Definition 2.1. Let V be a vector space, C a coalgebra, $\theta : C \to V$ a linear map. The pair (C, θ) is called a *free coalgebra on* V if, for any coalgebra D and a linear map $\varphi : D \to V$, there exists a unique coalgebra map $\Phi : D \to C$ completing the commutative diagram:

$$\begin{array}{ccc} D & \overset{\Phi}{\dashrightarrow} & C \\ & {\scriptstyle\varphi}\searrow \quad \swarrow{\scriptstyle\theta} & \\ & V & \end{array}$$

By a standard argument, if a free coalgebra on V exists, it is unique up to a uniquely defined isomorphism. We will denote it by $c\mathcal{F}(V)$. It is shown in [9] that $c\mathcal{F}(V)$ exists for any V. A more explicit construction was given by R. Block and P. Leroux [5]. First we introduce the generalized finite dual.

Definition 2.2. Let $A = \bigoplus_{n \geq 0} A_n$ be a graded algebra, V a vector space. Let $\mathrm{Hom}(A, T(V))$ denote the space of all graded linear functions of degree 0 from A to the tensor algebra $T(V) = \bigoplus_{n \geq 0} T^n(V)$, i.e. all linear functions $f : A \to T(V)$ such that if $a \in A_n$, then $f(a)$ is a tensor of degree n. We will call $f \in \mathrm{Hom}(A, T(V))$ *representative* if there exists a finite family $\{g_i, h_i\}$ of elements of $\mathrm{Hom}(A, T(V))$ such that

$$f(ab) = \sum_i g_i(a) h_i(b), \qquad \forall a, b \in A, \tag{1}$$

where the multiplication on the right-hand side is the tensor product in $T(V)$. Since we will later consider elements of $T(V) \otimes T(V)$, we reserve the symbol $\otimes$ for the "outer" tensor product and simply write $v_1 \ldots v_n$ for the element $v_1 \otimes \ldots \otimes v_n \in T(V)$. The set of all representative functions $A \to T(V)$ will be denoted by A_V°.

It follows that if $f \in A_V^\circ$, then the tensor $\sum_i g_i \otimes h_i$ is uniquely determined by (1). We define $\Delta f = \sum_i g_i \otimes h_i$, and it turns out that $\Delta f \in A_V^\circ \otimes A_V^\circ$

and (A°_{V}, Δ) is a coalgebra with counit $\varepsilon(f) = f(1)$ [5, Lemma 1 and Theorem 1]. If $V = \mathbb{k}$, we recover the usual finite dual coalgebra A° of the (underlying ungraded) algebra A.

If we now specify $A = T(W)$ (graded by degree), for some vector space W, then there is a natural linear map $\theta : T(W)^{\circ}_{V} \to \mathrm{Hom}(W, V)$ which sends $f \in T(W)^{\circ}_{V}$ to its restriction to $W = T^1(W)$.

Theorem 2.3 (Theorem 2′ in [5]). *Let V and W be vector spaces. Then $(T(W)^{\circ}_{V}, \theta)$ defined above is (a realization of) the free coalgebra on the space $\mathrm{Hom}(W, V)$. Moreover, if D is a coalgebra and $\varphi : D \to \mathrm{Hom}(W, V)$ is a linear map, then the lifting of φ to a coalgebra map $\Phi : D \to T(W)^{\circ}_{V}$ is given by*

$$\begin{aligned} \Phi(d)1 &= \varepsilon(d), & \\ \Phi(d)w &= \varphi(d)w, & \forall w \in W = T^1(W), \text{ and} \\ \Phi(d)z &= \left(\textstyle\sum \varphi(d_{(1)}) \otimes \ldots \otimes \varphi(d_{(n)})\right) z, & \forall z \in T^n(W),\ n > 1. \end{aligned}$$

■

In particular, if we set $V = \mathbb{k}$, we see that $T(W)^{\circ}$ is the free coalgebra on W^*. This is a result of M. Sweedler originally used to prove the existence of free coalgebras. It also sheds light on the nature of identities of a coalgebra. Let $F(X_1, \ldots, X_n)$ be an associative polynomial in n variables, set $W = \langle X_1 \ldots, X_n \rangle$, so $F \in T(W)$. Then the free coalgebra $c\mathcal{F}(W^*) = T(W)^{\circ}$ is a subspace of $T(W)^*$ containing $T(W^*)$ (see Remark 2.4 below). Moreover, $T(W)^*$ has a natural topology of a dual space and $T(W)$ can be recovered as the space of all continuous linear functions on $T(W)^*$. Since $T(W^*)$ is dense in $T(W)^*$, so is $c\mathcal{F}(W^*)$, and thus the spaces of continuous linear functions on $T(W)^*$ and on $c\mathcal{F}(W^*)$ (with topology inherited from $T(W)^*$) are in one-to-one correspondence. Thus we conclude that $T(W)$ is the space of continuous linear functions on $c\mathcal{F}(W^*)$ and so polynomial identities in $X_1, \ldots, X_n$ can be viewed as continuous linear functions on the free coalgebra on the space $\langle X_1, \ldots, X_n \rangle^*$.

On the other hand, if we set $W = \mathbb{k}$ in Theorem 2.3, we obtain that $\mathbb{k}[t]^{\circ}_{V}$ is the free coalgebra on V. This gives a rather explicit construction of $c\mathcal{F}(V)$ as follows. Denote $\hat{T}(V)$ the completion of the tensor algebra $T(V)$, i.e. the algebra of all infinite formal sums $z = z_0 + z_1 + \ldots$, where $z_i \in T^i(V)$. The topology on $\hat{T}(V)$ is defined by a fundamental system of neighbourhoods of 0 consisting of the sets

$$F^n\hat{T}(V) = \{z \in \hat{T}(V) \mid z_i = 0 \quad \forall i < n\}.$$

Then an element $f \in \mathrm{Hom}(\mathbf{k}[t], T(V))$ can be identified with the formal sum $f_0 + f_1 + \ldots$, where $f_i = f(t^i) \in T^i(V)$, and so $c\mathcal{F}(V)$ becomes a subspace of $\hat{T}(V)$. Upon this identification, the canonical map $\theta : c\mathcal{F}(V) \to V$ just sends the sum $f_0 + f_1 + \ldots$ to its degree 1 component $f_1 \in T^1(V) = V$, and the formulas of Theorem 2.3 for the lifting of a linear map $\varphi : D \to V$ to a coalgebra map $\Phi : D \to c\mathcal{F}(V)$ become:

$$\begin{aligned}\Phi(d)_0 &= \varepsilon(d), \\ \Phi(d)_1 &= \varphi(d), \qquad \text{and} \\ \Phi(d)_n &= \sum \varphi(d_{(1)}) \otimes \ldots \otimes \varphi(d_{(n)}) \qquad \text{for } n > 1.\end{aligned}$$

Moreover, an explicit formula for the comultiplication of $c\mathcal{F}(V)$ can be obtained as follows (see [4, Sections 1 and 2]). Let $\mathcal{D}$ denote the continuous linear map from $\hat{T}(V)$ to the completed tensor product $\hat{T}(V)\hat{\otimes}\hat{T}(V)$ defined by its action on the monomials:

$$\mathcal{D}(v_1 \ldots v_n) = \sum_{i=0}^{n} v_1 \ldots v_i \otimes v_{i+1} \ldots v_n, \qquad \forall v_1, \ldots, v_n \in V,\ n = 0, 1, \ldots \tag{2}$$

Then an element $f \in \hat{T}(V)$ belongs to $c\mathcal{F}(V)$ iff $\mathcal{D}f$ lies in the subspace $\hat{T}(V) \otimes \hat{T}(V) \subset \hat{T}(V)\hat{\otimes}\hat{T}(V)$, in which case $\Delta f = \mathcal{D}f$, i.e. the comultiplication of $c\mathcal{F}(V)$ is just the restriction of $\mathcal{D}$ on $c\mathcal{F}(V)$. We also see from here that $T(V) \subset c\mathcal{F}(V)$. In particular, this implies that the canonical map θ is surjective. The counit of $c\mathcal{F}(V)$ just sends the sum $f_0 + f_1 + \ldots$ to its degree 0 component $f_0 \in T^0(V) = \mathbf{k}$.

Remark 2.4. Assuming the space V finite-dimensional, set $W = V^*$. Then the above construction of $c\mathcal{F}(V) \subset \hat{T}(V)$ agrees with the construction of M. Sweedler which realizes $c\mathcal{F}(W^*)$ as the subspace $T(W)^\circ$ of $T(W)^* = \hat{T}(V)$. In particular, $T(W)^\circ$ contains $T(W^*)$.

Remark 2.5. R. Block also proves in [4] a number of interesting properties of $c\mathcal{F}(V)$ which we will not use here. But one thing should be mentioned, since it illustrates the duality with algebras. Namely, there is a natural multiplication $Sh : c\mathcal{F}(V) \otimes c\mathcal{F}(V) \to c\mathcal{F}(V)$ which is the lifting of the linear map $\theta \otimes \varepsilon + \varepsilon \otimes \theta : c\mathcal{F}(V) \otimes c\mathcal{F}(V) \to V$, so $c\mathcal{F}(V)$ has a structure of a commutative Hopf algebra, with the antipode induced by $-\theta : T(V) \to V$ — dually to the fact the free algebra $T(V)$ has a natural structure of a cocommutative Hopf algebra defined by $V \to T(V) \otimes T(V) : v \to v \otimes 1 + 1 \otimes v$, with antipode induced by $V \to T(V) : v \to -v$. In terms of the realization of $cF(V)$ in $\hat{T}(V)$, the multiplication Sh coincides with the so called "shuffle product".

As was already mentioned, one can also define a free non-coassociative coalgebra on a vector space. The existence of such objects and their representation by infinite formal sums is obtained in [6].

To conclude this section, let us introduce the notion of a cogenerating map for coalgebras, which is the dual of a generating set (or, more precisely, space) for algebras. Let A be an algebra, V a vector space. Suppose we have a linear map $\varphi : V \to A$, then the image $\varphi(V)$ generates A as an algebra iff

$$\sum_{n\geq 0} \left(m_n \circ \varphi^{\otimes n}\right) V^{\otimes n} = A,$$

where $m_n : A^{\otimes n} \to A$ is the multiplication of A (with $m_1 = id_A$ and $m_0 = u$, the unit map). The formal dual of this statement is the following:

Definition 2.6. Let C be a coalgebra, V a vector space. We will call a linear map $\varphi : C \to V$ *cogenerating* if

$$\bigcap_{n\geq 0} \operatorname{Ker}\left(\varphi^{\otimes n} \circ \Delta_n\right) = 0.$$

A generating set in an algebra A allows us to represent A as a factor of a free algebra. Dually, cogenerating maps for a coalgebra C correspond to the imbeddings of C into free coalgebras.

Proposition 2.7. *Let C be a coalgebra, V a vector space, $\varphi : C \to V$ a linear map. Then the induced coalgebra map $\Phi : C \to c\mathcal{F}(V)$ is injective iff φ is cogenerating.*

Proof. Recall from the construction of $c\mathcal{F}(V)$ that

$$\Phi(d)_n = \sum \varphi(d_{(1)}) \otimes \ldots \otimes \varphi(d_{(n)}), \qquad \forall d \in C$$

(with the convention that the right-hand side means $\varphi(d)$ for $n = 1$ and $\varepsilon(d)$ for $n = 0$). In other words, $\Phi(d)_n = (\varphi^{\otimes n} \circ \Delta_n)d$, hence $d \in \operatorname{Ker}\Phi$ iff $(\varphi^{\otimes n} \circ \Delta_n)d = 0$, for all n. ∎

In particular, any coalgebra can be imbedded into a free coalgebra (take $V = C$, then $id : C \to V$ is obviously a cogenerating map).

Remark 2.8. Our definition of a cogenerating map agrees with the one given in [1], namely, $\varphi : C \to V$ is cogenerating if $\operatorname{Ker}\varphi$ contains no nonzero coideals of C. This latter definition also works in the non-coassociative case.

3 An Explicit Construction of Relatively Free Coalgebras

Dualizing the universal property of a free algebra of a variety (expressed in terms of a vector space rather than a set of generators) gives the following relative version of Definition 2.1.

Definition 3.1. Let $\mathfrak{C}$ be a pseudo-variety of coalgebras. Let V be a vector space, C a coalgebra, $\theta : C \to V$ a linear map. The pair (C, θ) is called a *$\mathfrak{C}$-free coalgebra on V* if C belongs to $\mathfrak{C}$ and, for any coalgebra $D \in \mathfrak{C}$ and a linear map $\varphi : D \to V$, there exists a unique coalgebra map $\Phi : D \to C$ completing the commutative diagram:

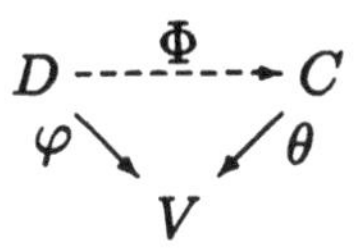

Such a $\mathfrak{C}$-free coalgebra is automatically unique, and we will denote it by $c\mathcal{F}_{\mathfrak{C}}(V)$. The existence is also immediate: the largest subcoalgebra of the free coalgebra $c\mathcal{F}(V)$ that belongs to $\mathfrak{C}$ obviously satisfies the universal property of $c\mathcal{F}_{\mathfrak{C}}(V)$. Any coalgebra of the pseudo-variety $\mathfrak{C}$ can be imbedded in a suitable $\mathfrak{C}$-free coalgebra (just take any imbedding into a free coalgebra, the image will automatically lie in the corresponding $\mathfrak{C}$-free coalgebra).

Example 3.2. Recall the pseudo-variety Grp of Example 1.8. The relatively free coalgebra $c\mathcal{F}_{Grp}(V)$ is spanned by the set $G(c\mathcal{F}(V))$ of all group-like elements of $c\mathcal{F}(V)$, which is in one-to-one correspondence with V by virtue of the map

$$e : V \to c\mathcal{F}(V) : v \to e(v) = 1 + v + v^2 + v^3 + \ldots,$$

where by our convention v^2 is the monomial $v \otimes v$, etc.

Denote $c\mathcal{F} = c\mathcal{F}(V)$, where $V = \langle X_1, X_2, \ldots \rangle$. Loosely speaking, $c\mathcal{F}$ is the free coalgebra "with countably many cogenerators".

Free coalgebras of pseudo-varieties inside $c\mathcal{F}$ can be characterized as T-subcoalgebras, whose definition is dual to the one of a T-ideal.

Definition 3.3. A subcoalgebra $L \subset c\mathcal{F}$ is called a *T-subcoalgebra* if $\alpha(L) \subset L$, for any endomorphism α of $c\mathcal{F}$, or, equivalently, if $\beta(L) \subset L$, for any coalgebra map $\beta : L \to c\mathcal{F}$ (the equivalence follows from the universal property of $c\mathcal{F}$).

Then pseudo-varieties of coalgebras are in a one-to-one correspondence with T-subcoalgebras as follows. We associate with a pseudo-variety $\mathfrak{C}$ the largest subcoalgebra $L \subset c\mathcal{F}$ belonging to $\mathfrak{C}$ (the sum of all such subcoalgebras, which belongs to $\mathfrak{C}$ because $\mathfrak{C}$ is closed under direct sums and factors). Since $\mathfrak{C}$ is closed under homomorphic images, L will be a T-subcoalgebra. Clearly, L is the $\mathfrak{C}$-free coalgebra on the space $\langle X_1, X_2, \ldots \rangle$.

Conversely, we associate with a T-subcoalgebra $L \subset c\mathcal{F}$ the class $\mathfrak{C}$ consisting of all coalgebras D such that, for any coalgebra map $\gamma : D \to c\mathcal{F}$, $\gamma(D) \subset L$. Obviously, $\mathfrak{C}$ is closed under isomorphisms, factors and direct sums. It is also closed under subcoalgebras, because coalgebra maps to $c\mathcal{F}$ can always be extended from subcoalgebras by the universal property of $c\mathcal{F}$. Two more checks are necessary.

Firstly, let L be a T-subcoalgebra, $\mathfrak{C}$ the pseudo-variety associated with L, and L' the T-subcoalgebra associated with $\mathfrak{C}$. Since L' belongs to $\mathfrak{C}$, then considering the inclusion map $L' \hookrightarrow \mathcal{F}$, we see that $L' \subset L$ by the construction of $\mathfrak{C}$. Conversely, using the definition of a T-coalgebra, we conclude that L also belongs to $\mathfrak{C}$, but then $L \subset L'$ since L' is the largest subcoalgebra with this property. So $L = L'$.

Secondly, let $\mathfrak{C}$ be a pseudo-variety, L the T-subcoalgebra associated with $\mathfrak{C}$, and $\mathfrak{C}'$ the pseudo-variety associated with L. If a coalgebra D belongs to $\mathfrak{C}$, then for any coalgebra map $\gamma : D \to c\mathcal{F}$, $\gamma(D) \subset L$ since $\mathfrak{C}$ is closed under homomorphic images and L is the largest subcoalgebra of $c\mathcal{F}$ belonging to $\mathfrak{C}$. Therefore, D is in $\mathfrak{C}'$ and we proved that $\mathfrak{C} \subset \mathfrak{C}'$. Conversely, if a coalgebra D belongs to $\mathfrak{C}'$, we want to prove that D must be in $\mathfrak{C}$ and so $\mathfrak{C}' \subset \mathfrak{C}$. To this end, observe that it suffices to prove that any finite-dimensional subcoalgebra of D lies in $\mathfrak{C}$, because D is a sum of such subcoalgebras and $\mathfrak{C}$ is closed under sums. So we may assume D finite-dimensional. Then there is an injective linear map $\varphi : D \to V = \langle X_1, X_2, \ldots \rangle$, which can be lifted to a coalgebra map $\Phi : D \to c\mathcal{F}$, necessarily also injective. Since $\Phi(D) \subset L$ by the definition of the class $\mathfrak{C}'$, we conclude that D is isomorphic to a subcoalgebra of L, hence D is in $\mathfrak{C}$. This completes the proof of the desired one-to-one correspondence.

From now on, we assume that $\mathfrak{C}$ is in fact a variety. One realization of $\mathfrak{C}$-free coalgebras can be obtained by extending M. Sweedler's result that $c\mathcal{F}(V^*) = T(V)^\circ$, as shown in [1, Theorem 2.3]. This result holds even in the non-(co)associative case.

Theorem 3.4. *Let $\mathfrak{C}$ be a variety of coalgebras and $\mathfrak{A}$ a variety of algebras defined by the same set of identities S. Then for any vector space V, the $\mathfrak{C}$-free coalgebra $c\mathcal{F}_{\mathfrak{C}}(V^*)$ on V^* is naturally isomorphic to $\mathcal{F}_{\mathfrak{A}}(V)^\circ$, where*

$\mathcal{F}_{\mathfrak{A}}(V)$ is the relatively free algebra of the variety $\mathfrak{A}$ generated by a basis of V. ■

Here we will give a more explicit construction inspired by the ideas of R. Block and P. Leroux [5]. Recall that, for any graded algebra

$$A = \bigoplus_{n \geq 0} A_n$$

and a vector space V, we defined the generalized finite dual coalgebra A_V° (Definition 2.2). We will need the following criterion [5, Corollary 4] for a subspace of A_V° to be a subcoalgebra. First we introduce some notation.

Fix $b \in A_n$, then, for any graded linear map $f : A \to T(V)$ of degree 0, we can define the right translate $R_b f : A \to T(V)$ by $(R_b f)a = f(ab)$, for all $a \in A$. Similarly, the left translate $L_b f : A \to T(V)$ is defined by $(L_b f)a = f(ba)$, for all $a \in A$. Obviously, $R_b f$ and $L_b f$ are graded linear maps of degree n.

Now, fix a multilinear map $\varphi : V \times \ldots \times V \to \mathbb{k}$ in n variables. We can also view it as a linear map $\varphi : T^n(V) \to \mathbb{k}$, i.e. an element of $T^n(V)^*$. Using φ, we can "truncate" tensors from $T(V)$ in the following way. Define $Rtrunc_{\varphi} : T(V) \to T(V)$ by

$$Rtrunc_{\varphi}(v_1 \ldots v_m) = \begin{cases} 0 & \text{if } m < n, \\ v_1 \ldots v_{m-n}\varphi(v_{m-n+1}, \ldots, v_m) & \text{if } m \geq n. \end{cases}$$

Similarly,

$$Ltrunc_{\varphi}(v_1 \ldots v_m) = \begin{cases} 0 & \text{if } m < n, \\ \varphi(v_1, \ldots, v_n)v_{n+1} \ldots v_m & \text{if } m \geq n. \end{cases}$$

Clearly, $Rtrunc_{\varphi}$ and $Ltrunc_{\varphi}$ are graded linear maps of degree $-n$.

Definition 3.5. Fix $b \in A_n$ and $\varphi \in T^n(V)^*$. Then, for any graded linear map $f : A \to T(V)$ of degree 0, the composite maps $Rtrunc_{\varphi} \circ R_b f$ and $Ltrunc_{\varphi} \circ L_b f$ are again graded of degree 0. We will call them the *right* and *left truncated translates of* f, respectively, and denote $R(b, \varphi)f$ and $L(b, \varphi)f$.

Recall from Definition 2.2 that the coalgebra A_V° consists of all graded linear functions $A \to T(V)$ of degree 0 that are representative.

Lemma 3.6. *Suppose $D \subset A_V^{\circ}$ is a subspace. Then D is a subcoalgebra iff $R(b, \varphi)D \subset D$ and $L(b, \varphi)D \subset D$, for all $b \in A_n$, $\varphi \in T^n(V)^*$, and $n \geq 0$.* ■

We are now ready for our construction. Recall $\mathcal{F} = T\langle X_1, X_2, \ldots \rangle$. We will denote by P_n the space of all multilinear polynomials in the first n variables, i.e.

$$P_n = \langle X_{\pi(1)} \ldots X_{\pi(n)} \mid \pi \in S_n \rangle.$$

As before, we can view $F \in P_n$ as an element of $\Bbbk S_n$, so $F = \sum_{\pi \in S_n} \lambda_\pi \pi$ acts on the tensors from $T^n(V)$ by the formula:

$$F \cdot (v_1 \ldots v_n) = \sum_{\pi \in S_n} \lambda_\pi v_{\pi^{-1}(1)} \ldots v_{\pi^{-1}(n)}.$$

Definition 3.7. Let $J \subset \mathcal{F}$ be a T-ideal generated by multilinear identities. Let A be a graded algebra and V a vector space. We will denote by $A_V^\circ(J)$ the space of all representative functions $f : A \to T(V)$ that satisfy, for all $n \geq 0$ and $a \in A_n$,

$$F \cdot f(a) = 0, \qquad \forall F \in J \cap P_n. \tag{3}$$

Specifying $J = 0$, we recover the whole A_V°. If J is generated by the identity $X_1X_2 - X_2X_1$, then a representative function $f : A \to T(V)$ belongs to $A_V^\circ(J)$ iff $f(a)$ is a symmetric tensor, for all $a \in A_n$, $n \geq 0$ (such "symmetric-valued" representative functions were used in [5] to construct free cocommutative coalgebras). As is well-known, in the case $\operatorname{char} \mathbf{k} = 0$ any T-ideal J is generated by multilinear identities.

Theorem 3.8. *Let J be a T-ideal generated by multilinear identities, A a graded algebra, and V a vector space. Then $A_V^\circ(J)$ is a subcoalgebra of A_V°. Moreover, if A is commutative (in the ordinary, non-graded sense), then $A_V^\circ(J)$ satisfies all the identities from J.*

Proof. By Lemma 3.6, we have to prove that if $f \in A_V^\circ(J)$, $b \in A_m$, $\varphi \in T^m(V)^*$ $(m \geq 0)$, then $R(b,\varphi)f$, $L(b,\varphi)f \in A_V^\circ(J)$. For the right truncated translate, we have to show that, for any $n \geq 0$ and $a \in A_n$, $(R(b,\varphi)f)a$ satisfies (3), i.e. $F \cdot Rtrunc_\varphi(f(ab)) = 0$, for all $F \in J \cap P_n$. Fix $F = \sum_{\pi \in S_n} \lambda_\pi \pi \in J \cap P_n$.

Set $z = f(ab)$, it is a tensor of degree $n + m$, so we can write:

$$z = \sum_{(i)} \mu_{i_1,\ldots,i_{n+m}} v_{i_1} \ldots v_{i_{n+m}},$$

where $\{v_i\}$ is a basis of V. Thus we can compute:

$$Rtrunc_\varphi(z) = \sum_{(i)} \mu_{i_1,\ldots,i_{n+m}} v_{i_1} \ldots v_{i_n} \varphi(v_{i_{n+1}}, \ldots, v_{i_{n+m}}),$$

hence

$$\begin{aligned}
F \cdot Rtrunc_\varphi(z) \\
&= \sum_{\pi\in S_n}\sum_{(i)} \lambda_\pi \mu_{i_1,\dots,i_{n+m}} v_{i_{\pi^{-1}(1)}} \cdots v_{i_{\pi^{-1}(n)}} \varphi(v_{i_{n+1}},\dots,v_{i_{n+m}}) \\
&= Rtrunc_\varphi\left(\sum_{\pi\in S_n}\sum_{(i)} \lambda_\pi \mu_{i_1,\dots,i_{n+m}} v_{i_{\pi^{-1}(1)}} \cdots v_{i_{\pi^{-1}(n)}} v_{i_{n+1}} \cdots v_{i_{n+m}}\right) \\
&= Rtrunc_\varphi(\tilde{F}\cdot z),
\end{aligned}$$

where $\tilde{F} = \sum_{\pi\in S_n} \lambda_\pi \tilde{\pi}$ and $\tilde{\pi}$ is the permutation of $1,\dots n+m$ that acts as π on $1,\dots,n$ and leaves $n+1,\dots,m$ intact.

Clearly, the identity $\sum_{\pi\in S_n} \lambda_\pi X_{\pi(1)}\dots X_{\pi(n)} X_{n+1}\dots X_{n+m} = 0$, corresponding to $\tilde{F}$, is a corollary of the identity $\sum_{\pi\in S_n} \lambda_\pi X_{\pi(1)}\dots X_{\pi(n)} = 0$, corresponding to F. Therefore, $\tilde{F} \in J$. But since $f \in A_V^\circ(J)$, $z = f(ab)$ must satisfy $G\cdot f(ab) = 0$, for all $G \in J \cap P_{n+m}$, hence $\tilde{F}\cdot z = 0$, and we have proved that $R(b,\varphi)f$ is in $A_V^\circ(J)$. The proof for the left truncated translate is similar. Therefore, $A_V^\circ(J)$ is a subcoalgebra.

Now assume that A is commutative. We want to prove that $A_V^\circ(J)$ satisfies all the identities from J. Since J is generated by multilinear identities, it suffices to show that $A_V^\circ(J)$ satisfies all $F \in J \cap P_n$, for all n. Fix $F = \sum_{\pi\in S_n} \lambda_\pi \pi \in J \cap P_n$.

Recalling Definition 1.2, we have to prove that, for all $f \in A_V^\circ(J)$,

$$F \cdot \sum f_{(1)} \otimes \dots \otimes f_{(n)} = 0. \tag{4}$$

Since $f_{(1)},\dots,f_{(n)}$ are linear functions from A to $T(V)$, the left-hand side of (4) can be viewed as a linear function from $A^{\otimes n}$ to $T(V)^{\otimes n}$. Therefore, (4) is equivalent to the following:

$$\left\langle F \cdot \sum f_{(1)} \otimes \dots \otimes f_{(n)}, a_1 \otimes \dots \otimes a_n \right\rangle = 0, \qquad \forall a_1,\dots,a_n \in A. \tag{5}$$

Clearly, it suffices to verify (5) only for homogeneous $a_1,\dots,a_n$, say, of degrees $m_1,\dots,m_n$, respectively. Then the left-hand side of (5) is an element of the space $T^{m_1}(V)\otimes\dots\otimes T^{m_n}(V)$, which is naturally imbedded in $T^m(V)$, where $m = m_1 + \dots + m_n$. Hence we can write, omitting, by our convention, the symbol $\otimes$ in the monomials from $T(V)$:

$$\begin{aligned}
LHS &= \sum_{\pi\in S_n} \lambda_\pi \left\langle \sum f_{(\pi^{-1}(1))} \otimes \dots \otimes f_{(\pi^{-1}(n))}, a_1 \otimes \dots \otimes a_n \right\rangle \\
&= \sum_{\pi\in S_n} \lambda_\pi \sum \langle f_{(\pi^{-1}(1))}, a_1\rangle \dots \langle f_{(\pi^{-1}(n))}, a_n\rangle.
\end{aligned} \tag{6}$$

Now, for any permutation π of $1,\ldots,n$, we define a permutation $\tilde{\pi}$ of $1,\ldots,m$ in the following way:

$$\begin{array}{rclcrcl} \tilde{\pi}(1) & = & M_{\pi(1)-1}+1, & \ldots, & \tilde{\pi}(m-m'_n+1) & = & M_{\pi(n)-1}+1, \\ \tilde{\pi}(2) & = & M_{\pi(1)-1}+2, & \ldots, & \tilde{\pi}(m-m'_n+2) & = & M_{\pi(n)-1}+2, \\ & \ldots, & & \ldots, & & \ldots, & \\ \tilde{\pi}(m'_1) & = & M_{\pi(1)-1}+m'_1, & \ldots, & \tilde{\pi}(m) & = & M_{\pi(n)-1}+m'_n, \end{array}$$

where $m'_i = m_{\pi(i)}$ and $M_i = m_1 + \ldots + m_i$, for $i = 1,\ldots,n$. In other words, $\tilde{\pi}$ permutes the blocks of sizes $m_1,\ldots,m_n$ according to the action of π on $1,\ldots,n$. Then we can continue with (6) as follows:

$$LHS = \sum_{\pi\in S_n} \lambda_\pi \sum \tilde{\pi} \cdot \left(\langle f_{(1)}, a_{\pi(1)}\rangle \ldots \langle f_{(n)}, a_{\pi(n)}\rangle\right). \tag{7}$$

By the iterated (1), for all $b_1,\ldots,b_n \in A$, we have:

$$\langle f, b_1 \ldots b_n\rangle = \sum \langle f_{(1)}, b_1\rangle \ldots \langle f_{(n)}, b_n\rangle,$$

hence (7) gives, with $b_i = a_{\pi(i)}$:

$$LHS = \sum_{\pi\in S_n} \lambda_\pi \tilde{\pi} \cdot \langle f, a_{\pi(1)} \ldots a_{\pi(n)}\rangle = \sum_{\pi\in S_n} \lambda_\pi \tilde{\pi} \cdot \langle f, a_1 \ldots a_n\rangle \tag{8}$$

by commutativity of A. Thus we have rewritten the left-hand side of (4) as $\tilde{F} \cdot f(a_1 \ldots a_n)$, where $\tilde{F} = \sum_{\pi\in S_n} \lambda_\pi \tilde{\pi} \in \mathbb{k}S_m$.

It remains to observe that the identity $\sum_{\pi\in S_n} \lambda_\pi X_{\tilde{\pi}(1)} \ldots X_{\tilde{\pi}(m)} = 0$, corresponding to $\tilde{F}$, follows from the identity $\sum_{\pi\in S_n} \lambda_\pi X_{\pi(1)} \ldots X_{\pi(n)} = 0$, corresponding to F, by substitution of $X_1 \ldots X_{m_1}$ for X_1, and so on, $X_{m-m_n+1} \ldots X_m$ for X_n (here we substitute 1 for X_i in the case $m_i = 0$). Hence $\tilde{F} \in J$ and $\tilde{F} \cdot f(a_1 \ldots a_n) = 0$ since $f \in A_V^\circ(J)$. ■

Corollary 3.9. *Let J be a T-ideal generated by multilinear identities and V a vector space. Then $\mathbb{k}[t]_V^\circ(J)$ is the* cVar(J)*-free coalgebra of V.*

Proof. By Theorem 3.8, $\mathbb{k}[t]_V^\circ(J)$ is a subcoalgebra of $\mathbb{k}[t]_V^\circ$, which belongs to cVar(J). It is the largest such subcoalgebra since if $f \in \mathbb{k}[t]_V^\circ$ satisfies

$$F \cdot \sum f_{(1)} \otimes \ldots \otimes f_{(n)} = 0, \tag{9}$$

for some $F \in P_n$, then we have:

$$F \cdot f(t^n) = F \cdot \left(\sum f_{(1)}(t) \ldots f_{(n)}(t)\right) = 0,$$

whence if f satisfies (9), for all $n \geq 0$ and $F \in J \cap P_n$, then f is in $\Bbbk[t]^\circ_V(J)$. It remains to recall that $\Bbbk[t]^\circ_V$ is the free coassociative coalgebra of V. ∎

Using the natural imbedding of the free coassociative coalgebra $c\mathcal{F}(V) = \Bbbk[t]^\circ_V$ into $\hat{T}(V)$, we obtain a realization of the cVar(J)-free coalgebra

$$c\mathcal{F}_{\mathrm{cVar}(J)}(V) = \Bbbk[t]^\circ_V(J) = \hat{T}_J(V) \cap c\mathcal{F}(V),$$

where $\hat{T}_J(V)$ consists of all formal sums $f_0 + f_1 + \ldots$ such that $f_n \in T^n(V)$ satisfies the "symmetry conditions" $F \cdot f_n = 0$, for all $F \in J \cap P_n$.

Remark 3.10. In the case of the so-called *A-homogeneous* non-coassociative varieties, another realization of free coalgebras by infinite formal sums is given in [1]. It should be mentioned that the condition of A-homogeneity is rather restrictive. For instance, the variety of cocommutative coassociative coalgebras is not A-homogeneous.

ACKNOWLEDGEMENTS: This paper is based on the Ph.D. thesis written by the author at Memorial University of Newfoundland under the supervision of Dr. Yu. Bahturin. While preparing the text for publication, the author was supported by a post-doctoral fellowship from the NSERC Discovery Grants of Drs. S. Kuhlmann, M. Marshall, and K. Taylor.

References

[1] J. Anquela and T. Cortés, *Cofree coalgebras*, Comm. Algebra, **24** (1996), 1, pp. 357–371

[2] J. Anquela, T. Cortés, and F. Montaner, *Nonassociative Coalgebras*, Comm. Algebra, **22** (1994), 12, pp.4693–4716

[3] G. Birkhoff, *On the structure of abstract algebras*, Proc. Cambridge Phil. Soc., **31** (1935), pp. 433–454

[4] R. Block, *Commutative Hopf algebras, Lie coalgebras, and divided powers*, J. Algebra, **96** (1985), pp.275–306

[5] R. Block and P. Leroux, *Generalized dual coalgebras of algebras, with applications to cofree coalgebras*, J. Pure and Applied Algebra, **36** (1985), pp.15–21

[6] G. Griffing, *The cofree nonassociative coalgebra*, Comm. Algebra, **16** (1988), 11, pp.2387–2414

[7] M. Kochetov, *On identities for coalgebras and Hopf algebras*, Comm. Algebra, **28** (2000), 3, pp.1211–1221

[8] S. Montgomery, *Hopf algebras and their actions on rings*, CMBS Regional Conference Series in Mathematics, **82**, Providence, 1993

[9] M. Sweedler, *Hopf algebras*, Benjamin, New York, 1969

[10] E. Taft, *The order of the antipode of a finite-dimensional Hopf algebra*, Proc. Nat. Acad. Sci. USA **68** (1971), pp.2631–2633

EXAMPLE OF ALMOST COMMUTATIVE HOPF ALGEBRAS WHICH ARE NOT COQUASITRIANGULAR

AKIRA MASUOKA

INSTITUTE OF MATHEMATICS, UNIVERSITY OF TSUKUBA, IBARAKI 305-8571, JAPAN

E-mail: akira@math.tsukuba.ac.jp.

INTRODUCTION

Drinfeld defined the important notions of quasitriangular (or QT) bialgebra and almost cocommutative bialgebra. A QT bialgebra is necessarily almost cocommutative. But, there seem to be known only few examples of bialgebras which are almost cocommutative, but are not QT. In this paper we will give an example of finite-dimensional Hopf algebras with such properties, including the selfdual Hopf algebras $A_{\zeta,g}$ constructed in [M1]. Since we prefer to consider comodules rather than modules, we will actually work with the dual notions: coquasitriangular (or CQT) bialgebra and almost commutative bialgebra.

1. THE MAIN THEOREM

We work over a field k. A bialgebra H is *CQT* if the monoidal category H-Comod of left (or equivalently right) H-comodules is braided. A braiding

$$\tau_{V,W} : V \otimes W \xrightarrow{\cong} W \otimes V$$

for H-Comod is given uniquely by a (convolution-)invertible linear map $\tau : H \otimes H \to k$, which we call a *CQT structure*, so that

$$\tau_{V,W}(v \otimes w) = \sum \tau(v_{-1}, w_{-1}) w_0 \otimes v_0.$$

The conditions which τ should satisfy are given by (B1'), (B3') and (B4') in [LT, Thm. 2.7]. In particular, (B1') is given by

$$hg = \sum \tau^{-1}(g_1, h_1) g_2 h_2 \tau(g_3, h_3), \tag{1}$$

where $g, h \in H$. For a CQT structure τ, its transpose ${}^t\tau$ defined by ${}^t\tau(h,g) = \tau(g,h)$ gives a braiding for right H-comodules in a similar way as above; see [LT, Thm. 2.12].

A bialgebra H is *almost commutative*, if there is an invertible linear map $\tau : H \otimes H \to k$ which only satisfies (1). This condition (1) is equivalent to

$$H^{\mathrm{op}} = H^{\tau},$$

where H^{op} denotes the bialgebra with the opposite product, and H^{τ} is the coalgebra H with the deformed product $g \cdot h$ given by the right-hand side of (1).

A finite-dimensional bialgebra H is CQT (resp., almost commutative) if and only if the dual bialgebra H^* is QT (resp., almost cocommutative) in Drinfeld's sense; see [Mo, Defs. 10.1.1 and 5].

Let n, m be integers > 1 such that n divides m. We denote by $\mathbb{Z}_n$ the cyclic group of order n. Define two groups by

$$\begin{aligned} F &= \mathbb{Z}_n \oplus \mathbb{Z}_m = \langle a, b \mid a^n = 1 = b^m,\ ab = ba \rangle, \\ G &= \mathbb{Z}_n = \langle x \mid x^n = 1 \rangle. \end{aligned}$$

Let (F, G) be the matched pair of groups with the structure maps

$$G \xleftarrow{\text{triv}} G \times F \xrightarrow{\triangleright} F,$$

where triv is the trivial action, and $\triangleright$ is the action of group automorphisms of F determined by

$$x \triangleright a = a, \quad x \triangleright b = ab.$$

(Notice that $(x \triangleright b)^m = 1$ since $n \mid m$.) The matched pair (F, G) naturally gives rise to a Singer pair, or an abelian matched pair, (kF, k^G) of Hopf algebras. See [M3, Sect. 1a; M4, Sect. 2].

Theorem 1. *Let H be a Hopf algebra of dimension mn^2 which fits into an extension $k^G \rightarrowtail H \twoheadrightarrow kF$ associated to the Singer pair just obtained.*

(1) *Suppose that n is odd and the characteristic* $\operatorname{ch} k$ *does not divide n. Then H is not CQT unless it is commutative.*
(2) *Suppose that $(k^\times)^m = k^\times$, where $k^\times = k \backslash \{0\}$. Then H is almost commutative. Moreover, we can choose such a structure $\tau : H \otimes H \to k$ that factors through the natural map $H \otimes H \twoheadrightarrow kF \otimes kF$.*

Suppose $n = m$. In the proof of [M2, Thm. 3.1], we constructed Hopf algebras, $A_{\zeta,\eta}$, of dimension n^3 whose dual Hopf algebra $(A_{\zeta,\eta})^*$ fits into an extension associated the Singer pair given above, where ζ and η are n-th roots of 1. One sees that $(A_{\zeta,\eta})^*$ is commutative if and only if $\zeta = 1$.

Let p be an odd prime. Suppose that $\operatorname{ch} k \neq p$, and k contains a primitive p-th root of 1. In [M1, Sect. 2], we constructed $p^2 - p$ Hopf algebras, $A_{\zeta,g}$, of dimension p^3, where ζ are primitive p-th roots of 1, and $g : \mathbb{Z}_p \oplus \mathbb{Z}_p \to k^\times$ are certain group homomorphisms. The Hopf algebra $A_{\zeta,g}$ is semisimple and cosemisimple, but is neither commutative nor cocommutative. Conversely, any Hopf algebra of dimension p^3 with these properties is of the form $A_{\zeta,g}$, if k is algebraically closed and $\operatorname{ch} k = 0$; see [M1, Thm. 3.1].

Corollary 2. *Suppose $(k^\times)^p = k^\times$. Then $A_{\zeta,g}$ is almost commutative and almost cocommutative, but is neither QT nor CQT.*

Proof. Since $(k^\times)^p = k^\times$, it follows from [M1, Remark 2.21] that $H = A_{\zeta,g}$ is selfdual. By construction the dual Hopf algebra H^* fits into an extension associated to the Singer pair (kF, k^G) given as above in the special case $n = m = p$. The corollary now follows from the theorem. □

The corollary improves [M3, Prop. 2.5], which states that $A_{\zeta,g}$ is neither triangular nor cotriangular.

2. Proof of Part 1 of the Theorem

Let n, H, C and V be as given in the beginning of [M3, Sect. 2]. Thus, $n > 1$ is an integer such that $\operatorname{ch} k \not| n$, H is a finite-dimensional Hopf algebra, $C \subset H$ is an $n \times n$ comatric subcoalgebra, and V is a simple left C-comodule. Let $a \in H$ be a grouplike such that $aC = C$.

Lemma 3. *The order $o(a)$ of a divides n, and so in particular $o(a) \leq n$.*

Proof. We may suppose k is algebraically closed.

Since $a \otimes V$ is still a simple left C-comodule, there is a C-colinear isomorphism $\alpha : a \otimes V \xrightarrow{\cong} V$.

Since C is a left $(H, k\langle a \rangle)$-Hopf module, it follows by the Nichols-Zöller theorem that C is free as a left $k\langle a \rangle$-module, and so that $o(a) = \dim k\langle a \rangle$ divides $n^2 = \dim C$. Since this implies $\operatorname{ch} k \not| o(a)$, the same argument as in the proof of [M3, Lemma 2.1] shows that α is diagonalizable, regarded as a linear endomorphism of V $(= a \otimes V)$, and that there exist scalars $\lambda_i \neq 0$ and a comatric basis (c_{ij}) of C such that

$$ac_{ij} = \frac{\lambda_i}{\lambda_j} c_{ij} \quad (1 \leq i, j \leq n).$$

Since this implies that the subspace of C spanned by $c_{11}, c_{21}, \cdots, c_{n1}$ is a left $(H, k\langle a \rangle)$-Hopf module, it follows again by the Nichols-Zöller theorem that $o(a)$ divides n. □

Suppose in addition that $Ca = C$, which follows from $aC = C$ provided H-Comod is braided. Then there exist C-colinear isomorphisms

$$\alpha : a \otimes V \xrightarrow{\cong} V, \quad \alpha' : V \otimes a \xrightarrow{\cong} V.$$

Define a C-colinear isomorphism $V \xrightarrow{\cong} V$, which is a scalar multiplication, say, by $\chi \neq 0$, by the following commutative diagram.

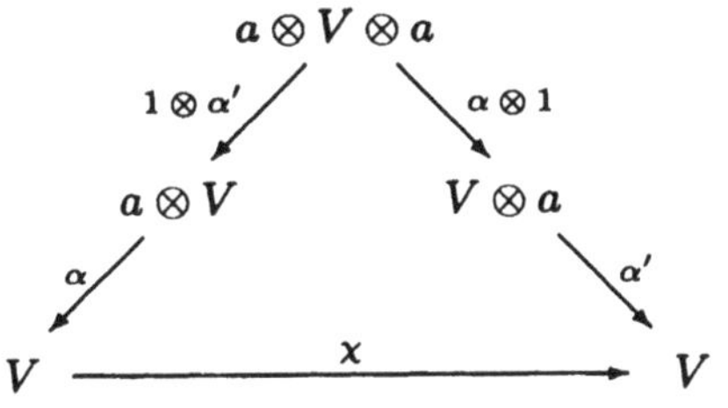

Notice that χ is independent of choice of α, α', since they are unique up to scalar multiplication.

Lemma 4. *If H has a CQT structure $\tau : H \otimes H \to k$, then*

$$\chi = \tau(a, a) = \pm 1.$$

Proof. Let $\tau_{V,W}$ denote the braiding given by τ. Then, $\tau_{a,a} = \tau(a, a)$. For a C-colinear isomorphism $\alpha : a \otimes V \xrightarrow{\cong} V$, take as α' the composite

$$V \otimes a \xrightarrow{\tau_{V,a}} a \otimes V \xrightarrow{\alpha} V.$$

The condition for $\tau_{V,W}$ given by (B3) in [LT, Def. 2.2] makes the following diagram commute.

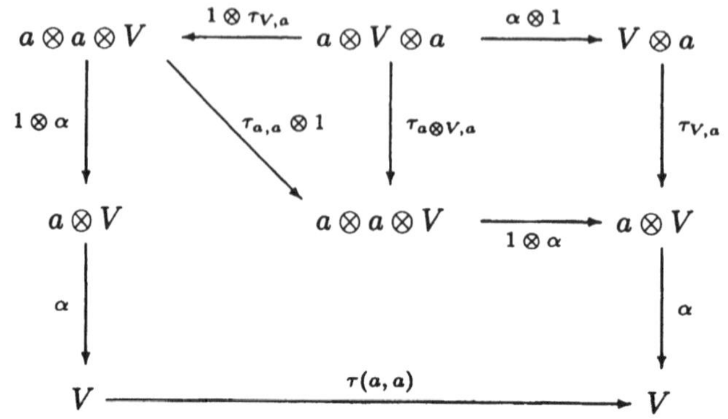

It follows that $\chi = \tau(a,a)$. Since $({}^t\tau)^{-1}$ is also a CQT structure, we have

$$\chi = ({}^t\tau)^{-1}(a,a) = \tau(a,a)^{-1},$$

which implies $\tau(a,a) = \pm 1$. □

Corollary 5. *Suppose $o(a) \geq n$, which implies by Lemma 3 that $o(a) = n$. If n is odd and H is CQT, then $ec = ce$ for all $c \in C$, where $e = (1/n)\sum_{i=0}^{n-1} a^i$.*

Proof. Let τ be a CQT structure for H. By Lemma 4, $\tau(a,a) = \pm 1$. Since $\tau(a,-)$ is multiplicative, we have that

$$\tau(a,a)^n = \tau(a,a^n) = \tau(a,1) = 1,$$

and so that

$$\chi = \tau(a,a) = 1$$

since n is odd. The corollary follows from [M3, Lemma 2.1]. □

Proof of Part 1 of Theorem 1. Let H be as in the theorem. Then we have such an identification

$$H = k^G \otimes kF \tag{2}$$

that preserves at least unit, counit and structures of left k^G-module and right kF-comodule.

By base extension we may suppose that k is algebraically closed, and so that the dual algebra H^* is the semi-direct product, or the smash product, $G \ltimes k^F$ since G is cyclic. Hence we can suppose that (2) identifies the coalgebra H with the smash coproduct $k^G \otimes kF$ which is given by the comodule-algebra structure $\rho : kF \to kF \otimes k^G$ determined by

$$\rho(a) = a \otimes 1, \quad \rho(b) = \sum_{i=1}^{n-1} a^i b \otimes e_i,$$

where (e_i) in k^G is the dual basis of (x^i) in kG. Consequently, $1 \otimes a$ is a grouplike in H and the subspace

$$C = \sum_{i=0}^{n-1} k^G \otimes a^i b$$

is a comatric subcoalgebra of H with a comatric basis $(e_{i-j}a^j b)_{i,j}$. Notice that (2) identifies H also with the F-crossed product $k^G \otimes kF$ over $k^G = k^G \otimes k$ given by the trivial action $F \times k^G \to k^G$ together with some 2-cocycle $F \times F \to k^G$. Hence, k^G is central in H, and $(1 \otimes a)C = C = C(1 \otimes a)$. Since we see that $(1 \otimes a)^i \in k^G \otimes a^i$ for $1 \leq i \leq n$, the order $o(1 \otimes a) \geq n$.

Suppose H is CQT. Since we also suppose that n is odd and $\operatorname{ch} k \not| n$, it follows by Corollary 5 that $e = (1/n) \sum_{i=0}^{n-1} (1 \otimes a)^i$ commutes with $1 \otimes b$. Since $(1 \otimes a)(1 \otimes b)$ and $(1 \otimes b)(1 \otimes a)$ are units of the same degree $ab = ba$ in the F-crossed product, there exists a unit u in k^G such that

$$(1 \otimes b)(1 \otimes a) = (u \otimes 1)(1 \otimes a)(1 \otimes b)$$

and so that

$$(1 \otimes b)e = \frac{1}{n} \sum_{n=0}^{n-1} (u^i \otimes 1)(1 \otimes a)^i (1 \otimes b).$$

Since we see that $(1 \otimes b)e = e(1 \otimes b)$ requires $u = 1$, it follows that $1 \otimes a$ and $1 \otimes b$ commute with each other. Since they generate the k^G-algebra H, it is commutative. □

Remark 6. The result just proved does not hold true if n is even. If k is algebraically closed and $\operatorname{ch} k \neq 2$, there is uniquely a semisimple Hopf algebra H_8 of dimension 8 which is neither commutative nor cocommutative. Although H_8 fits into an extension $k^G \rightarrowtail H_8 \twoheadrightarrow kF$ associated with the Singer pair (kF, k^G) given as above in the special case $n = m = 2$, H_8 has precisely 8 CQT structures as was proved by S. Suzuki [S, Prop. 3.10], in which $H_8 = A_{1,2}^{(+-)} \simeq A_{1,2}^{(--)}$.

3. Proof of Part 2 of the Theorem

In general, given a matched pair (F, G) of finite groups, we have a double complex [M4, Sect. 1] essentially due to George Kac,

$$\begin{array}{ccccc}
 & \vdots & & \vdots & \\
 & \uparrow & & \uparrow & \\
A^{\cdot\cdot} = & \operatorname{Map}_+(G^2 \times F, k^\times) & \xrightarrow{\partial} & \operatorname{Map}_+(G^2 \times F^2, k^\times) & \longrightarrow \cdots \\
 & \uparrow \partial' & & \uparrow \partial' & \\
 & \operatorname{Map}_+(G \times F, k^\times) & \xrightarrow{\partial} & \operatorname{Map}_+(G \times F^2, k^\times) & \longrightarrow \cdots,
\end{array}$$

such that the total 1-cohomology group $H^1(\operatorname{Tot} A^{\cdot\cdot})$ is isomorphic to the group $\operatorname{Opext}(kF, k^G)$ of the equivalence classes of Hopf algebra extensions associated to the Singer pair (kF, k^G) which arises from the given matched pair. The isomorphism is induced by

$$(\sigma, \theta) \mapsto k^G \#_{\sigma,\theta} kF,$$

where

$$\sigma : G \times F \times F \to k^\times, \quad \theta : G \times G \times F \to k^\times$$

are supposed to form a total 1-cocycle in $A^{\cdot\cdot}$, and $k^G \#_{\sigma,\theta} kF$ then denotes the Hopf algebra of bicrossed product which is constructed from σ, θ on the vector space

$k^G \otimes kF$. In particular as an algebra, $k^G \#_{\sigma,\theta} kF$ is the F-crossed product over k^G given by the action $F \times k^G \to k^G$ which arises from $\triangleleft : G \times F \to G$, together with the 2-cocycle $(u, v) \mapsto \sigma(-; u, v)$, $F \times F \to k^G$.

Proof of Part 2 of Theorem 1. Suppose that (F, G) is the special matched pair given just above the theorem. Let H be as in the theorem, and suppose it is given by a total 1-cocycle (σ, θ). Since F is abelian and the action of F on G is trivial, H^{op} fits into an extension associated to the same Singer pair, and is given by $({}^t\sigma, \theta)$, where ${}^t\sigma(x^r; u, v) = \sigma(x^r; v, u)$.

Suppose $(k^\times)^m = k^\times$. Then the argument which gave $\bar{a}^n = \bar{b}^n = 1$ in [M2, p. 577, line 19] allows us to suppose

$$\sigma(x; a^i b^j, a^k b^l) = \zeta^{jk},$$

where ζ is an n-th root of 1 in k. Set $\omega = {}^t\sigma\sigma^{-1}$. Then this together with the trivial map $G \times G \times F \to k^\times$ form a total 1-cocycle in $A^{\cdot\cdot}$. Let G act on $F \times F$ diagonally. One sees that $\omega(x; -, -)$ is constant on each G-orbit in $F \times F$. Since $\partial' \omega$ is trivial, it follows that

$$\omega(x^r; a^i b^j, a^k b^l) = \omega(x; a^i b^j, a^k b^l)^r = \zeta^{r(il-jk)}.$$

The order of the G-orbit containing $(a^i b^j, a^k b^l)$ is the least common multiple of the orders in $\mathbb{Z}_n$ of j and l. Since this is divided by the order of ζ^{il-jk}, we have a map $\nu : F \times F \to k^\times$ such that

$$\frac{\nu(u, v)}{\nu(x \triangleright u, x \triangleright v)} = \omega(x; u, v)$$

and necessarily

$$\frac{\nu(u, v)}{\nu(x^r \triangleright u, x^r \triangleright v)} = \omega(x^r; u, v), \tag{3}$$

where $u, v \in F$. This implies that $H^{\mathrm{op}} = H^\tau$, where $\tau : H \otimes H \to k$ is the composite of ν with the natural map $H \otimes H \twoheadrightarrow kF \otimes kF$. For the product of H^τ is given by $\sigma\partial'\nu$, if we define the value $\partial'\nu(x^r; u, v)$ to be the left-hand side of (3); cf. [M4, Prop. 3.1]. □

Remark 7. For a split cosemisimple bialgebra H, the following are equivalent.

(i) H is almost commutative;

(ii) The K_0-ring $K_0(H)$ of the monoidal category of finite-dimensional left (or right) H-comodules is commutative;

(iii) There exist an invertible linear map $\tau : H \otimes H \to k$ and a coassociator $\phi : H \otimes H \otimes H \to k$ [P, p. 331] such that $(H^\tau, \phi^\tau) = H^{\mathrm{op}}$. Here (H^τ, ϕ^τ) denotes the coquasi-bialgebra deformed from (H, ϕ) via τ [P, p. 332], and H^{op} is regarded as a coquasi-bialgebra with the trivial coassociator.

Notice that (ii) is equivalent to $K_0(H) = K_0(H^{\mathrm{op}})$. Then the equivalence (i) $\Leftrightarrow$ (ii) follows from the proof of [N, Thm. 4.1], while (ii) $\Leftrightarrow$ (iii) follows from [M5, Prop. 5.8].

Let H, ν and τ be as in the proof of Part 2 of Theorem 1. Let $\partial\nu : F \times F \times F \to k^\times$ denote the coboundary of ν in the standard complex for computing the group cohomology $H^\cdot(F, k^\times)$ with coefficients in the trivial left F-module $k^\times$. Let $\phi : H \otimes H \otimes H \to k$ be the composite of $\partial\nu$ with the natural map $H^{\otimes 3} \twoheadrightarrow (kF)^{\otimes 3}$. Then ϕ is a coassociator for H. We see $(H^\tau, \phi^\tau) = H^{\mathrm{op}}$; see [M5, Sect. 4].

References

[LT] R.G. Larson and J. Towber, *Two dual classes of bialgebras related to the concept of "quantum groups" and "quantum Lie algebras"*, Comm. Algebra 19 (1991), 3295–3345.

[M1] A. Masuoka, *Selfdual Hopf algebras of dimension p^3 obtained by extension*, J. Algebra 178 (1995), 791–806.

[M2] A. Masuoka, *Calculations of some groups of Hopf algebra extensions*, J. Algebra 191 (1997), 568–588.

[M3] A. Masuoka, *Cocycle deformations and Galois objects for some cosemisimple Hopf algebras of finite dimension*, Contemp. Math. 267 (2000), 195–214.

[M4] A. Masuoka, *Hopf algebra extensions and cohomology*, In: S. Montgomery and H.-J. Schneider (eds.), "New Directions in Hopf Algebras", MSRI Publ. 43, Cambridge Univ. Press, Cambridge, 2002, pp. 167–209.

[M5] A. Masuoka, *Cohomology and coquasi-bialgebra extensions associated to a matched pair of bialgebras*, Adv. Math. 173 (2003), 262–315.

[Mo] S. Montgomery, *"Hopf Algebras and Their Actions on Rings"*, CBMS Reg. Conf. Ser. in Math. 82, Amer. Math. Soc., Providence, 1993.

[N] D. Nikshych, *K_0-rings and twistings of finite-dimensional semisimple Hopf algebras*, Comm. Algebra 26 (1998), 321–342; Corrigendum, Comm. Algebra 26 (1998), 2019.

[P] Peter Schauenburg, *Hopf algebra extensions and monoidal categories*, In: S. Montgomery and H.-J. Schneider (eds.), "New Directions in Hopf Algebras", MSRI Publ. 43, Cambridge Univ. Press, Cambridge, 2002, pp. 321–381.

[S] Satoshi Suzuki, *A family of braided cosemisimple Hopf algebras of finite dimension*, Tsukuba J. Math. 22 (1998), 1–29.

Hopf Algebras of Dimension p^2

Siu-Hung Ng
Mathematics Department, Towson University, Baltimore, MD 21252

Abstract

Let p be a prime number. It is known that any non-semisimple Hopf algebra of dimension p^2 over an algebraically closed field of characteristic 0 is isomorphic to a Taft algebra. In this exposition, we will give a more direct alternative proof to this result.

0 Introduction

Let p be an odd prime and k an algebraically closed field of characteristic 0. It was shown independently in [AS98b], [BDG99] and [Gel98] that there exist infinitely many isomorphism classes of Hopf algebras of dimension p^4 over k. These results give counterexamples to a Kaplansky conjecture that there were only finitely many types in each dimension [Kap75]. However, for some special dimensions, there are, indeed, only finitely many isomorphism classes of Hopf algebras. For example, Hopf algebras of dimension p over k are the group algebra $k[\mathbf{Z}_p]$ and hence there is only one type of Hopf algebras of dimension p [Zhu94].

If H is a semisimple Hopf algebra of dimension p^2 over k, then H is isomorphic to $k[\mathbf{Z}_{p^2}]$ or $k[\mathbf{Z}_p \times \mathbf{Z}_p]$ by [Mas96]. A more general result for semisimple Hopf algebras of dimension pq, where p, q are odd prime, is obtained by [EG98]. However, there are very few examples of non-semisimple Hopf algebras of these dimensions. In fact, there are no non-semisimple Hopf algebras over k of dimension 15, 21 or 35 (cf. [AN01]).

For non-semisimple Hopf algebras of dimension p^2, the only known example during the last three decades is the Taft algebras (cf. [Mon98, Section 5] and [Taf71]). Let $\omega \in k$ be a primitive nth root of unity. The Taft algebra associated with ω, denoted by $T(\omega)$, is the Hopf algebra generated by x and a, as a k-algebra, subject to the relations

$$a^n = 1, \quad ax = \omega xa, \quad x^n = 0,$$

with the coalgebra structure, and the antipode of $T(\omega)$ given by

$$\Delta(a) = a \otimes a, \quad S(a) = a^{-1}, \quad \varepsilon(a) = 1,$$
$$\Delta(x) = x \otimes a + 1 \otimes x, \quad S(x) = -xa^{-1}, \quad \varepsilon(x) = 0.$$

The Hopf algebra $T(\omega)$ is not semisimple and $\dim T(\omega) = n^2$. It is also known that $T(\omega) \cong T(\omega)^*$ as Hopf algebras, and that $T(\omega) \cong T(\omega')$ only if $\omega = \omega'$.

The question whether the Taft algebras are the only non-semisimple Hopf algebras of dimension p^2 remained open for the last three decades, and it was repeatedly asked by Susan Montgomery in several conferences. However, there were some partial answers to the question. Andruskiewitsch and Chin proved that non-semisimple *pointed* Hopf algebras of dimension p^2 are indeed Taft algebras (cf. [Şte97]). Using this result, Andruskiewitsch and Schneider proved in [AS98a] that if H is non-semisimple Hopf algebra with antipode S of dimension p^2 such that the order of S^2 is p, then H is isomorphic to a Taft algebra. In the course of studying Hopf algebras of dimension pq where p, q are odd primes, the author proved in [Ng02], using the later results, that non-semisimple Hopf algebras of dimension p^2 are indeed Taft algebras. Hence, the classification of Hopf algebras of dimension p^2 is completed and they are the following list of $p+1$ non-isomorphic Hopf algebras over k of dimension p^2

(a) $k[\mathbf{Z}_{p^2}]$;

(b) $k[\mathbf{Z}_p \times \mathbf{Z}_p]$;

(c) $T(\omega)$, $\omega \in k$ a primitive pth of unity.

In this paper, we will give a more direct proof to the result that if H is a non-semisimple Hopf algebra with antipode S of dimension p^2 over k, then $o(S^2) = p$ and hence by [AS98a], H is isomorphic to a Taft algebra. The statement for the case when $p = 2$ is proven in [Kap75].

1 Notation and Preliminaries

Throughout this paper, p is an odd prime, k is an algebraically closed field of characteristic 0 and H is a finite-dimensional Hopf algebra over k with antipode S. Its comultiplication and counit are, respectively, denoted by Δ and ε. A non-zero element $a \in H$ is called group-like if $\Delta(a) = a \otimes a$. The set of all group-like elements $G(H)$ of H is a linearly independent set, and it forms a group under the multiplication of H. For the details of elementary aspects for finite-dimensional Hopf algebras, readers are referred to the references [Swe69] and [Mon93].

The order of the antipode is of fundamental importance to the semisimplicity of H. We recall some important results on the antipodes of finite-dimensional Hopf algebras.

Theorem 1.1 [LR87, LR88] *Let H be a finite-dimensional Hopf algebra with antipode S over a field of characteristic 0. Then the following statements are equivalent:*

(i) *H is semisimple.*

(ii) *H^* is semisimple.*

(iii) $\mathrm{Tr}(S^2) \neq 0$.

(iv) $S^2 = id_H$.

Let $\lambda \in H^*$ be a non-zero right integral of H^* and let $\Lambda \in H$ be a non-zero left integral of H. There exists $\alpha \in \mathrm{Alg}(H, k) = G(H^*)$, independent of the choice of Λ, such that $\Lambda a = \alpha(a)\Lambda$ for $a \in H$. Likewise, there is a group-like element $g \in H$, independent of the choice of λ, such that $\beta\lambda = \beta(g)\lambda$ for $\beta \in H^*$. We call g the distinguished group-like element of H and α the distinguished group-like element of H^*. Then we have a formula for S^4 in terms of α and g [Rad76]:

$$S^4(a) = g(\alpha \rightharpoonup a \leftharpoonup \alpha^{-1})g^{-1} \quad \text{for } a \in H, \tag{1.1}$$

where $\rightharpoonup$ and $\leftharpoonup$ denote the natural actions of the Hopf algebra H^* on H described by

$$\beta \rightharpoonup a = \sum a_{(1)}\beta(a_{(2)}) \quad \text{and} \quad a \leftharpoonup \beta = \sum \beta(a_{(1)})a_{(2)}$$

for $\beta \in H^*$ and $a \in H$. Hence, we have

$$o(S^4) \mid \mathrm{lcm}(o(g), o(\alpha)). \tag{1.2}$$

The Hopf algebra H (respectively H^*) is said be *unimodular* if α (resp. g) is trivial. If both H and H^* are unimodular, then $S^4 = id_H$.

Lemma 1.2 [AS98a] [LR95] *Let H be a finite-dimensional non-semisimple Hopf algebra with antipode S of odd dimension over k. Then, H and H^* can not be both unimodular and $S^4 \neq id_H$.*

Proof. It follows immediately from [LR95, Theorem 2.1] or [AS98a, Lemma 2.5]. ■

Following [Ng02], we define the index of H to be the least positive integer n such that

$$g^n = 1 \quad \text{and} \quad S^{4n} = id_H \,.$$

Lemma 1.3 *Let H be a finite-dimensional non-semisimple Hopf algebra with antipode S of dimension p^2 where p is an odd prime. Suppose that g and α are the distinguished group-like elements of H and H^* respectively. Then*

$$o(S^4) = \operatorname{lcm}(o(g), o(\alpha)) = p \,.$$

In particular, H is of index p.

Proof. Since H is not semisimple and $\dim H$ is odd, by Lemma 1.2,

$$S^4 \neq id_H \tag{1.3}$$

and H, H^* cannot be both unimodular. Hence $\operatorname{lcm}(o(g), o(\alpha)) > 1$. By Nichols-Zoeller Theorem [NZ89],

$$o(\alpha) \mid p^2 \quad \text{and} \quad o(g) \mid p^2 \,.$$

If $o(g) = p^2$ or $o(\alpha) = p^2$, then

$$H \cong k[\mathbf{Z}_{p^2}] \quad \text{or} \quad H \cong k[\mathbf{Z}_{p^2}]^*$$

and hence H is semisimple. Therefore,

$$o(g) \mid p \quad \text{and} \quad o(\alpha) \mid p$$

and hence $\operatorname{lcm}(o(g), o(\alpha)) = p$. Since $o(S^4) \neq 1$ and $o(S^4) \mid \operatorname{lcm}(o(g), o(\alpha)) = p$, we have $o(S^4) = p$. ■

2 Eigenspace Decompositions for Hopf algebras of Odd Index

Let H be a finite-dimensional Hopf algebra with antipode S of odd index $n > 1$. Suppose that g and α are the distinguished group-like elements of H and H^* respectively. The element g defines a coalgebra automorphism $r(g)$ on H as follows:

$$r(g)(a) = ag \quad \text{for} \quad a \in H \,.$$

Since S^2 is an algebra automorphism on H and $S^2(g) = g$, we have

$$S^2 \circ r(g) = r(g) \circ S^2$$

and so $r(g)$ and S^2 are simultaneously diagonalizable. Let $\omega \in k$ be a primitive nth root of unity. The eigenvalues of S^2 are of the form $(-1)^a\omega^i$ and the eigenvalues of $r(g)$ are of the form ω^j. Define

$$H^\omega_{a,i,j} = \{u \in H \mid S^2(u) = (-1)^a\omega^i u\,, ug = \omega^j u\} \text{ for any } (a,i,j) \in \mathbf{Z}_2\times\mathbf{Z}_n\times\mathbf{Z}_n$$

We will simply write $\mathcal{K}_n$ for the group $\mathbf{Z}_2 \times \mathbf{Z}_n \times \mathbf{Z}_n$, write $H^\omega_{(a,i,j)}$ for $H^\omega_{a,i,j}$ and $\mathbf{a}$ for (a, i, j). Since S^2 and $r(g)$ are simultaneously diagonalizable, we have the decomposition

$$H = \bigoplus_{\mathbf{a}\in\mathcal{K}_n} H^\omega_{\mathbf{a}}\,. \tag{2.1}$$

Note that $H^\omega_{\mathbf{a}}$ could be zero.

Since $g^n = 1$ and α is an algebra map, $\alpha(g)$ is a nth root of 1. There exists a unique element $x(\omega, H) \in \mathbf{Z}_n$ such that

$$\omega^{x(\omega,H)} = \alpha(g)\,.$$

Using the eigenspace decomposition of H in (2.1), the diagonalization of a left integral Λ of H admits an interesting form.

Lemma 2.1 [Ng02] *Let H be a finite-dimensional Hopf algebra with antipode S of odd index $n > 1$, $\omega \in k$ primitive nth root of 1, and $x = x(\omega, H)$. Suppose Λ is a non-zero left integral of H. Then we have*

$$\Delta(\Lambda) = \sum_{\mathbf{a}\in\mathcal{K}_n} \left(\sum u_{\mathbf{a}} \otimes v_{-\mathbf{a}+\mathbf{x}}\right) \tag{2.2}$$

where $\mathbf{x} = (0, -x, x)$ and $\sum u_{\mathbf{a}} \otimes v_{-\mathbf{a}+\mathbf{x}} \in H^\omega_{\mathbf{a}} \otimes H^\omega_{-\mathbf{a}+\mathbf{x}}$. Moreover, if λ is a right integral of H^* such that $\lambda(\Lambda) = 1$, then

$$\dim H^\omega_{\mathbf{a}} = \sum \lambda(S(v_{\mathbf{x}-\mathbf{a}})u_{\mathbf{a}})$$

for any $\mathbf{a} \in \mathcal{K}_n$.

3 Antipodes of Hopf Algebras of Dimension p^2

In [Ng02], the author proved that if H is a non-semisimple Hopf algebra with antipode S of dimension p^2, then $o(S^2) = p$ and hence, by [AS98a], H is isomorphic to a Taft algebra. In this section, we will provide a more direct proof to the result using the following lemmas.

Lemma 3.1 [Ng02] Let H be a finite-dimensional Hopf algebra of index p where p is an odd prime, and let $\omega \in k$ be a primitive pth root of 1. If H^* is not unimodular, then for any $j \in \mathbf{Z}_p$,

(i) there exists an integer d_j such that

$$\dim H^{\omega}_{0,i,j} - \dim H^{\omega}_{1,i,j} = d_j \quad \text{for all } i \in \mathbf{Z}_p,$$

(ii) and

$$\sum_{(a,i)\in \mathbf{Z}_2\times\mathbf{Z}_p} \dim H^{\omega}_{a,i,j} = \frac{\dim H}{p}.$$

Lemma 3.2 *Let H be a finite-dimensional Hopf algebra with antipode S of index p where p is an odd prime, and $\omega \in k$ a primitive pth root of 1. If H^* is not unimodular, then*

$$\dim H^{\omega}_{0,-x,j} \neq 0$$

for all $j \in \mathbf{Z}_p$ where $x = x(\omega, H)$.

Proof. Let λ be a right integral of H^* and Λ a left integral of H such that $\lambda(\Lambda) = 1$. By Lemma 2.1,

$$\Delta(\Lambda) = \sum_{\mathbf{a}\in\mathcal{K}_p} \left(\sum u_{\mathbf{a}} \otimes v_{-\mathbf{a}+\mathbf{x}}\right)$$

where $\mathbf{x} = (0, -x, x)$ and $\sum u_{\mathbf{a}} \otimes v_{-\mathbf{a}+\mathbf{x}} \in H^{\omega}_{\mathbf{a}} \otimes H^{\omega}_{-\mathbf{a}+\mathbf{x}}$. Note that for any $(a, i, j) \in \mathcal{K}_p$,

$$\sum S(v_{a,-x-i,x-j})u_{a,i,j} \in H^{\omega}_{0,-x,j}\,. \tag{3.1}$$

It suffices to show that for any $j \in \mathbf{Z}_p$, there exists $(a, i) \in \mathbf{Z}_2 \times \mathbf{Z}_p$ such that

$$\sum S(v_{a,-x-i,x-j})u_{a,i,j} \neq 0.$$

Suppose not. Then there exists $j \in \mathbf{Z}_p$ such that for all $(a, i) \in \mathbf{Z}_2 \times \mathbf{Z}_p$,

$$\sum S(v_{a,-x-i,x-j})u_{a,i,j} = 0.$$

Then, by Lemma 2.1,

$$\dim H^{\omega}_{a,i,j} = \sum \lambda(S(v_{a,-x-i,x-j})u_{a,i,j}) = 0$$

for all $(a, i) \in \mathbf{Z}_2 \times \mathbf{Z}_p$. Thus,

$$\sum_{(a,i)\in\mathbf{Z}_2\times\mathbf{Z}_p} \dim H^{\omega}_{a,i,j} = 0$$

but this contradicts Lemma 3.1(ii). ∎

Lemma 3.3 *Let H be a non-semisimple Hopf algebra with antipode S of dimension p^2 where p is a odd prime, and let $\omega \in k$ be a primitive pth root of 1. Then for any $i, j \in \mathbf{Z}_p$,*

$$\dim H^{\omega}_{0,i,j} = 1 \quad \textit{and} \quad \dim H^{\omega}_{1,i,j} = 0 .$$

Proof. For $j \in \mathbf{Z}_p$, by Lemma 3.1(i),

$$\sum_{i \in \mathbf{Z}_p} \dim H^{\omega}_{0,i,j} - \dim H^{\omega}_{1,i,j} = pd_j .$$

Hence, by Lemma 3.1(ii), $|d_j| = 0$ or 1 for all $j \in \mathbf{Z}_p$. If $d_j = 0$, for any $i \in \mathbf{Z}_p$, we have

$$\dim H^{\omega}_{0,i,j} = \dim H^{\omega}_{1,i,j}$$

and hence

$$p = \sum_{i \in \mathbf{Z}_p} \dim H^{\omega}_{0,i,j} + \dim H^{\omega}_{1,i,j} = 2 \sum_{i \in \mathbf{Z}_p} \dim H^{\omega}_{0,i,j} .$$

This contradicts that p is odd. If $d_j = -1$, then

$$\sum_{i \in \mathbf{Z}_p} H^{\omega}_{0,i,j} - H^{\omega}_{1,i,j} = -p .$$

By Lemma 3.1(ii),

$$\sum_{i \in \mathbf{Z}_p} \dim H^{\omega}_{0,i,j} = 0 \quad \text{and} \quad \sum_{i \in \mathbf{Z}_p} \dim H^{\omega}_{1,i,j} = p$$

but the first equation contradicts Lemma 3.2. Therefore, $d_j = 1$ and hence, by Lemma 3.1, the result follows. ■

Theorem 3.4 *If H be a non-semisimple Hopf algebra with antipode S of dimension p^2 over k, then $o(S^2) = p$ and hence H is isomorphic to a Taft algebra.*

Proof. By Lemma 3.3,

$$H = \bigoplus_{i,j \in \mathbf{Z}_p} H^{\omega}_{0,i,j}$$

where $\omega \in k$ is a primitive pth root of 1. Therefore, $S^{2p}(u) = u$ for all $u \in H$. By Lemma 1.3, $o(S^2) = p$. It follows from [AS98a] that H is isomorphic to a Taft algebra. ■

Acknowledgement

The author would like to thank Susan Montgomery for bringing his attention to the question on the Hopf algebras of dimension p^2.

References

[AN01] Nicolás Andruskiewitsch and Sonia Natale. Counting arguments for Hopf algebras of low dimension. *Tsukuba J. Math.*, 25(1):187–201, 2001.

[AS98a] Nicolás Andruskiewitsch and Hans-Jürgen Schneider. Hopf algebras of order p^2 and braided Hopf algebras of order p. *J. Algebra*, 199(2):430–454, 1998.

[AS98b] N. Andruskiewitsch and H.-J. Schneider. Lifting of quantum linear spaces and pointed Hopf algebras of order p^3. *J. Algebra*, 209(2):658–691, 1998.

[BDG99] M. Beattie, S. Dăscălescu, and L. Grünenfelder. On the number of types of finite-dimensional Hopf algebras. *Invent. Math.*, 136(1):1–7, 1999.

[EG98] Pavel Etingof and Shlomo Gelaki. Semisimple Hopf algebras of dimension pq are trivial. *J. Algebra*, 210(2):664–669, 1998.

[Gel98] Shlomo Gelaki. Pointed Hopf algebras and Kaplansky's 10th conjecture. *J. Algebra*, 209(2):635–657, 1998.

[Kap75] Irving Kaplansky. *Bialgebras*. Department of Mathematics, University of Chicago, Chicago, Ill., 1975. Lecture Notes in Mathematics.

[LR87] Richard G. Larson and David E. Radford. Semisimple cosemisimple Hopf algebras. *Amer. J. Math.*, 109(1):187–195, 1987.

[LR88] Richard G. Larson and David E. Radford. Finite-dimensional cosemisimple Hopf algebras in characteristic 0 are semisimple. *J. Algebra*, 117(2):267–289, 1988.

[LR95] Richard G. Larson and David E. Radford. Semisimple Hopf algebras. *J. Algebra*, 171(1):5–35, 1995.

[Mas96] Akira Masuoka. The p^n theorem for semisimple Hopf algebras. *Proc. Amer. Math. Soc.*, 124(3):735–737, 1996.

[Mon93] Susan Montgomery. *Hopf algebras and their actions on rings*, volume 82 of *CBMS Regional Conference Series in Mathematics*. Published for the Conference Board of the Mathematical Sciences, Washington, DC, 1993.

[Mon98] Susan Montgomery. Classifying finite-dimensional semisimple Hopf algebras. In *Trends in the representation theory of finite-dimensional algebras (Seattle, WA, 1997)*, pages 265–279. Amer. Math. Soc., Providence, RI, 1998.

[Ng02] Siu-Hung Ng. Non-semisimple Hopf algebras of dimension p^2. *J. Algebra*, 255(1):182–197, 2002.

[NZ89] Warren D. Nichols and M. Bettina Zoeller. A Hopf algebra freeness theorem. *Amer. J. Math.*, 111(2):381–385, 1989.

[Rad76] David E. Radford. The order of the antipode of a finite dimensional Hopf algebra is finite. *Amer. J. Math.*, 98(2):333–355, 1976.

[Şte97] D. Ştefan. Hopf subalgebras of pointed Hopf algebras and applications. *Proc. Amer. Math. Soc.*, 125(11):3191–3193, 1997.

[Swe69] Moss E. Sweedler. *Hopf algebras*. W. A. Benjamin, Inc., New York, 1969. Mathematics Lecture Note Series.

[Taf71] Earl J. Taft. The order of the antipode of finite-dimensional Hopf algebra. *Proc. Nat. Acad. Sci. U.S.A.*, 68:2631–2633, 1971.

[Zhu94] Yongchang Zhu. Hopf algebras of prime dimension. *Internat. Math. Res. Notices*, (1):53–59, 1994.

SUPPORT CONES FOR INFINITESIMAL GROUP SCHEMES

JULIA PEVTSOVA*
IAS

SCHOOL OF MATHEMATICS, INSTITUTE FOR ADVANCED STUDY, PRINCETON, NJ 08540
E-mail address: julia@ias.edu

ABSTRACT. We verify that the construction of *support cone* for infinite dimensional modules, introduced in [12], extends to modules over any infinitesimal group scheme and satisfies all good properties of support varieties for finite dimensional modules, thereby extending the results of the author for infinite dimensional modules of Frobenius kernels [12]. We show, using an alternative description of support cones in terms of Rickard idempotents, that for an algebraic group G over an algebraically closed field k of positive characteristic p and a point s in the cohomological support variety of a Frobenius kernel $G_{(r)}$, the orbit $G \cdot s$ can be realized as a support cone of a rational G-module.

0. INTRODUCTION

The theory of support varieties for finite dimensional modules for finite groups ([1],[5],[2]) or restricted Lie algebras ([7],[8],[11]) has drawn considerable attention over the last twenty years. One of its most attractive features is the elegant connection it provides between the cohomological behaviour and intrinsic representation-theoretic properties of a module. This connection proves to be crucial in establishing basic properties of support varieties such as good behavior with respect to tensor products or the property of detecting projectivity of a module on its support variety.

Motivated by the work of Benson, Carlson and Rickard ([4]) for finite groups, we seek to associate a geometric object (the "support cone") to an arbitrary module M of an infinitesimal group scheme G. It turns out that the original, cohomological, approach to support varieties does not provide a good generalization for infinite dimensional modules. For this reason, in our study of geometric properties of infinite dimensional modules for infinitesimal group schemes we take the representation-theoretic approach developed in [15],[16] and define *support cone* of a module M in purely representation-theoretic terms. The lack of cohomological description, though, makes it more difficult to show that one of the most fundamental properties of support varieties, detection of projectivity, is satisfied by our construction. The proof of this fact for infinitesimal group schemes, built upon an earlier result for Frobenius kernels, occupies §2.

In the spirit of work of Benson, Carlson and Rickard for infinite dimensional modules for finite groups, we provide an equivalent description of support cones in terms of Rickard idempotents, the universal modules corresponding to tensor-ideal thick subcategories of the stable module category. In the last section we give an

* This material is based upon work supported by the NSF under agreement No. DMS-9729992. Any opinions, findings and conclusions or recommendations expressed in this material are those of the author and do not necessarily reflect the views of the NSF.

example of the interplay of these two approaches which allows us to prove certain "realization" results.

This short note is complementary to [12], where the author studied possible extensions of the notion of support variety to infinite dimensional modules for Frobenius kernels. We have tried to make this note self-contained by recalling all the main ingredients going into the definition of support cones and construction of Rickard idempotents. At the same time, many technical details closely mimic those provided in [12] and are simply omitted here.

The author gratefully acknowledges Andrei Suslin for providing the key idea in the proof of Proposition 2.1 and Eric Friedlander for many useful discussions and his interest in the subject.

Throughout the paper k will denote an algebraically closed field of positive characteristic p. All group schemes are assumed to be defined over k.

1. Support cones for Frobenius kernels

In this section we recall various definitions and results concerning infinitesimal group schemes, leading up to the construction of support cones.

Definition 1.1. A finite group scheme G over k is a functor $G : \{\text{k - alg}\} \to \{\text{groups}\}$ from the category of finitely generated commutative k-algebras to the category of groups which is represented by a finite-dimensional commutative k-algebra (denoted $k[G]$). A finite group scheme G is infinitesimal if $k[G]$ is a local ring.

Let I be the augmentation ideal of the coordinate algebra $k[G]$ of an infinitesimal group G. The *height* of G is the minimal integer r such that for any $x \in I$, $x^{p^r} = 0$.

Example 1.2. Let G be an affine algebraic group. We denote by $G_{(1)}$ the scheme-theoretic kernel of the Frobenius map

$$G_{(1)} = \ker\{F : G \to G^{(1)}\},$$

where $G^{(1)}$ is the base change of G via the Frobenius map (i.e., the p-th power map) on k. For example, $GL_{n(1)}$ is the group scheme given by $GL_{n(1)}(A) = \{(a_{ij}) \in M_n(A) : a_{ij}^p = \delta_{ij} \text{ for all } 1 \le i,j \le n\}$ for any finitely generated commutative k-algebra A. The r-th Frobenius kernel of G, denoted $G_{(r)}$, is the scheme-theoretic kernel of $F^r : G \to G^{(r)}$. The height of $G_{(r)}$ is precisely r. Moreover, any infinitesimal group scheme of height r can be embedded into the r-th Frobenius kernel of GL_n for an appropriate n.

Example 1.3. For the additive group $\mathbb{G}_{(a)}$, we have $k[\mathbb{G}_{a(r)}] = k[T]/T^{p^r}$. We fix notation for the dual algebra $k[\mathbb{G}_{a(r)}]^\#$ which will be used later in the text. Let $v_0, \dots, v_{p^r-1}$ be the basis of $k[\mathbb{G}_{a(r)}]^\# = (k[T]/T^{p^r})^\#$ dual to the basis of $k[T]/T^{p^r}$ consisting of powers of T. Denote v_{p^i} by u_i. Then

$$k[\mathbb{G}_{a(r)}]^\# = k[u_0, \dots, u_{r-1}]/(u_0^p, \dots, u_{r-1}^p).$$

A *1-parameter subgroup* of height r of an affine group scheme G is a homomorphism $\mathbb{G}_{a(r)} \to G$. We say that a 1-parameter subgroup is injective if this homomorphism is a closed embedding of group schemes.

For a field extension K/k we shall use the subscript $_K$ to denote the extension of scalars from k to K.

1-parameter subgroups constitute a detecting family of small subgroups for an infinitesimal group, analogous to the family of "shifted cyclic subgroups" of an elementary abelian p-group. We make this detecting property precise in the following theorem, which will be generalized to all infinitesimal group schemes in the next section.

Theorem 1.4. [12, 1.7] *Let $G_{(r)}$ be the r-th Frobenius kernel of an algebraic group G and let M be a $G_{(r)}$-module such that for any field extension K/k and any subgroup scheme $H_K \hookrightarrow G_{(r),K}$ isomorphic to $\mathbb{G}_{a(s),K}$, $s \leq r$, the restriction of M_K to H_K is projective. Then M is projective as a $G_{(r)}$-module.*

Remark 1.5. Considering field extensions is essential here when the module in question is allowed to be infinite dimensional. We refer the reader to [4] or [12] for examples of modules which are projective restricted to every 1-parameter subgroup (or cyclic shifted subgroup in the case of a module for a finite group) defined over the ground field k, but not projective as $G_{(r)}$-modules.

Following [15], we define a functor

$$V_r(G) : (\text{comm } k\text{-alg}) \to (\text{sets})$$

by setting

$$V_r(G)(A) = \text{Hom}_{Gr/A}(\mathbb{G}_{a(r)} \otimes_k A, G \otimes_k A).$$

This functor is representable by an affine scheme of finite type over k, which we will still denote $V_r(G)$. Indeed, the following holds:

Theorem 1.6. [15, 1.5] *The functor $V_r(G)$ is represented by an affine scheme of finite type over k. Moreover, $G \to V_r(G)$ is a covariant functor from the category of affine group schemes over k of height $\leq r$ to the category of affine schemes of finite type over k, which takes closed embeddings to closed embeddings.*

Thus, a point $s \in V_r(G)$ defines a canonical $k(s)$-rational point of $V_r(G)$ and the associated 1-parameter subgroup defined over $k(s)$:

$$\nu_s : \mathbb{G}_{a(r),k(s)} \to G_{k(s)}.$$

For G an infinitesimal group scheme of height r, we will write $V(G) = V_r(G)$. We can now define *support cone* of a G-module.

Definition 1.7. Let G be an infinitesimal k-group scheme of height r and let M be a G-module. The *support cone* of M is the following subset of $V(G)$:

$$V(G)_M = \{s \in V(G) : M_{k(s)} \text{ is not projective as a module for the subalgebra}$$
$$k(s)[u_{r-1}]/(u_{r-1}^p) \subset k(s)[u_0, \ldots, u_{r-1}]/(u_0^p, \ldots, u_{r-1}^p) = k(s)[\mathbb{G}_{a(r)}]^{\#}\}.$$

We remark that by a "subset" of an affine scheme $X = \operatorname{Spec} A$ we mean simply a set of prime ideals in A. We shall often use the same notation for a point in X and the corresponding prime ideal in A. The algebra $k[V(G)]$ is graded connected, thus, there is a well-defined map $V(G) - \{0\} \xrightarrow{\pi} \operatorname{Proj} V(G)$, where $\operatorname{Proj} V(G)$ denotes the projective spectrum of $k[V(G)]$. We call a subset of $V(G)$ *conical* if it coincides with a full preimage of a subset in $\operatorname{Proj} V(G)$ with added $\{0\}$.

Definition 1.7 was first introduced in [16] for finite dimensional modules where it was further shown that $V(G)_M$ is a closed subvariety of $V(G)$ and, furthermore, is naturally homeomorphic to the cohomological support variety of M, i.e. the variety of the ideal $\mathrm{Ann}_{H^{ev}(G,k)}(\mathrm{Ext}^*_G(M,M))$ in $\mathrm{Specm}\, H^{ev}(G,k)$ (respectively $\mathrm{Specm}\, H^*(G,k)$ if $p=2$), the maximal ideal spectrum of the cohomology algebra of G. In particular, $V(G)$ is naturally identified with $\mathrm{Spec}\, H^{ev}(G,k)$.

It is shown in [12] that support cones for Frobenius kernels satisfy most of the standard properties of support varieties, except for being closed.

Theorem 1.8. [12, 2.6] *Let $G_{(r)}$ be the r-th Frobenius kernel of an algebraic group G and M and N be $G_{(r)}$-modules. Support cones satisfy the following properties:*

(1) *$V(G)_M$ is a conical subset of $V(G)$.*
(2) *"Naturality." Let $f : H \to G$ be a homomorphism of infinitesimal group schemes of height $\leq r$. Denote by $f_* : V(H) \to V(G)$ the associated morphism of schemes. Then*
$$f_*^{-1}(V(G)_M) = V(H)_M,$$
where M is considered as an H-module via f.
(3) *$V(G)_M = 0$ if and only if M is projective.*
(4) *"Tensor product property." $V(G)_{(M\otimes N)} = V(G)_M \cap V(G)_N$.*
(5) *Let $0 \to M_1 \to M_2 \to M_3 \to 0$ be a short exact sequence of G-modules. Then for any permutation (ijk) of (123) we have*
$$V(G)_{M_i} \subset V(G)_{M_j} \cup V(G)_{M_k}.$$
(6) *For any collection of G-modules $\{M_\alpha\}$, we have*
$$V(G)_{\bigoplus_\alpha M_\alpha} = \bigcup_\alpha V(G)_{M_\alpha}.$$

2. Detection of projectivity.

In this section we prove that for any infinitesimal group scheme G and any G-module M, projectivity of M can be detected via restricting to 1-parameter subgroups of G, building upon Theorem 1.4.

Proposition 2.1. *Let G be an infinitesimal group scheme and let M be a G-module such that for any field extension K/k and any subgroup scheme $H_K \hookrightarrow G_K$ isomorphic to $\mathbb{G}_{a(r),K}$, the restriction of M_K to H_K is projective. Let $G \hookrightarrow G'$ be a closed embedding of G into some Frobenius kernel of the same height as G. Then $\mathrm{Ind}_G^{G'}(M)$ is projective as a G'-module.*

Proof. By Theorem 1.8.3, it suffices to show that $V(G')_{\mathrm{Ind}_G^{G'}(M)} = 0$. Let s be a point in $V(G')$, corresponding to a 1-parameter subgroup $\nu_s : \mathbb{G}_{a(r),K} \to G'_K$. By definition of the support cone, to show that $s \notin V(G')_{\mathrm{Ind}_G^{G'}(M)}$, we have to show that the restriction of $(\mathrm{Ind}_G^{G'}(M))_K$ to $K[u_{r-1}]/(u_{r-1}^p) \subset K[u_0,\ldots,u_{r-1}]/(u_0^p,\ldots,u_{r-1}^p) = K[\mathbb{G}_{a(r)}]^\#$ via ν_s is projective. Since Ind commutes with extension of scalars, we may assume that everything is defined over the ground field k, i.e. $K=k$. By lowering the height of $\mathbb{G}_{a(r)}$, if necessary, we can further assume that the map ν_s is an embedding. This will involve factoring through the "projection" map $p_{r,r'} : \mathbb{G}_{a(r)} \to \mathbb{G}_{a(r')}$, which takes generator u_{r-1} of $k[\mathbb{G}_{a(r)}]^\#$ to the generator $u_{r'-1}$ of $k[\mathbb{G}_{a(r')}]^\#$, and, thus, will not affect the projectivity of the restriction of $\mathrm{Ind}_G^{G'}(M)$ to the corresponding subalgebra.

Let $\mathbb{G}_{a(t)}$ be a 1-parameter subgroup of G defined as $G \cap \mathbb{G}_{a(r)} \subset G'$. Consider the following Cartesian square of group schemes:

$$
\begin{array}{ccc}
\mathbb{G}_{a(t)} & \xrightarrow{\hat{A}\ \ddot{A}} & G\ddot{A} \\[2pt]
\Big\downarrow{\scriptstyle p} & & \Big\downarrow{\scriptstyle p} \\[2pt]
\mathbb{G}_{a(r)} & \xrightarrow{\hat{A}\,\ddot{A}} & G'
\end{array}
$$

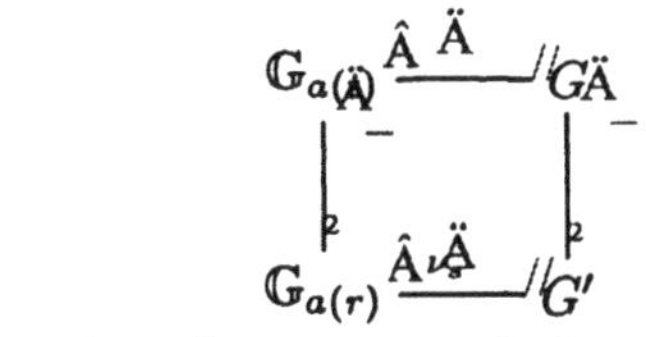

Let $\Lambda = \mathrm{End}_k(M, M)$. Then Λ is an associative unital G-algebra. Observe that $H^1(\mathbb{G}_{a(t)}, \Lambda) = \mathrm{Ext}^1_{\mathbb{G}_{a(t)}}(M, M) = 0$, since the restriction of M to any injective 1-parameter subgroup of G is projective. Since $\mathbb{G}_{a(t)}$ is unipotent, vanishing of the first cohomology group implies that Λ is projective as a $\mathbb{G}_{a(t)}$-module. Thus, $\mathrm{Ind}^{\mathbb{G}_{a(r)}}_{\mathbb{G}_{a(t)}}(\Lambda)$, which is again an associative unital $\mathbb{G}_{a(r)}$ - algebra, is projective as a $\mathbb{G}_{a(r)}$-module.

The natural map of $\mathbb{G}_{a(r)}$-algebras $\mathrm{Ind}^{G'}_G(\Lambda) \to \mathrm{Ind}^{\mathbb{G}_{a(r)}}_{\mathbb{G}_{a(t)}}(\Lambda)$ (determined by the adjointness of Induction and Restriction functors) is surjective and has a nilpotent kernel (cf. [16, 4.3]). Denote the kernel by I. Projectivity of $\mathrm{Ind}^{\mathbb{G}_{a(r)}}_{\mathbb{G}_{a(t)}}(\Lambda)$ as a $\mathbb{G}_{a(r)}$-module implies that it is projective restricted further to $k[u_{r-1}]/(u^p_{r-1})$. Thus,

$$
H^*(k[u_{r-1}]/(u^p_{r-1}), \mathrm{Ind}^{\mathbb{G}_{a(r)}}_{\mathbb{G}_{a(t)}}(\Lambda)) = 0 \text{ for } * > 0.
$$

Therefore, the long exact sequence in cohomology corresponding to the short exact sequence

$$
0 \to I \to \mathrm{Ind}^{G'}_G(\Lambda) \to \mathrm{Ind}^{\mathbb{G}_{a(r)}}_{\mathbb{G}_{a(t)}}(\Lambda) \to 0
$$

of modules gives an isomorphism

$$
H^*(k[u_{r-1}]/(u^p_{r-1}), \mathrm{Ind}^{G'}_G(\Lambda)) \cong H^*(k[u_{r-1}]/(u^p_{r-1}), I)
$$

in positive degrees. The ideal I is nilpotent, so the algebra without unit $H^*(k[u_{r-1}]/(u^p_{r-1}), I)$ is also nilpotent. The isomorphism above implies that the augmentation ideal $H^{*>0}(k[u_{r-1}]/(u^p_{r-1}), \mathrm{Ind}^{G'}_G(\Lambda))$ is nilpotent. On the other hand, the map of algebras $k \to \mathrm{Ind}^{G'}_G(\Lambda)$ induces an action of $H^{ev}(k[u_{r-1}]/(u^p_{r-1}), k) \cong k[x]$ (where x is a generator in degree two) on $H^*(k[u_{r-1}]/(u^p_{r-1}), \mathrm{Ind}^{G'}_G(\Lambda))$. The action of x, in particular, induces a periodicity isomorphism

$$
H^i(k[u_{r-1}]/(u^p_{r-1}), \mathrm{Ind}^{G'}_G(\Lambda)) \cong H^{i+2}(k[u_{r-1}]/(u^p_{r-1}), \mathrm{Ind}^{G'}_G(\Lambda)) \text{ for all } i > 0
$$

Since image of x in $H^2(k[u_{r-1}]/(u^p_{r-1}), \mathrm{Ind}^{G'}_G(\Lambda))$ under the map of algebras $H^{ev}(k[u_{r-1}]/(u^p_{r-1}), k) \to H^*(k[u_{r-1}]/(u^p_{r-1}), \mathrm{Ind}^{G'}_G(\Lambda))$ is nilpotent, the periodicity isomorphism induced by the action of x is trivial. Hence,

$$
H^{*>0}(k[u_{r-1}]/(u^p_{r-1}), \mathrm{Ind}^{G'}_G(\Lambda)) = 0.
$$

There is a natural action of Λ on M compatible with the G-structure. This induces an action of $\mathrm{Ind}^{G'}_G(\Lambda)$ on $\mathrm{Ind}^{G'}_G(M)$ compatible with their structure as G'-modules, and, therefore, $k[u_{r-1}]/(u^p_{r-1})$-modules. Hence, the action of $H^{ev}(k[u_{r-1}]/(u^p_{r-1}), k) \cong k[x]$ on $H^*(k[u_{r-1}]/(u^p_{r-1}), \mathrm{Ind}^{G'}_G(M))$ factors through the action of $H^*(k[u_{r-1}]/(u^p_{r-1}), \mathrm{Ind}^{G'}_G(\Lambda))$. Since the latter vanishes in positive degrees, the action of x on $H^*(k[u_{r-1}]/(u^p_{r-1}), \mathrm{Ind}^{G'}_G(M))$ is trivial. On the other hand, it induces a periodicity isomorphism. We conclude that $H^1(k[u_{r-1}]/(u_{r-1}^p), \mathrm{Ind}_G^{G'}(M)) = 0$ and, hence, $\mathrm{Ind}_G^{G'}(M)$ is projective as a $k[u_{r-1}]/(u_{r-1}^p)$-module. Hence, $s \notin V(G)_{\mathrm{Ind}_G^{G'}(M)}$. The statement follows.

□

Theorem 2.2. *Let G be an infinitesimal group scheme and let M be a G-module such that for any field extension K/k and any subgroup scheme $H_K \hookrightarrow G_K$ isomorphic to $\mathbb{G}_{a(r),K}$ the restriction of M_K to H_K is projective. Then M is projective as a G-module.*

Proof. Embed G into some Frobenius kernel G'. By Proposition 2.1, $\mathrm{Ind}_G^{G'}(M)$ is a projective G'-module. Therefore, $H^*(G, M) = H^*(G', \mathrm{Ind}_G^{G'}(M)) = 0$ for $* > 0$. Applying the same argument to all modules of the form $M \otimes S^{\#}$ for all simple G-modules S, we get $\mathrm{Ext}_G^*(S, M) = 0$ for $* > 0$. Applying Lemma 1.2 in [12], we conclude that M is projective.

□

Theorem 2.3. *Let G be an infinitesimal group scheme, and let M, N be G-modules. Support cones $V(G)_M, V(G)_N$ satisfy properties (1)-(6) of Theorem 1.8.*

We omit the proof of this theorem since (1),(2) and (4)-(6) were proved in [12] and (3) is a straightforward application of Theorem 2.2.

3. Rickard idempotents.

Theorem 2.2 allows us to extend the description of support cones in terms of Rickard idempotents given in [12] for Frobenius kernels to any infinitesimal group scheme. We begin by briefly recalling the notion of Rickard idempotent modules ([13]) and then state Theorem 3.3 which provides an alternative description of the support cones. This approach to supports of infinite dimensional modules for finite groups was introduced by Benson, Carlson and Rickard in [4] and it works equally well in our context of infinitesimal group schemes. This section does not have any proofs since the existence of Rickard idempotent modules is a general statement about Bousfield localization (cf. [13], [14] or [10]) and the proof of Theorem 3.3 goes exactly as in the case of Frobenius kernels which is presented in [12].

We shall denote by StMod(G) the stable category of all G-modules. Recall that objects of StMod(G) are G-modules and maps are equivalence classes of G-module homomorphisms where two maps are equivalent if their difference factors through a projective G-module.

The fact that in the category of G-modules projectives are injectives and vice versa (cf. [9]) implies the existence of a triangulated structure on StMod(G). The shift operator in StMod(G) is given by the Heller operator Ω^{-1} : StMod(G) $\rightarrow$ StMod(G) (cf., for example, [3] for the definition of Ω) and distinguished triangles come from short exact sequences in Mod(G).

We shall denote by stmod(G) the full triangulated subcategory of StMod(G) whose objects are represented by finite dimensional modules. A full triangulated subcategory C of stmod(G) (respectively StMod(G)) is called *thick* if it is closed under taking direct summands (respectively taking direct summands and arbitrary direct sums). It is called *tensor-ideal* if it is closed under taking tensor products with any G-module. We shall use the notation <u>Hom</u> for $\mathrm{Hom}_{\mathrm{StMod}}$ and "$\cong$" for stable isomorphisms.

Two modules are stably isomorphic (i.e. isomorphic in StMod(G)) if and only if they become isomorphic after adding projective summands to them. This implies that support cones are well-defined in StMod(G).

Let $\mathcal{C}$ be a thick subcategory of stmod(G). Denote by $\vec{\mathcal{C}}$ the full triangulated subcategory of StMod(G) whose objects are filtered colimits of objects in $\mathcal{C}$. ($\vec{\mathcal{C}}$ coincides with the smallest full triangulated subcategory of StMod(G) which contains $\mathcal{C}$ and is closed under taking direct summands and arbitrary direct sums (cf. [13]).)

The following theorem introduces the universal modules $E(W)$ and $F(W)$ and establishes some of their properties.

Theorem 3.1. *Let W be a subset of $V(G)$ and let $\mathcal{C}_W$ be the subcategory of stmod(G) consisting of all finite dimensional modules M such that $V(G)_M \subset W$. Then*

(1) *$\mathcal{C}_W$ is a tensor-ideal thick subcategory of stmod(G).*

(2) *There exists a distinguished triangle*

$$T(W) : E(W) \xrightarrow{\epsilon} k \xrightarrow{\eta} F(W) \to \Omega^{-1}E(W)$$

in StMod(G), satisfying the following universal properties for any G-module M:

(i) $E(W) \otimes M \in \vec{\mathcal{C}_W}$;

(ii) the map $\epsilon \otimes id_M$ is the universal map in StMod(G) from an object in $\vec{\mathcal{C}_W}$ to M, i.e. for any $C \in \vec{\mathcal{C}_W}$, $\epsilon \otimes id_M$ induces an isomorphism

$$\underline{Hom}(C, E(W) \otimes M) \simeq \underline{Hom}(C, M);$$

(iii) the map $\eta \otimes id_M : M \to F(W) \otimes M$ is the universal map in StMod(G) from M to a $\mathcal{C}_W$-local object (where N is called a $\mathcal{C}_W$-local object iff $\underline{Hom}(M, N) = 0$ for any $M \in \mathcal{C}_W$).

(3) *There are stable isomorphisms:*

$$E(W) \otimes E(W) \cong E(W) \text{ and } F(W) \otimes F(W) \cong F(W)$$

and $E(W) \otimes F(W)$ is projective;

(4) *For a G-module M, the following are equivalent:*

- *$M \in \vec{\mathcal{C}_W}$*
- *$M \otimes E(W)$ is stably isomorphic to M*
- *$M \otimes F(W)$ is projective.*

The modules $E(W)$ and $F(W)$ were introduced by J. Rickard ([13]) for finite groups and are thereby called *Rickard idempotent modules.* It is not hard to see that the universal properties (2) determine $E(W), F(W)$ uniquely up to a stable isomorphism.

Let V be a closed conical subset of $V(G)$. Denote by V' the subset of V consisting of all points of V except for generic points of irreducible components of V. Define

$$\kappa(V) \stackrel{def}{=} E(V) \otimes F(V').$$

As a tensor product of idempotent modules, $\kappa(V)$ is again idempotent, i.e. $\kappa(V) \otimes \kappa(V) \cong \kappa(V)$.

Note that the generic point of an irreducible closed conical subvariety is a homogeneous prime ideal, so that there is a natural 1-1 correspondence between homogeneous prime ideals of $k[V(G)]$ and closed irreducible conical subvarieties of $V(G)$.

For an irreducible closed conical set V with the generic point s we shall use $\kappa(s)$ to denote $\kappa(V)$.

We conclude the review of the properties of Rickard idempotents with the following lemma establishing their good behaviour with respect to restriction to a subgroup scheme (cf. [4], [12]).

Lemma 3.2. *Let G be an infinitesimal group scheme, H be a closed subgroup scheme of G and W be a subset of $V(G)$. Let $i_* : V(H) \hookrightarrow V(G)$ be the embedding of schemes induced by the inclusion $i : H \hookrightarrow G$. Then the following two distinguished triangles in StMod(H) are stably isomorphic:*

$$T(i_*^{-1}(W)) : E(i_*^{-1}(W)) \to k \to F(i_*^{-1}(W)) \to \Omega^{-1}E(i_*^{-1}(W))$$

and

$$T(W)\downarrow_H : E(W)\downarrow_H \to k \to F(W)\downarrow_H \to \Omega^{-1}E(W)\downarrow_H .$$

For a conical subset W in $V(G)$ we denote by **Proj W**, the "projectivization" of W, the set of points in W which correspond to homogeneous prime ideals of $k[V(G)]$ excluding the augmentation ideal. Proj W can be viewed as a subset of the scheme Proj $k[V(G)]$. There is 1-1 correspondence between conical subsets of $V(G)$ and their "projectivizations", i.e. a conical subset is completely determined by its homogeneous ideals. Therefore, the standard properties of support cones, described in Theorem 1.8, apply to their "projectivizations".

The proof of the following theorem can be found in [12].

Theorem 3.3. *Let G be an infinitesimal group scheme.*

(1) *Let s be a point in $V(G)$ corresponding to a homogeneous prime ideal. Then $\mathrm{Proj}\, V(G)_{\kappa(s)} = \{s\}$.*

(2) *Let M be a G-module. Then*

$$\mathrm{Proj}\ V(G)_M = \{s \in \mathrm{Proj}\ V(G) : M \otimes \kappa(s) \textit{ is not projective as a } G\textit{-module}\}.$$

The following "realization" statement is an immediate application of the first part of Theorem 3.3 and Theorem 2.3.6.

Corollary 3.4. *Let W be any subset of $\mathrm{Proj}\, V(G)$. Then there exists a G-module M such that $\mathrm{Proj}\, V(G)_M = W$.*

4. Support Cones and Induction.

A group scheme is called unipotent if it can be embedded in $U_N \subset GL_N$, the subgroup scheme of upper triangular matrices in the general linear group. Recall that a module M of a unipotent group scheme U is injective if and only if $H^1(U, M) = 0$.

Proposition 4.1. *Let $i : H \hookrightarrow G$ be a closed embedding of infinitesimal group schemes and assume further that H is unipotent. Then for any H-module M,*

$$V(G)_{\mathrm{Ind}_H^G(M)} = V(H)_M.$$

Proof. Since H is unipotent, the isomorphism $H^*(G, \mathrm{Ind}_H^G(N)) \simeq H^*(H, N)$ implies that projectivity of $\mathrm{Ind}_H^G(N)$ as a G-module implies projectivity of N as an H-module. Conversely, since Induction takes injectives to injectives, projectivity of N implies projectivity of $\mathrm{Ind}_H^G(N)$. Thus, for any H-module N, N is projective if and only if $\mathrm{Ind}_H^G(N)$ is projective.

Now we can prove the equality of support cones in four easy steps: $s \in V_G(Ind_H^G M) \overset{Th.\,3.3}{\Longleftrightarrow} Ind_H^G M \otimes \kappa(s)$ is not projective $\Longleftrightarrow Ind_H^G(M \otimes \kappa(s))$ is not projective (tensor identity and Lemma 3.2) $\Longleftrightarrow M \otimes \kappa(s)$ is not projective over $H \overset{Th.\,3.3}{\Longleftrightarrow} s \in V_H(M)$.

□

To proceed, we need the following algebraic lemma.

Lemma 4.2. *Let A be a regular ring of finite Krull dimension d, $k(\mu) = A/\mu A$ be the residue field of A at a prime ideal μ, and $J^\bullet$ be a cochain complex of flat A-modules acyclic in positive degrees. Then $J^\bullet \otimes_A k(\mu)$ is also acyclic in positive degrees.*

Proof. We proceed by induction on $d = \dim A$.

Since $J^\bullet \otimes_A k(\mu) \simeq J^\bullet_\mu \otimes_{A_\mu} k(\mu)$ and localization is exact, it suffices to assume that A is a regular local ring with maximal ideal μ.

Let $d = 1$. Then A is a discrete valuation ring. Denote by π a generator of the maximal ideal of A. Since $J^\bullet$ is flat, tensoring $J^\bullet$ with the short exact sequence $0 \to A \to A \to A/\pi A \to 0$ gives a long exact sequence of complexes

$$0 \to J^\bullet \to J^\bullet \to J^\bullet \otimes_A k((\pi)) \to 0$$

and, thus, a long exact sequence in cohomology

$$\ldots H^{n-1}(J^\bullet \otimes_A k((\pi))) \to H^n(J^\bullet) \to H^n(J^\bullet) \to H^n(J^\bullet \otimes_A k((\pi))) \to \ldots.$$

Since $H^n(J^\bullet) = 0$ for $n > 0$, we conclude that $H^n(J^\bullet \otimes_A k((\pi))) = 0$ for $n > 0$.

Let A be a regular local ring of dimension d. Since A is regular, we can find an element t in the maximal ideal of A such that A/tA is a regular ring of dimension strictly less than dimension of A. Applying the same argument as above with $\pi = t$, we conclude that $J^\bullet/tJ^\bullet$ is acyclic in positive degrees. Since tensoring preserves flatness (cf. [6, 6.6a]), $J^\bullet/tJ^\bullet = J^\bullet \otimes_A A/tA$ is a cochain complex of flat A/tA-modules. Applying induction hypothesis, we conclude that

$$J^\bullet \otimes_A k(\mu) = J^\bullet/tJ^\bullet \otimes_{A/tA} k(\mu/t\mu)$$

is acyclic in positive degrees. □

Let G be an algebraic group over k. The action of G on $G_{(r)}$ by conjugation induces a natural action on the scheme $V(G_{(r)})$. For a subset W in $V(G_{(r)})$, we denote by $G \cdot W$ the G-orbit of W in $V(G_{(r)})$. The following result is a refinement of Proposition 1.4 in [12].

Proposition 4.3. *Let G be a connected reductive algebraic group, and M be a $G_{(r)}$-module. Then $V(G_{(r)})_{Ind_{G_{(r)}}^G(M)} = G \cdot V(G_{(r)})_M$.*

Proof. Let K/k be a field extension and $H_K \to G_{(r),K}$ be a 1-parameter subgroup. Since M is a G-module, the support cone of M is stable under the action of G. Hence, to prove the theorem it suffices to check the following:

(I). If M_K restricted to H_K is not projective, then so is $(\mathrm{Ind}_{G_{(r)}}^G(M))_K \simeq \mathrm{Ind}_{G_{(r),K}}^{G_K}(M_K)$.

(II). If M_K is projective restricted to all conjugates of H_K under the action of $G(K)$, then $(\mathrm{Ind}_{G_{(r)}}^G(M))_K$ is projective as an H_K-module.

Since Induction commutes with extension of scalars, we can assume that $K = k$ in both cases listed above. By taking the image of H in $G_{(r)}$, we can also assume that $H \to G_{(r)}$ is an embedding.

Let $M \to I^\bullet$ be the standard $G_{(r)}$-injective resolution of M and let $J^\bullet = (\mathrm{Ind}_{G_{(r)}}^G I^\bullet)^H$. We have $H^*(H, \mathrm{Ind}_{G_{(r)}}^G M) = H^*(J^\bullet)$. The complex $J^\bullet$ is naturally a complex of flat $k[G^{(r)}] = k[G/G_{(r)}]$-modules and, moreover, for any $g \in G$ there is an isomorphism:

$$J^\bullet \otimes_{k[G^{(r)}]} k(g) \cong (I^\bullet \otimes k(g))^{g^{-1}(H \otimes k(g))g}. \qquad (*)$$

(cf. [12], p.6)

Suppose $M \downarrow_H$ is not projective. Then $J^\bullet \otimes_{k[G^{(r)}]} k \cong (I^\bullet)^H$ has non-trivial cohomology in positive degrees. Since the scheme $G^{(r)}$ is smooth, the coordinate ring $k[G^{(r)}]$ is regular, and, thus, applying Lemma 4.2, we conclude that $J^\bullet$ is not acyclic in positive degrees. Thus, $\mathrm{Ind}_{G_{(r)}}^G M$ has non-trivial cohomology in positive degrees which implies that it is not projective. We have then proved (I).

In the case described in (II), the same isomorphism (*) shows that $J^\bullet \otimes_{k[G^{(r)}]} k(g)$ is acyclic in positive degrees for all $g \in G$. Since the projection $F^r : G \to G/G_{(r)} \simeq G^{(r)}$ is a bijection on points, and the extension of scalars from $k(F^r(g))$ to $k(g)$ gives an injective map in cohomology of $J^\bullet \otimes_{k[G^{(r)}]} k(F^r(g))$, we get that for any $x \in G^{(r)}$, $J^\bullet \otimes_{k[G^{(r)}]} k(x)$ has trivial cohomology in positive degrees. Lemma 1.3 in [12] now implies that $J^\bullet$ is acyclic and, thus, $\mathrm{Ind}_{G_{(r)}}^G M$ is injective (and, hence, projective) as an H-module. We have now verified (II), which completes the proof. □

Corollary 4.4. *Let G be a connected reductive algebraic group.*

1. *For any $s \in \mathrm{Proj}V(G_{(r)})$, there exists a G-rational module M such that $\mathrm{Proj}V(G_{(r)})_M = G \cdot s$.*
2. *For any conical subset W of $V(G_{(r)})$, there exists a G-rational module M such that $V(G_{(r)})_M = G \cdot W$. In particular, any conical subset stable under the G-action can be realized as a support cone of a rational G-module.*

Proof. Proposition 4.3 and Theorem 3.3.1 immediately imply that the module $M = \mathrm{Ind}_{G_{(r)}}^G(\kappa(s))$ has the desired support cone. The second statement now follows by applying Theorem 2.3.6. □

For a k-rational point $s \in V(G_{(r)})$, we denote by L_s the line through s in $V(G)$. By using Proposition 4.1, we can realize the orbit of $L_s \subset V(G_{(r)})$ in a more explicit way than the one described in Corollary 4.4.

Corollary 4.5. *Let G be a connected reductive algebraic group and s be a k-rational point of $V(G_{(r)})$. Let further $\nu_s : \mathbb{G}_{a(r)} \to G_{(r)}$ be the 1-parameter subgroup corresponding to s and $M_s = k[u_0, \dots u_{r-2}]/(u_0^p, \dots u_{r-2}^p)$, where we identify $k[\mathbb{G}_{a(r)}]^\#$ with $k[u_0, \dots u_{r-1}]/(u_0^p, \dots u_{r-1}^p)$. Then M_s has a natural structure of a $\mathbb{G}_{a(r)}$-module as a quotient of $k[\mathbb{G}_{a(r)}]^\#$. We have*

$$V(G_{(r)})_{Ind_{\mathbb{G}_{a(r)}}^G(M_s)} = G \cdot L_s.$$

Proof. It is immediate from the definition that the support cone of M_s as a $\mathbb{G}_{a(r)}$-module is the line in $V(\mathbb{G}_{a(r)}) \simeq \mathbb{A}^r$ through the origin and the point corresponding to the 1-parameter subgroup $id : \mathbb{G}_{a(r)} \to \mathbb{G}_{a(r)}$. This line maps to L_s under $\nu_{s,*} : V(\mathbb{G}_{a(r)}) \to V(G_{(r)})$. Thus, by Proposition 4.1, $V(G_{(r)})_{\mathrm{Ind}_{\mathbb{G}_{a(r)}}^{G_{(r)}} M_s} = L_s$. The statement now follows from Proposition 4.3 and transitivity of induction.

□

Remark 4.6. For $r = 1$, the category of $G_{(1)}$ modules is equivalent to the category of restricted $g = \mathrm{Lie}\, G$-modules and the support variety $V(G)_{(1)}$ can be identified with the restricted nullcone $\mathcal{N}_p(g)$ of g ([8],[16]). Then the corollary above implies that for any $x \in \mathcal{N}_p(g)$ we can realize the orbit of kx, $G \cdot kx \subset \mathcal{N}_p(g)$, as the support cone of $\mathrm{Ind}_{\mathbb{G}_{a(1)}}^{G} k$, where $\mathbb{G}_{a(1)} \to G_{(1)} \to G$ is the 1-parameter subgroup corresponding to x.

REFERENCES

[1] J.L. Alperin, L. Evens, *Representations, resolutions, and Quillen's dimension theorem*, J. Pure & Applied Algebra 22 (1981) 1-9.

[2] G. Avrunin, L. Scott, *Quillen stratification for modules*, Inventiones Math. 66 (1982) 277-286.

[3] D.J. Benson, *Representations and cohomology*, Volume I and II, Cambridge University Press, (1991).

[4] D.J. Benson, J.F. Carlson, J. Rickard, *Complexity and varieties for infinitely generated modules II*, Math. Proc. Camb. Phil. Soc. 120 (1996) 597-615.

[5] J. Carlson, *The varieties and cohomology ring of a module*, J. of Algebra 85 (1983) 104-143.

[6] D. Eisenbud, *Commutative Algebra with a View Towards Algebraic Geometry*, (1995) Springer-Verlag, New York.

[7] E. Friedlander, B. Parshall, *Geometry of p-unipotent Lie Algebras*, J. Algebra 109 (1987) 25-45.

[8] E. Friedlander, B. Parshall, *Support varieties for restricted Lie algebras*, Invent. Math. 86 (1986) 553-562.

[9] C.G. Faith, E.A. Walker, *Direct sum representations of injective modules*, J. of Algebra 5 (1967) 203-221.

[10] M. Hovey, J. Palmieri, N. Strickland, *Axiomatic stable homotopy theory*, Mem. Amer. Math. Soc. 128 (1997) no 610, x+114.

[11] J.C. Jantzen, *Kohomologie von p-Lie Algebren und nilpotente Elemente*, Abh. Math. Sem. Univ. Hamburg 56 (1986) 191-219.

[12] J. Pevtsova, *Infinite dimensional modules for Frobenius kernels*, J. Pure & Applied Algebra 173 (2002) 59-86.

[13] J. Rickard, *Idempotent modules in the stable category*, J. London Math. Society (2), 56 (1997) no. 1, 149-170.

[14] J. Rickard, *Bousfield localization for representation theorists*, Trends. Math. (2000) 273-283, Birkhäuser, Basel.

[15] A. Suslin, E. Friedlander, C. Bendel, *Infinitesimal 1-parameter subgroups and cohomology*, J. Amer. Math. Soc. 10 (1997) 693-728.

[16] A. Suslin, E. Friedlander, C. Bendel, *Support varieties for infinitesimal group schemes*, J. Amer. Math. Soc. 10 (1997) 729-759.

Coalgebras from Formulas

Serban Raianu
California State University Dominguez Hills
Department of Mathematics
1000 E Victoria St
Carson, CA 90747
e-mail:sraianu@csudh.edu

Abstract

Nichols and Sweedler showed in [5] that generic formulas for sums may be used for producing examples of coalgebras. We adopt a slightly different point of view, and show that the reason why all these constructions work is the presence of certain representative functions on some (semi)group. In particular, the indeterminate in a polynomial ring is a primitive element because the identity function is representative.

Introduction

The title of this note is borrowed from the title of the second section of [5]. There it is explained how each generic addition formula naturally gives a formula for the action of the comultiplication in a coalgebra. Among the examples chosen in [5], this situation is probably best illustrated by the following two:
Let C be a k-space with basis $\{s, c\}$. We define $\Delta : C \longrightarrow C \otimes C$ and $\varepsilon : C \longrightarrow k$ by

$$\begin{aligned} \Delta(s) &= s \otimes c + c \otimes s \\ \Delta(c) &= c \otimes c - s \otimes s \\ \varepsilon(s) &= 0 \\ \varepsilon(c) &= 1. \end{aligned}$$

Then (C, Δ, ε) is a coalgebra called the trigonometric coalgebra.
Now let H be a k-vector space with basis $\{c_m \mid m \in \mathbf{N}\}$. Then H is a coalgebra with comultiplication Δ and counit ε defined by

$$\Delta(c_m) = \sum_{i=0,m} c_i \otimes c_{m-i}, \quad \varepsilon(c_m) = \delta_{0,m}.$$

This coalgebra is called the divided power coalgebra.
Identifying the "formulas" in the above examples is not hard: the formulas for sin and cos applied to a sum in the first example, and the binomial formula in the second one. Nevertheless, some more questions still need to be answered: Can one associate a coalgebra to any formula? If not, how can one characterize formulas leading to coalgebra structures? Do all formulas of that kind need to be formulas for sums? If not, are the formulas for sums special in any way?
The aim of this note is to answer all these questions. First, we remark that a formula will define a coalgebra structure precisely when it represents an equality showing that a certain function is a "representative function". The easiest, and perhaps the most striking example, is the coalgebra (or bialgebra) structure defined on the polynomial ring over an infinite field k, and it may be summarized as follows: the identity function from k to itself is a primitive representative function on the additive group of k. This in turn explains why "addition" plays a privileged role among the other operations: any polynomial function (or a function represented as the sum of a power series) is a representative function on the additive group of k, and the comultiplication applied to that function is in fact the function applied to a sum of variables.
Other examples include numerous coalgebra or bialgebra structures used in combinatorics.
All the above will be explained in the second section. Before doing this, we briefly recall in the first section the construction of the representative bialgebra of a semigroup (we include enough detail so that the exposition becomes self contained, but the reader is referred to [1] or [2] for unexplained notions or notation).

1 The representative bialgebra of a semigroup

Let k be a field, and G a monoid. Denote by kG the semigroup algebra (kG has basis G as a k-vector space and multiplication given by

$(ax)(by) = (ab)(xy)$ for $a, b \in k$, $x, y \in G$), and by $(kG)^*$ its linear dual, $(kG)^* = Hom_k(kG, k)$. Put

$$(kG)^\circ =$$

$$= \{f \in (kG)^* \mid \exists f_i, g_i \in (kG)^* \ : \ f(xy) = \sum f_i(x)g_i(y),\ \forall x, y \in kG\}$$

It is clear that $(kG)^\circ$ is a kG-subbimodule of $(kG)^*$ with respect to $\rightharpoonup$ and $\leftharpoonup$, given by

$$(x \rightharpoonup f)(y) = f(yx), \quad (f \leftharpoonup x)(y) = f(xy), \quad \forall x, y \in kG, \quad f \in (kG)^*.$$

If we take $f \in (kG)^\circ$, we can assume the f_i's and g_i's are linearly independent.

(This can be seen as follows: let n be the least positive integer for which there exist f_i, g_i with $f(xy) = \sum_{i=1}^{n} f_i(x)g_i(y)$, for all $x, y \in kG$, with $(f_i)_{i=1,n}$ linearly independent. Then $(g_i)_{i=1,n}$ are also linearly independent, because if g_n is a linear combination of the others, say $g_n = \sum_{i=1}^{n-1} \alpha_i g_i$, then $f(xy) = \sum_{i=1}^{n-1} (f_i + \alpha_i f_n)(x)g_i(y)$, and $\{f_1 + \alpha_1 f_n, \ldots, f_{n-1} + \alpha_{n-1} f_n\}$ are linearly independent, contradicting the minimality of n).

Then there exist linearly independent $v_1, \ldots, v_n \in kG$ with $g_i(v_j) = 0$ for any $i, j = 1, \ldots, n$, $i \neq j$, while $g_i(v_i) \neq 0$. (This follows by induction on n. For $n = 1$ the result is clear. Let now $g_1, \ldots, g_{n+1}$ be linearly independent, and applying the induction hypothesis we find $v_1, \ldots, v_n \in kG$ satisfying the conditions for $g_1, \ldots, g_n$, and pick $v \in kG$ with $g_{n+1}(v) \neq \sum_{i=1,n} g_{n+1}(v_i)g_i(v)$. Then put $w_{n+1} = v - \sum_{i=1,n}(g_i(v_i))^{-1} v_i g_i(v)$, and $w_j = v_j - g_{n+1}(v_j)(g_{n+1}(w_{n+1}))^{-1} w_{n+1}$ for $j = 1, \ldots, n$. These $w_1, \ldots, w_{n+1}$ satisfy the required conditions).

Thus we obtain that $f_i \in kG \rightharpoonup f \subset (kG)^\circ$ and $g_i \in f \leftharpoonup kG \subset (kG)^\circ$. Hence we can define

$$\Delta : (kG)^\circ \longrightarrow (kG)^\circ \otimes (kG)^\circ,$$

$$\Delta(f) = \sum f_i \otimes g_i \quad \Leftrightarrow \quad f(xy) = \sum f_i(x)g_i(x), \quad \forall x, y \in kG,$$

and

$$\varepsilon : (kG)^\circ \longrightarrow k, \quad \varepsilon(f) = f(1_G)$$

The above maps define a coalgebra structure on $(kG)^\circ$, which turns it into a bialgebra.

Also note that for all $f \in (kG)^\circ$, $kG \rightharpoonup f \leftharpoonup kG$ is a subcoalgebra of $(kG)^\circ$

(the smallest one containing f), and it is finite dimensional.
Now there exists an isomorphism of vector spaces

$$\phi : k^G \longrightarrow (kG)^* = Hom(kG, k), \quad \phi(f)(\sum_i a_i x_i) = \sum_i a_i f(x_i).$$

Consequently, k^G becomes a kG-bimodule by transport of structures via ϕ:

$$(xf)(y) = f(yx), \quad (fx)(y) = f(xy), \quad \forall x, y \in G, \quad f \in k^G.$$

Definition 1.1 *If G is a monoid, we call*

$$R_k(G) := \phi^{-1}((kG)^\circ)$$

the representative bialgebra *of the monoid G.* ∎

Note that the bialgebra structure on $R_k(G)$ is also transported via ϕ. $R_k(G)$ is a kG-subbimodule of k^G, and consists of the functions (which are called representative) generating a finite dimensional kG-subbimodule (or, equivalently, a left or right kG-submodule). We have

$$R_k(G) = \{f \in k^G \mid \exists f_i, g_i \in k^G, \ f(xy) = \sum f_i(x) g_i(y) \ \forall x, y \in G\},$$

and the comultiplication map on $R_k(G)$ is given as follows: for $f \in R_k(G)$,

$$\Delta(f) = \sum f_i \otimes g_i \Leftrightarrow f_i, g_i \in k^G \text{ are such that } f(xy) = \sum f_i(x) g_i(y). \tag{1}$$

If G is a group, the $R_k(G)$ is a Hopf algebra with antipode $S(f)(x) = f(x^{-1})$.

Remarks 1.2 *1) The following is a partial explanation for the name of representative functions. Let G be a group, and $\rho : G \longrightarrow GL_n(k)$ a representation of G. Put $\rho(x) = (f_{ij}(x))_{i,j}$, and let $V(\rho)$ be the k-subspace of k^G spanned by the $\{f_{ij}\}_{i,j}$. Then $R_k(G) = \sum_\rho V(\rho)$, where ρ ranges over all finite dimensional representations of G. Indeed, let $f \in V(\rho)$, and $y \in G$. If $f = \sum a_{ij} f_{ij}$, then $(yf)(x) = \sum a_{ij} f_{ij}(xy) = \sum a_{ij} f_{ik}(x) g_{kj}(y)$, and thus $yf = \sum a_{ij} g_{kj} f_{ik} \in V(\rho)$. Similarly, $fy \in V(\rho)$, and so we have ($\supseteq$). For the reverse inclusion, let $f \in R_k(G)$. Then the left kG-submodule generated by f is finite dimensional, say with basis $\{f_1, \ldots, f_n\}$, and thus $xf_i = \sum g_{ij}(x) f_j$. Then $\rho : G \longrightarrow GL_n(k)$, $\rho(x) = (g_{ij}(x))_{i,j}$, is a representation of G, $V(\rho)$ is*

spanned by the g_{ij}'s, and we obtain that $f_i = \sum f_j(1_G)g_{ij} \in V(\rho)$, thus $f \in V(\rho)$.
2) Let $\theta : (kG)^* \otimes (kG)^* \longrightarrow (kG \otimes kG)^*$, $\theta(f \otimes g)(x \otimes y) = f(x)g(y)$. Then θ is an injective linear map, which is an isomorphism when G is finite. If we denote by $M : kG \otimes kG \longrightarrow kG$ the multiplication map, then $(kG)^\circ$ consists of those $f \in (kG)^*$ with the property that $M^*(f) \in Im(\theta)$. Consequently, $(kG)^\circ = (kG)^*$ when G is finite.
3) We give now a second explanation for the name of representative functions. Assume that G is a finite group. Then consider the representable functor from commutative k-algebras to groups, given by $F(R) = Alg_k((kG)^*, R)$ ($Alg_k((kG)^*, R)$ is a group under convolution, and the inverse of a map is the composition of that map with the antipode). For commutative k-algebras R without other idempotents than zero and one, we have that $F(R) \simeq G$, so in this case $(kG)^\circ = (kG)^*$ "almost represents" the functor associating the group G to any commutative k-algebra R. ∎

2 Representative Functions and Coalgebra Structures

We start this section by looking at some examples of representative functions:
1) Let G be a monoid. Then $f \in R_k(G)$ is a grouplike element if and only if $\Delta(f) = f \otimes f$ and $\varepsilon(f) = 1$, which will happen if and only if, by (1), $f(xy) = f(x)f(y)$ and $f(1_G) = 1$, meaning that f is a monoid morphism from G to $(k, \cdot)$.
An example is the exponential function $\exp : \mathbf{R} \longrightarrow \mathbf{R}$, because $e^{x+y} = e^x e^y$, hence $\Delta(\exp) = \exp \otimes \exp$, and the subspace spanned by exp in $R_{\mathbf{R}}((\mathbf{R}, +))$ is a one-dimensional subcoalgebra.
Another example is $\mathbf{1}$, the constant function taking the value 1: we have $\mathbf{1}(x+y) = 1 = \mathbf{1}(x)\mathbf{1}(y)$.
2) If G is a monoid, then $f \in R_k(G)$ is primitive if and only if $\Delta(f) = f \otimes \mathbf{1} + \mathbf{1} \otimes f$, where $\mathbf{1}$ is the constant function taking the value 1. By (1), this will happen if and only if $f(xy) = f(x)\mathbf{1}(y) + \mathbf{1}(x)f(y) = f(x) + f(y)$, i.e. if and only if f is a semigroup morphism from G to $(k, +)$.
An example is the logarithmic function $\lg : (0, \infty) \longrightarrow \mathbf{R}$, because $\lg(xy) = \lg(x) + \lg(y)$, hence $\Delta(\lg) = \lg \otimes \mathbf{1} + \mathbf{1} \otimes \lg$, and thus $\mathbf{1}$ and lg span a two-dimensional subcoalgebra of $R_{\mathbf{R}}((0, \infty), \cdot))$.

3) Let $d_n : \mathbf{R} \longrightarrow \mathbf{R}$ be defined by $d_n(x) = \frac{x^n}{n!}$. Since $d_n(x+y) = \sum_i d_i(x)d_{n-i}(y)$ (by the binomial formula), it follows that the d_n's are representative functions on the group $(\mathbf{R}, +)$, and the subspace they span is a subcoalgebra of $R_{\mathbf{R}}((\mathbf{R}, +))$, isomorphic to the divided power coalgebra from the introduction. This explains the name of this coalgebra.
The binomial formula for $n = 0$ just says that $\mathbf{1}$, the constant function 1, is a grouplike, because $(x+y)^0 = \mathbf{1}(x+y) = 1 = \mathbf{1}(x)\mathbf{1}(y)$.
For $n = 1$, it shows that the identity function is primitive: $x+y = x+y$ may be seen as $Id(x+y) = Id(x) + Id(y) = Id(x)\mathbf{1}(y) + \mathbf{1}(x)Id(y)$.
In general, if we replace $\mathbf{R}$ by any infinite field k (so that polynomial functions on k correspond one-to-one with polynomials in $k[X]$), this explains the definition of the comultiplication defined on the polynomial ring $k[X]$: the indeterminate X corresponds to the identity function, which is primitive. Consequently, in this case all polynomial functions are representative.

Divided power coalgebras are behind other examples of coalgebras appearing in combinatorics. We briefly list three of their possible generalizations (see [3] for more details), along with some other appearances of representative functions.
4) Incidence coalgebras for partially ordered sets may be viewed as generalizations of the divided powers coalgebra, since the latter is just the standard reduced incidence coalgebra spanned by segments of nonnegative integers under natural ordering.
5) Divided powers coalgebras can also be extended in a different direction, by generalizing the binomial coefficients as follows: Let G be a commutative semigroup (written additively). Then *section coefficients* on G are a mapping $(i, j, k) \mapsto (i \mid j, k) \in \mathbf{Z}$ such that for any i the number of ordered pairs j, k such that $(i \mid j, k) \neq 0$ is finite, and

$$\sum_k (i \mid j, k)(k \mid p, q) = \sum_s (i \mid s, q)(s \mid j, p) \tag{2}$$

The section coefficients are called *bisection coefficients* if

$$(i + j \mid p, q) = \sum_{p_1+p_2=p, q_1+q_2=q} (i \mid p_1, q_1)(j \mid p_2, q_2) \tag{3}$$

Therefore, the bisection coefficients may be viewed as representative functions on G (over a field k of characteristic zero). To associate a coalgebra C to G, associate an x_i to each $i \in G$, take the span C of all the x_i's, and define

$$\Delta(x_i) = \sum_{j,k}(i \mid j,k)x_j \otimes x_k.$$

Also put $\varepsilon(x_i) = \delta_{i,0}$, where $0 \in G$ is unique such that $(i \mid 0, j) = (i \mid j, 0) = \delta_{i,j}$. Then the coassociativity of Δ follows from (2), while putting $x_j x_j = x_{i+j}$ turns C into a bialgebra because of (3). This is an illustration of how useful is the fact that being "representative" ensures both coassociativity of the comultiplication and compatibility with the multiplication.

6) Another possible extension of divided powers coalgebras uses polynomial sequences of binomial type as a replacement for the sequence d_n from 3). By definition, these are representative functions on the additive group of k (see [3, (5.4)]): the ploynomial sequence $p_n(x)$ is said to be of *binomial type* if deg $p_n = n$ for all n, and

$$p_n(x+y) = \sum_{k=0}^{n} \binom{n}{k} p_k(x)p_{n-k}(y).$$

Similar examples of representative functions are provided by polynomial sequences of Boolean type (see [3, (5.8)]): a polynomial sequence indexed by the finite subsets of a set $\{p_A(x)\}$ is said to be of *Boolean type* if

$$p_A(x+y) = \sum_{A_1+A_2=A} p_{A_1}(x)p_{A_2}(y).$$

The motivating examples behind them are chromatic polynomials of graphs (see [3, (5.9)]), which are also examples of representative functions.

We now go back to the first example in the introduction, and give an explanation for the name "trigonometric coalgebra". The functions sin and cos : $\mathbf{R} \longrightarrow \mathbf{R}$ satisfy the equalities

$$\sin(x+y) = \sin(x)\cos(y) + \cos(x)\sin(y),$$

and

$$\cos(x+y) = \cos(x)\cos(y) - \sin(x)\sin(y).$$

These equalities are both of the type (1), and show that sin and cos are representative functions on the group $(\mathbf{R}, +)$. The subspace generated by them in the space of the real functions is then a subcoalgebra of $R_{\mathbf{R}}((\mathbf{R}, +))$, isomorphic to the trigonometric coalgebra.

As in Remark 1.2 3), the fact that sin and cos are representative functions is also suitable for a different interpretation. Instead of taking the subspace of $R_{\mathbf{R}}((\mathbf{R},+))$ spanned by sin and cos, take the subalgebra generated by them, $\mathbf{R}[\sin,\cos]$, factor it by the ideal generated by $\sin^2+\cos^2-1$, and denote the quotient algebra by H. Then H is a commutative Hopf algebra, with comultiplication defined on sin and cos as above. Moreover, as shown in [6, Section 2.2], the Hopf algebra H represents (as in "representable functor represented by the commutative $\mathbf{R}$-algebra H") the affine group scheme $\mathcal{C}:\mathbf{R}-\mathbf{Alg_c}\longrightarrow\mathbf{Gr}$, defined by

$$\mathcal{C}(R)=\{(a,b)\in R\times R\mid a^2+b^2=1\},$$

on which the group structure is defined by

$$(a,b)\cdot(c,d)=(ac-bd,ad+bc).$$

Finally, we have a brief look at what makes addition special among other operations (the fact that addition is not the only operation providing formulas leading to coalgebras is illustrated in 2) above).
Let k be an infinite field. Then we have the k-algebra isomorphism

$$k[X,Y]\longrightarrow k[X]\otimes k[X],\quad X\mapsto X\otimes 1,\quad Y\mapsto 1\otimes X.$$

Let P be a polynomial function. Since P is representative, there exist (polynomial) functions F_i and G_i, $i=1,\ldots,n$ such that

$$P(x+y)=\sum_{i=1}^{n}F_i(x)G_i(y). \tag{4}$$

(Note that (4) may also be derived simply by applying the binomial theorem several times and collecting like terms.) Consequently, by (1) we get that $\Delta(P)=\sum F_i\otimes G_i$. The same thing may be obtained directly from (4) by applying Δ to P and using the fact that Δ is multiplicative:

$$\begin{aligned}\Delta(P(x)) &= P(\Delta(x))\\ &= P(x\otimes 1+1\otimes x)\\ &= P(x+y)\\ &= \sum_{i=1}^{n}F_i(x)G_i(y)\\ &= \sum_{i=1}^{n}F_i(x\otimes 1)G_i(1\otimes x)\end{aligned}$$

$$= \sum_{i=1}^{n}(F_i(x) \otimes 1)(1 \otimes G_i(x))$$
$$= \sum_{i=1}^{n} F_i(x) \otimes G_i(x).$$

This way of obtaining a formula for Δ from the formula (4) without mentioning representative functions or (1) was used in [5].

We end by remarking that the point of view exhibiting various representative functions behind coalgebra structures may be expanded by looking at Hopf algebras acting on algebras (or fields), because the measuring condition is also a way of saying that acting by an element of the Hopf algebra is a representative function (i.e. it is a formula of type (1). In this way, "acting as homomorphisms" produces grouplikes, while "acting as derivations" produces primitives.

All of the above come as no surprise: since it is well known that any coalgebra may be viewed as a subcoalgebra of the representative coalgebra of the multiplicative monoid of its linear dual [2, Exercise 1.5.15], wherever there is a coalgebra there are also representative functions around.

References

[1] E. Abe, *Hopf Algebras*, Cambridge Univ. Press, 1977.

[2] S. Dăscălescu, C. Năstăsescu, S. Raianu, *Hopf Algebras: an Introduction*, Monographs and Textbooks in Pure and Applied Mathematics, 235, Marcel Dekker, Inc., New York, 2001.

[3] S.A. Joni, G.-C. Rota, Coalgebras and Bialgebras in Combinatorics, in *Umbral calculus and Hopf algebras* (Norman, Okla., 1978), pp. 1–47, Contemp. Math., 6, Amer. Math. Soc., Providence, R.I., 1982.

[4] S. Montgomery, *Hopf algebras and their actions on rings*, CBMS Regional Conference Series in Mathematics, 82, Amer. Math. Soc., Providence, RI, 1993.

[5] W.D. Nichols, M.E. Sweedler, Hopf algebras and combinatorics, in *Umbral calculus and Hopf algebras* (Norman, Okla., 1978), pp. 49–84, Contemp. Math., 6, Amer. Math. Soc., Providence, R.I., 1982.

[6] B. Pareigis, *Four lectures on Hopf algebras*, Centre de Recerca Matematica, Insitut d'Estudis Catalans, No. 6, 1984.

[7] M.E. Sweedler, *Hopf Algebras*, Benjamin, New York, 1969.

[8] W.C. Waterhouse, *Introduction to Affine Group Schemes*, Graduate Texts in Mathematics 66, Springer-Verlag, New York 1979.

FOURIER THEORY FOR COALGEBRAS, BICOINTEGRALS AND INJECTIVITY FOR BICOMODULES

WALTER RICARDO FERRER SANTOS

CENTRO DE MATEMÁTICA. FACULTAD DE CIENCIAS. UNIVERSIDAD DE LA REPÚBLICA

IGUÁ 4225, CP 11400. MONTEVIDEO. URUGUAY

Walter Ricardo Ferrer Santos: wrferrer@cmat.edu.uy.

ABSTRACT. In this paper we introduce the basic concepts of Fourier theory in the minimal setup of coalgebras. We let the basic concepts of this theory –as well as concepts concerning integrals– interact with injectivity conditions on special objects in the category of bicomodules. In particular we prove that if C is a coalgebra, the category of C–bicomodules is semisimple if and olny if C admits a normalized Fourier transform. We show that in the case of Hopf algebras the situation becomes more symmetric than for general coalgebras. We also consider the case of biFrobenius algebras as defined by Doi-Takeuchi.

1. INTRODUCTION

In this section we describe briefly the content of this paper. In Section 2 for a given coalgebra C we review the basic results about injective C–comodules and injective C–bicomodules. We also relate the injectivity of the base field k as a C–comodule or C–bicomodule, with the existence of total cointegrals or total bicointegrals respectively. In Section 3 we define the concept of Fourier transform and Fourier product for general coalgebras defining along the way the concept of Fourier bialgebra. A Fourier bialgebra is simultaneously an algebra and a coalgebra satisfying the following compatibility condition : $\sum(c \cdot d)_1 \otimes (c \cdot d)_2 = \sum c \cdot d_1 \otimes d_2 = \sum c_1 \otimes c_2 \cdot d$. We show that the basic coalgebra C is injective as a bicomodule if and only if C is a Fourier bialgebra and the product and coproduct satisfy the additional compatibility condition $\sum c_1 \cdot c_2 = c$. We also show that

C is injective as a bicomodule if and only if the category of all bicomodules is semisimple. In Section 4 we treat the case of Hopf algebras and reinterpret some well known results –for example the result that guarantees that the existence of a total cointegral is equivalent to the semisimplicity of the category– in terms of the considerations of Section 3 about Fourier bialgebras. In Section 5 we deal with biFrobenius algebras as defined by Doi and Takeuchi in [4] and prove that the semisimplicity of the category is related to the existence of a convolution inverse for the antipode.

We work over a fixed base field k and assume all the objects and maps to be k–linear. Some of the results of this paper could be obtained in a more general setup. For example working with the projective modules over a commutative ring R, or assuming that the maps are R–split. We will not deal with these possible generalizations that do not seem to add any relevant insight or application.

Next we describe briefly some of the notations we use. If C is a k–coalgebra, Δ_C and ε_C will denote its comultiplication and counit and when there is no danger of confusion the subscript will be omitted. We call $\mathcal{M}^C$ the category of right C–comodules and ${}^C\mathcal{M}^C$ the category of C–bicomodules. We adopt Sweedler's notation and write $\Delta(c) = \sum c_1 \otimes c_2$, $\chi(n) = \sum n_0 \otimes n_1 \in N \otimes C$ if $(N, \chi) \in \mathcal{M}^C$ and $\chi(m) = \sum m_{-1} \otimes m_0 \otimes m_1 \in C \otimes M \otimes C$ if $(M, \chi) \in {}^C\mathcal{M}^C$.

A coaugmentation for a coalgebra C is an element $1 \in C$, $1 \neq 0$ such that $\Delta(1) = 1 \otimes 1$, the pair $(C, 1)$ is called a coaugmented coalgebra. In this case $\varepsilon(1) = 1$. If C is coaugmented we view the base field k as an object in $\mathcal{M}^C$ or ${}^C\mathcal{M}^C$ in the evident manner endowed with the trivial coactions, i.e., with the maps $\lambda \to \lambda \otimes 1 : k \to k \otimes C$ and $\lambda \to 1 \otimes \lambda \otimes 1 : k \to C \otimes k \otimes C$ respectively.

As it is customary we call $\Delta^{op} : C \to C \otimes C$ the comultiplication opposite to Δ, i.e., $\Delta^{op}(c) = \sum c_2 \otimes c_1$, we abbreviate (C, Δ^{op}) as C^{cop}.

If V is a k–space, $(C \otimes V \otimes C, \Delta \otimes id \otimes \Delta) \in {}^C\mathcal{M}^C$. If $(M, \chi) \in {}^C\mathcal{M}^C$, then $\chi : M \to C \otimes M \otimes C$ is a morphism of bicomodules when M is equipped with the structure χ and $C \otimes M \otimes C$ with $\Delta \otimes id \otimes \Delta$. Similarly for $(M, \chi) \in \mathcal{M}^C$. We call

$\Delta^2 : C \to C \otimes C \otimes C$ the map $\Delta^2 = (\Delta \otimes id)\Delta = (id \otimes \Delta)\Delta$. Clearly Δ^2 endows C with a structure of C–bicomodule, i.e., $(C, \Delta^2) \in {}^C\mathcal{M}^C$.

In the case where C is a Hopf algebra its antipode will be denoted as $S_C : C \to C$, or simply as $S : C \to C$.

The author would like to thank G. Hochschild for allowing him to use in this paper –in particular in Section 4– some results appearing in an unpublished manuscript of his authorship.

2. Injective comodules and bicomodules

It is easy to deal with injective objects in ${}^C\mathcal{M}^C$ or in $\mathcal{M}^C$. For example the following criterion is well known, we sketch below its very easy proof.

Lemma 2.1. *An object $M \in \mathcal{M}^C$ is injective if and only if the structure morphism $\chi : M \to M \otimes C$ splits in $\mathcal{M}^C$–here $M \otimes C$ is equipped with the structure $id \otimes \Delta$–. Similarly an object $N \in {}^C\mathcal{M}^C$ is injective if and only if $\chi : N \to C \otimes N \otimes C$ splits in ${}^C\mathcal{M}^C$–as before $C \otimes N \otimes C$ is equipped with the structure $\Delta \otimes id \otimes \Delta$–.*

Proof: As the situations of comodules and bicomodules are similar we treat the case of $M \in \mathcal{M}^C$. We only prove that if $\chi : M \to M \otimes C$ splits, then M is injective.

Let γ be a map that splits χ, i.e., assume that $\gamma : M \otimes C \to M$ satisfies: a) $\gamma\chi = id_M$, b) the diagram below is commutative

$$\begin{array}{ccc} M \otimes C & \xrightarrow{\gamma} & M \\ {\scriptstyle id\otimes\Delta}\downarrow & & \downarrow{\scriptstyle \chi} \\ M \otimes C \otimes C & \xrightarrow[\gamma\otimes id]{} & M \otimes C \end{array}$$

Let $\alpha : S \to T$ be an injective map in $\mathcal{M}^C$ and consider a diagram of the form

$$\begin{array}{ccc} S & \xrightarrow{\alpha} & T \\ {\scriptstyle \beta}\downarrow & & \\ M & & \end{array}$$

To prove our assertion we need to complete it with a C–morphism $\widehat{\beta} : T \to M$ such that $\widehat{\beta}\alpha = \beta$. To construct $\widehat{\beta}$ we first take $\sigma : T \to S$ a k–linear map that splits α, i.e., such that $\sigma\alpha = id_S$. Define $\widehat{\beta} = \gamma(\beta \otimes id)(\sigma \otimes id)\chi_T : T \to M$. Then $\widehat{\beta}\alpha = \gamma(\beta \otimes id)(\sigma \otimes id)\chi_T\alpha = \gamma(\beta \otimes id)(\sigma \otimes id)(\alpha \otimes id)\chi_S = \gamma(\beta \otimes id)\chi_S = \gamma\chi\beta = \beta$. Moreover $\widehat{\beta}$ is a C–morphism. Indeed: $\chi\widehat{\beta} = \chi\gamma(\beta \otimes id)(\sigma \otimes id)\chi_T = (\gamma \otimes id)(id \otimes \Delta)(\beta \otimes id)(\sigma \otimes id)\chi_T = (\gamma \otimes id)(\beta \otimes id \otimes id)(\sigma \otimes id \otimes id)(id \otimes \Delta)\chi_T = (\gamma \otimes id)(\beta \otimes id \otimes id)(\sigma \otimes id \otimes id)(\chi_T \otimes id)\chi_T = (\widehat{\beta} \otimes id)\chi_T$. □

It is interesting to notice that the above result is the dual version of the following well known assertion: let R be a k-algebra and M an R–module, then M is projective if and only if the multiplication morphism $R \otimes M \to M$ splits in the category of R–modules. In this manner M is a direct summand of the free module $R \otimes M$ in a canonical way.

Corollary 2.1. *The objects $(C, \Delta) \in \mathcal{M}^C$ and $(C \otimes C, \Delta \otimes \Delta) \in {}^C\mathcal{M}^C$ are injective in the corresponding categories.*

PROOF: We treat the case of $C \otimes C$. Consider $\gamma : C \otimes C \otimes C \otimes C \to C \otimes C$, $\gamma = id \otimes \varepsilon \otimes \varepsilon \otimes id$. Clearly $\gamma(\Delta \otimes \Delta) = id_{C \otimes C}$ and the diagram

$$\begin{array}{ccc} C \otimes C \otimes C \otimes C & \xrightarrow{\quad id\otimes\varepsilon\otimes\varepsilon\otimes id \quad} & C \otimes C \\ {\scriptstyle \Delta\otimes id\otimes id\otimes\Delta}\big\downarrow & & \big\downarrow{\scriptstyle \Delta\otimes\Delta} \\ C \otimes C \otimes C \otimes C \otimes C \otimes C & \xrightarrow[\quad id\otimes id\otimes\varepsilon\otimes\varepsilon\otimes id\otimes id \quad]{} & C \otimes C \otimes C \otimes C \end{array}$$

commutes.

□

The first of the definitions that follow is well known and some authors call it an integral instead of a cointegral –see for example [1]– but the name cointegral seems to be more appropriate.

Definition 2.1. A right cointegral for a coaugmented coalgebra $(C, \Delta, \varepsilon, 1)$ is a k–linear map $\phi : C \to k$ that satisfies $\sum \phi(c_1)c_2 = \phi(c)1$. If $\phi(1) = 1$ we say that the cointegral is total. A bicointegral for $(C, \Delta, \varepsilon, 1)$ is a map $\tau : C \otimes C \to k$ that satisfies that $\sum \tau(c, d_1)d_2 = \sum c_1\tau(c_2, d) = \tau(c, d)1$. If moreover $\tau(1,1) = 1$ we say that the bicointegral is total.

Observation 2.1. It follows immediately from Lemma 2.1 that the augmented coalgebra C admits a total bicointegral if and only if k is injective in ${}^C\mathcal{M}^C$ and C admits a total right cointegral if and only if k is injective in $\mathcal{M}^C$.

In some situations, for example for group coalgebras or more generally for the coalgebra part of a Hopf algebra –see [1][Th. 3.3.2]– the injectivity of $k \in \mathcal{M}^C$ implies the injectivity of all objects $M \in \mathcal{M}^C$, i.e., the semisimplicity of the category. This is not so in the case of arbitrary coalgebras as the following example shows. The case of Hopf algebras will be treated in Section 4.

Observation 2.2. Consider the three dimensional coalgebra $C = k1 + kg + kt$ being 1 and g group like elements, $\Delta(t) = g \otimes t + t \otimes 1$ and $\varepsilon(t) = 0$, $\varepsilon(g) = \varepsilon(1) = 1$. Define $\phi : C \to k$ as $\phi(c) = \alpha$ for $c = \alpha 1 + \beta g + \gamma t$. Clearly $\sum \phi(c_1)c_2 = \alpha 1 = \phi(c)1$. If we consider $\chi_1, \chi_g : k \to k \otimes C$ defined as $\chi_1(\lambda) = \lambda \otimes 1$ and $\chi_g(\lambda) = \lambda \otimes g$ respectively , then $(k, \chi_1) \in \mathcal{M}^C$ is injective and $(k, \chi_g) \in \mathcal{M}^C$ is not injective.

3. Fourier bialgebras and the injectivity of C in ${}^C\mathcal{M}^C$

In this section we present in the minimal set up of coalgebras some basic concepts concerning Fourier theory.

Definition 3.1. Let C be a k–coalgebra and call C^* its dual, a Fourier transform in C is a linear map $\mathcal{F} : C \to C^*$ such that for all $c, d \in C$, $\sum \mathcal{F}(c)(d_1)d_2 = \sum c_1\mathcal{F}(c_2)(d)$. A Fourier transform is said to be normalized if for all $c \in C$, $\sum \mathcal{F}(c_1)(c_2) = \varepsilon(c)$. A Fourier product on C is a bilinear map $f : C \otimes C \to C$, verifying for all $c, d \in C$, $\Delta(c \cdot d) = \sum c \cdot d_1 \otimes d_2 = \sum c_1 \otimes c_2 \cdot d$ –we denote $f(c \otimes d) = c \cdot d$–. A Fourier product is said to be normalized if for all $c \in C$, $\sum c_1 \cdot c_2 = c$. A triple $(C, \Delta, \varepsilon, f)$ where (C, Δ, ε) is a counital coalgebra and $f : C \otimes C \to C$ is a Fourier product is called a Fourier bialgebra.

We have chosen to call a product as in Definition 3.1 a Fourier product –instead of convolution product that would be more consistent with the usage in Fourier analysis– in order to avoid confusion with the already established nomenclature in Hopf theory concerning the use of the word "convolution".

Observation 3.1. From the equality $\sum(c{\cdot}d)_1\otimes(c{\cdot}d)_2 = \sum c{\cdot}d_1\otimes d_2 = \sum c_1\otimes c_2{\cdot}d$ we deduce applying $\varepsilon\otimes id$ and $id\otimes\varepsilon$ respectively that $\sum\varepsilon(c\cdot d_1)d_2 = c\cdot d = \sum c_1\varepsilon(c_2\cdot d)$

Next we show that a Fourier product is automatically associative.

Lemma 3.1. *Assume that C is an arbitrary coalgebra and that $f : C\otimes C \to C$ is a Fourier product in C. Then f is associative.*

PROOF: Let $c, d, e \in C$, as $c\cdot d = \sum\varepsilon(c\cdot d_1)d_2$ we have that $(c\cdot d)\cdot e = \sum\varepsilon(c\cdot d_1)d_2\cdot e$. Similarly, $c\cdot(d\cdot e) = \sum\varepsilon(c\cdot(d\cdot e)_1)(d\cdot e)_2 = \sum\varepsilon(c\cdot d_1)d_2\cdot e$. Hence $(c\cdot d)\cdot e = c\cdot(d\cdot e)$. □

Observation 3.2. If G is a finite group and C the k–space of all functions from G into k, the so called –non-normalized– convolution product used in Fourier analysis is defined as follows, if $f, g\in C$ then $(f\cdot g)(z) = \sum_{y\in G} f(y)g(y^{-1}z)$. In terms of the basis of C consisting of the characteristic functions $\{\delta_x : x\in G\}$, this product is just $\delta_x\cdot\delta_y = \delta_{xy}$. If we consider the standard comultiplication in C given as $\Delta(\delta_x) = \sum_{yz=x}\delta_y\otimes\delta_z$, it is easy to verify that $(C,\Delta,\varepsilon,\cdot)$ is a Fourier bialgebra.

Definition 3.2. Given an arbitrary Fourier transform $\mathcal{F}$ in C define $f_{\mathcal{F}} : C\otimes C \to C$ as $f_{\mathcal{F}}(c\otimes d) = \sum\mathcal{F}(c)(d_1)d_2 = \sum c_1\mathcal{F}(c_2)(d)$. Conversely, given a Fourier product $f : C\otimes C\to C$, we define $\mathcal{F}_f : C\to C^*$ as $\mathcal{F}_f(c)(d) = \varepsilon(f(c\otimes d))$.

Observation 3.3. It is not hard to prove that in this manner we produce in the first case a Fourier product and in the second case a Fourier transform.

The main properties of Fourier transforms and products are summarized in the theorem that follows.

Theorem 3.1. *Let C be a coalgebra. The correspondences considered in Definition 3.2 are inverses of each other. For $\mathcal{F}$ and f related as above, the map $\mathcal{F}: C \to C^{cop*}$ is a morphism of algebras when C is equipped with the associative product f and C^* with the convolution product corresponding to $C^{cop} = (C, \Delta^{op})$. Moreover the Fourier transform $\mathcal{F}$ is normalized if and only if the Fourier product f is normalized.*

PROOF: Let $\mathcal{F}$ be a Fourier transform and consider $f_{\mathcal{F}}$, then $\mathcal{F} = \mathcal{F}_{f_{\mathcal{F}}}$. Indeed, for arbitrary $c, d \in C$, $\mathcal{F}_{f_{\mathcal{F}}}(c)(d) = \varepsilon(f_{\mathcal{F}}(c \otimes d)) = \varepsilon(\sum \mathcal{F}(c)(d_1)d_2) = \mathcal{F}(c)(d)$. Conversely if f is a Fourier product then $f = f_{\mathcal{F}_f}$. This follows from the fact that for $c, d \in C$, $f_{\mathcal{F}_f}(c \otimes d) = \sum c_1 \mathcal{F}_f(c_2)(d) = \sum c_1 \varepsilon(f(c_2 \otimes d)) = f(c \otimes d)$ –see Observation 3.1–.

To prove the assertion concerning the multiplicativity of the Fourier transform we first observe that $\mathcal{F}(c \cdot d)(e) = \varepsilon((c \cdot d) \cdot e) = \varepsilon(c \cdot d \cdot e)$. Hence –denoting the convolution product with the symbol $\star$– we have that $(\mathcal{F}(d) \star \mathcal{F}(c))(e) = \sum \mathcal{F}(d)(e_1)\mathcal{F}(c)(e_2) = \sum \varepsilon(d \cdot e_1)\varepsilon(c \cdot e_2)$. From the equality $\sum(d \cdot e_1) \otimes e_2 = \sum(d \cdot e)_1 \otimes (d \cdot e)_2$ we deduce that $\sum(d \cdot e_1) \otimes (c \cdot e_2) = \sum(d \cdot e)_1 \otimes c \cdot (d \cdot e)_2$ and applying $\varepsilon \otimes \varepsilon$ we deduce that $\sum \varepsilon(d \cdot e_1)\varepsilon(c \cdot e_2) = \varepsilon(c \cdot d \cdot e) = \mathcal{F}(c \cdot d)(e)$. □

Observation 3.4. In recent papers, vector spaces equipped with a coassociative comultiplication and an associative multiplication satisfying a compatibility condition different from the pentagonal axiom that characterizes bialgebras, have been considered in different contexts. Important examples are: infinitesimal bialgebras (see [2]), biFrobenius algebras (see [4]) and braided bialgebras (see [7]). Fourier bialgebras belong to this realm of structures.

Theorem 3.2. *Let C be an arbitrary coalgebra, $(C, \Delta^2) \in {}^C\mathcal{M}^C$ is injective if and only if C admits a normalized Fourier product.*

PROOF: Suppose that (C, Δ^2) is injective. In accordance with Lemma 2.1 there exists a linear map $\gamma : C \otimes C \otimes C \to C$ such that $\sum \gamma(c_1 \otimes c_2 \otimes c_3) = c$ and $\sum(\gamma(c \otimes d \otimes e))_1 \otimes (\gamma(c \otimes d \otimes e))_2 \otimes (\gamma(c \otimes d \otimes e))^3 = \sum c_1 \otimes \gamma(c_2 \otimes d \otimes e_1) \otimes e_2$

for all $c, d, e \in C$. If we define for $c, d \in C$, $c \cdot d = \sum \gamma(c \otimes d_1 \otimes d_2)$ it is clear that $\sum c_1 \cdot c_2 = \sum \gamma(c_1 \otimes c_2 \otimes c_3) = c$. Moreover, we have that $\sum (c \cdot d)_1 \otimes (c \cdot d)_2 \otimes (c \cdot d)_3 = \sum c_1 \otimes \gamma(c_2 \otimes d_1 \otimes d_2) \otimes d_3 = \sum c_1 \otimes c_2 \cdot d_1 \otimes d_2$. Applying $\varepsilon \otimes id \otimes id$ and $id \otimes id \otimes \varepsilon$ we obtain that $\sum (c \cdot d)_1 \otimes (c \cdot d)_2 = \sum c \cdot d_1 \otimes d_2 = \sum c_1 \otimes c_2 \cdot d$. Conversely, assuming the existence of the Fourier product we define $\gamma : C \otimes C \otimes C \to C$ as $\gamma(c \otimes d \otimes e) = c \cdot d \cdot e$. Then $\sum \gamma(c_1 \otimes c_2 \otimes c_3) = \sum (c_1 \cdot c_2) \cdot c_3 = \sum (c_1 \cdot c_2) = c$. Also $\sum (c \cdot d \cdot e)_1 \otimes (c \cdot d \cdot e)_2 = \sum c_1 \otimes c_2 \cdot (d \cdot e) = \sum c_1 \otimes (c_2 \cdot d) \cdot e$. Applying $id \otimes \Delta$ we obtain that $\sum (c \cdot d \cdot e)_1 \otimes (c \cdot d \cdot e)_2 \otimes (c \cdot d \cdot e)_3 = \sum c_1 \otimes c_2 \cdot d \cdot e_1 \otimes e_2$. □

Observation 3.5. Concerning injectivity properties it frequently happens that algebraic operations on a given basic object imply injectivity of certain special objects in the category, for example: if C has a coproduct, then $C \in \mathcal{M}^C$ is injective. The situation described in Theorem 3.2 is rather peculiar, in that an injectivity condition is shown to imply the existence of a certain new algebraic operation – the Fourier product–.

Corollary 3.3. *Let C be an arbitrary coalgebra, then $(C, \Delta^2) \in {}^C\mathcal{M}^C$ is injective if and only if all objects $(M, \chi_M) \in {}^C\mathcal{M}^C$ are injective, i.e., if the category of bicomodules is semisimple.*

PROOF: We use the Fourier product in order to construct the maps $\gamma_N : C \otimes N \otimes C \to N$ that split in ${}^C\mathcal{M}^C$ the structure morphism. Define $\gamma_N(c \otimes n \otimes d) = \sum \varepsilon(c \cdot n_{-1}) n_0 \varepsilon(n_1 \cdot d)$. In this situation we have that $\gamma_N \chi_N(n) = \gamma_N(\sum n_{-1} \otimes n_0 \otimes n_1) = \sum \varepsilon(n_{-2} \cdot n_{-1}) n_0 \varepsilon(n_1 \cdot n_2) = \sum \varepsilon(n_{-1}) n_0 \varepsilon(n_1) = n$. The computation that follows shows that γ_N is a morphism in the appropriate category: $\chi_N \gamma_N(c \otimes n \otimes d) = \sum \varepsilon(c \cdot n_{-2}) n_{-1} \otimes n_0 \otimes n_1 \varepsilon(n_2 \cdot d) = \sum c \cdot n_{-1} \otimes n_0 \otimes n_1 \cdot d$ –see Observation 3.1–; $(id \otimes \gamma_N \otimes id)(\Delta \otimes id \otimes \Delta)(c \otimes n \otimes d) = \sum c_1 \otimes \gamma_N(c_2 \otimes n \otimes d_1) \otimes d_2 = \sum c_1 \varepsilon(c_2 \cdot n_{-1}) \otimes n_0 \varepsilon(n_1 \cdot d_1) d_2 = \sum c \cdot n_{-1} \otimes n_0 \otimes n_1 \cdot d$. □

Next we present another characterization of the injectivity of $(C, \Delta^2) \in {}^C\mathcal{M}^C$.

Theorem 3.4. *Let C be a coalgebra, then $(C, \Delta^2) \in {}^C\mathcal{M}^C$ is injective if and only if there exists a k–linear map $\omega_0 : C \otimes C \to k$ such that $\omega_0 \Delta = \varepsilon$ and $\sum c_1 \omega_0(c_2 \otimes d) = \sum \omega_0(c \otimes d_1) d_2$.*

PROOF: Consider the corresponding Fourier product and define $\omega_0(c\otimes d) = \varepsilon(c\cdot d)$. Then $\sum \omega_0(c_1\otimes c_2) = \sum \varepsilon(c_1\cdot c_2) = \varepsilon(c)$. From the equality $\sum c\cdot d_1\otimes d_2 = \sum c_1 \otimes c_2\cdot d$ we deduce –applying $id\otimes\varepsilon$– that $c\cdot d = \sum c_1\omega_0(c_2\otimes d)$, similarly –applying $\varepsilon\otimes id$– we deduce that $c\cdot d = \sum \omega_0(c\otimes d_1)d_2$. Conversely, assuming the existence of ω_0 we define $c\cdot d = \sum c_1\omega_0(c_2\otimes d) = \sum \omega_0(c\otimes d_1)d_2$. Then $\Delta(c\cdot d) = \sum c_1\otimes c_2\omega_0(c_3\otimes d) = \sum c_1\otimes c_2\cdot d$, and also $\Delta(c\cdot d) = \sum \omega_0(c\otimes d_1)d_2\otimes d_3 = \sum c\cdot d_1\otimes d_2$. □

Observation 3.6. The following formulae express the Fourier product and Fourier transform in terms of ω_0. $\mathcal{F}(c)(d) = \varepsilon(c\cdot d) = \sum \omega_0(c\otimes d_1)\varepsilon(d_2) = \omega_0(c\otimes d)$; $c\cdot d = \sum c_1\omega_0(c_2\otimes d) = \sum \omega_0(c\otimes d_1)d_2$.

4. THE CASE OF HOPF ALGEBRAS

In the case where the coalgebra C is part of a Hopf algebra structure, the situation is far more symmetric. The Hopf algebras will be considered as coaugmented cannonically with the neutral element of the multiplication.

Theorem 4.1. *Let H be a Hopf algebra, then k is injective in $\mathcal{M}^H$ if and only if it is injective in ${}^H\mathcal{M}^H$.*

PROOF: Given a bicointegral $\tau : H\otimes H \to k$ we construct a right cointegral $\phi : H \to k$ simply as $\phi(h) = \sum \tau(h_1\otimes h_2)$. If τ is total then ϕ is also total and in order to check that ϕ is a right cointegral we proceed as follows: $\sum \phi(h_1)h_2 = \sum \tau(h_1\otimes h_2)h_3 = \sum \tau(h_1\otimes h_2)1 = \phi(h)1$. Conversely given ϕ we define $\tau : H\otimes H \to k$ as $\tau(x\otimes y) = \phi(Sx)\phi(y)$. If ϕ is a total right cointegral, then τ is a total bicointegral. Indeed, for $x, y \in H$: $\sum \tau(x\otimes y_1)y_2 = \sum \phi(Sx)\phi(y_1)y_2 = \phi(Sx)\phi(y)1 = \tau(x\otimes y)1$ and $\sum x_1\tau(x_2\otimes y) = \sum x_1\phi(Sx_2)\phi(y) = \phi(Sx)\phi(y)1 = \tau(x\otimes y)1$. □

Observation 4.1. In the argument above we used that if ϕ is a right cointegral then for all $x\in H$, $\sum x_1\phi(Sx_2) = \phi(Sx)$. This equality will be deduced from the general formula that follows: if ϕ is a right cointegral and $x, y \in H$ are arbitrary elements of H then that $\sum \phi((Sx)y_1)y_2 = \sum x_1\phi((Sx_2)y)$.

Indeed $\sum \phi((Sx)y_1)y_2 = \sum \phi((Sx_2)y_1)\varepsilon(x_1)y_2 = \sum x_1\phi((Sx_3)y_1)(Sx_2)y_2 = \sum x_1\phi((S(x_2)y)_1)(S(x_2)y)_2 = \sum x_1\phi((Sx_2)y)$. In particular is $y = 1$ we deduce that $\sum x_1\phi(Sx_2) = \phi(Sx)1$.

Theorem 4.2. *Assume that H is a Hopf algebra, then the injectivity of k is equivalent to the injectivity of $(H, \Delta^2) \in {}^H\mathcal{M}^H$.*

PROOF: In accordance with the results of Section 3 we have to produce from a total right cointegral $\phi : H \to k$ a normalized Fourier product in H. We do this in the usual fashion defining $x \cdot y = \sum \phi((Sx)y_1)y_2 = \sum x_1\phi((Sx_2)y)$. Clearly $\sum x_1 \cdot x_2 = \sum \phi((Sx_1)x_2)x_3 = \phi(1)\sum \varepsilon(x_1)x_2 = x$. Also $\Delta(x \cdot y) = \sum \phi((Sx)y_1)y_2 \otimes y_3 = \sum x \cdot y_1 \otimes y_2$ and $\Delta(x \cdot y) = \sum x_1 \otimes x_2\phi((Sx_3)y) = \sum x_1 \otimes x_2 \cdot y$. □

Observation 4.2. The map ω_0 constructed in Theorem 3.4 is just $\omega_0(x \otimes y) = \phi((Sx)y)$ and the Fourier transform in the context of Hopf algebras $\mathcal{F} : H \to H^*$ is $\mathcal{F}(x)(y) = \phi((Sx)y)$. This is the usual Fourier transform for Hopf algebras equipped with a cointegral (see for example [3]). In this case the Fourier product is $x \cdot y = \sum \phi((Sx)y_1)y_2 = \sum x_1\phi((Sx_2)y)$.

5. The case of biFrobenius algebras

In this section we endow a biFrobenius algebra with a structure of Fourier bialgebra. In this context the relationship between the new product and injectivity is not as symmetric as for the case of Hopf algebras.

The concept of biFrobenius algebra was introduced in [4] where [5] and [6] are cited as predecessors.

We start by briefly reviewing the basic definitions concerning biFrobenius algebras, the reader should consult [4] for the details we skip in our presentation.

Let A be a finite dimensional unital k–algebra. If $\phi \in A^*$, we say that the pair (A, ϕ) is a Frobenius algebra if $A^* = \phi \leftharpoonup A$. Recall that if $f \in A^*$ and $a \in A$, $f \leftharpoonup a \in A^*$ is the element defined by the formula $(f \leftharpoonup a)(b) = f(ab)$. The map $c_\phi : A \to A^*$ given as $c_\phi(x) = \phi \leftharpoonup x$ is an isomorphism and $c_\phi(xy) = c_\phi(x) \leftharpoonup y$.

In other words when we consider A and A^* with the A–module structures given respectively by the right multiplication and $\leftharpoonup$, c_ϕ is an isomorphism of right A–modules.

The map $id \otimes c_\phi : A \otimes A \to A \otimes A^*$ is an isomorphism and an element $\xi = \sum a_i \otimes b_i$: $(id \otimes c_\phi)(\xi) = \sum e_j \otimes e_j^*$ is called a dual basis. We call $\{e_j : j = 1 \cdots n\}$ a k–basis of A and $\{e_j^* : j = 1 \cdots n\}$ the corresponding basis in the dual space. In accordance with the definitions above, the basic property of ξ is that for all $x \in A$, $x = \sum a_i \phi(b_i x)$. Then $(\phi \leftharpoonup \sum \phi(xa_i)b_i)(z) = \sum \phi(xa_i)\phi(b_i z) = \phi(\sum xa_i\phi(b_i z)) = \phi(xz) = (\phi \leftharpoonup x)(z)$. Hence $x = \sum \phi(xa_i)b_i = \sum(a_i \rightharpoonup \phi)(x)b_i$. The action $a \rightharpoonup f$ for $a \in A$ and $f \in A^*$ is defined as $(a \rightharpoonup f)(b) = f(ba)$. If $f \in A^*$ is an arbitrary linear functional, by applying it to the equality $x = \sum(a_i \rightharpoonup \phi)(x)b_i$ we conclude that $f = \sum a_i f(b_i) \rightharpoonup \phi$ and then $A^* = A \rightharpoonup \phi$.

Suppose that A is augmented with $\varepsilon : A \to k$. If V is a right A–module we denote as V^A the submodule of the invariantes, i.e., $V^A = \{v \in V : v \cdot a = \varepsilon(a)v\}$.

In our particular situation if A is Frobenius and augmented, the map $c_\phi : A \to A^*$ induces an isomorphism from A^A onto A^{*A}. The following computation shows that $A^{*A} = k\varepsilon$. If for all $x \in A$, $f \leftharpoonup x = \varepsilon(x)f$, then $f(x) = (f \leftharpoonup x)(1) = \varepsilon(x)f(1)$, i.e. , $f = f(1)\varepsilon$. As $\varepsilon \in {}^*A$ our assertion follows. Using the isomorphism c_ϕ restricted to A^A –that will be abbreviated as $\mathcal{I}_r(A)$– we conclude that the vector space $\mathcal{I}_r(A)$ is one-dimensional and that there exists an element $t \in \mathcal{I}_r(A)$ such that $\phi \leftharpoonup t = \varepsilon$. If follows also that the element $t \in \mathcal{I}_r(A)$ is unique.

Dually, we say that a pair (C, t) where C is a finite dimensional coalgebra and $t \in C$, is a coFrobenius coalgebra if $C = t \leftharpoonup C^*$. The action of C^* on C is defined by the rule: if $c \in C$ and $f \in C^*$, $c \leftharpoonup f = \sum f(c_1)c_2$. If C is coFrobenius, for an arbitrary $c \in C$ there exists a unique element $f \in C^*$ such that $c = \sum f(t_1)t_2$.

We consider a map $\Theta : C^* \otimes C^* \to C \otimes C^*$ defined as $\Theta(\sum \alpha_i \otimes \beta_i) = \sum(t \leftharpoonup \alpha_i) \otimes \beta_i$. If we take a k–basis of C , $\{e_j : j = 1 \cdots n\}$ of C and $\{e_j^* : j = 1 \cdots n\}$ the corresponding basis on the dual, we can find an element $\sum f_i \otimes g_i$ such that

$\sum(t \leftharpoondown f_i) \otimes g_i = \sum e_j \otimes e_j^*$. Let $g \in C^*$ be an arbitrary linear functional, we deduce from the above that $g = \sum g(t \leftharpoondown f_i)g_i = \sum f_i(t_1)g(t_2)g_i$. Hence $g = \sum(f_i g)(t)g_i$ or equivalently $g = \sum(gg_i)(t)f_i$. The element $\widehat{\xi} = \sum f_i \otimes g_i$ is called a dual basis.

In the case where C is coaugmented with $1 \in C$ we define $\mathcal{I}_r(C^*) = \{\phi : \sum \phi(c_1)c_2 = \phi(c)1 \ \forall c \in C\}$ and considering the map $c_t : C^* \to C$: $c_t(f) = t \leftharpoondown f$, it follows immediately that c_t is a C^*– morphism and that $c_t(\mathcal{I}_r(C^*)) = k1$. We choose $\phi \in \mathcal{I}_r(C^*)$ such that $t \leftharpoondown \phi = 1$. Clearly ϕ is unique.

Definition 5.1. Let $(H, \Delta, \varepsilon, m, 1)$ be a finite dimensional k–space such that (H, Δ, ε) is a counital coalgebra and $(H, m, 1)$ is a unital algebra. Assume that $1 \in H$ is a group like element and that $\varepsilon : H \to k$ is a morphism of algebras. Let $t \in H$ and $\phi \in H^*$ be such that (H, t) is a coFrobenius coalgebra and (H, ϕ) a Frobenius algebra and define $S : H \to H$ by the formula $S(h) = \sum \phi(t_1 h)t_2$. We say that (H, t, ϕ) is a biFrobenius algebra if S is an antimorphism of algebras and of coalgebras.

In the situation above it is not hard to prove –see [4]– that S is bijective and satisfies that $\varepsilon \circ S = \varepsilon$ and $S(1) = 1$. Then it follows that $1 = S(1) = \sum \phi(t_1)t_2 = t \leftharpoondown \phi$ and also $\varepsilon(h) = \varepsilon(S(h)) = \sum \phi(t_1 h)\varepsilon(t_2) = \phi(th) = (\phi \leftharpoondown t)(h)$, i.e., $\varepsilon = \phi \leftharpoondown t$. Using the uniqueness of t and ϕ we deduce that $\phi \in \mathcal{I}_r(C^*)$ and $t \in \mathcal{I}_r(C)$.

The map S is called the antipode of the biFrobenius algebra and it is worth noticing that in general it is not the convolution inverse of id –see the Example 5.3 below–.

Example 5.1. It is very easy to show that an arbitrary finite dimensional Hopf algebra is a biFrobenius algebra.

Example 5.2. If (H, t, ϕ) is a biFrobenius algebra then (H^*, ϕ, t) is also a biFrobenius algebra and in particular we deduce that for all $f \in H^*$, $f \circ S = \sum(\phi_1 f)(t)\phi_2$.

Example 5.3. ([4]) Consider the quotient algebra $B_4 = k[X]/(X^4)$ and call $x = X + (X^4)$ with the the coalgebra structure: $\Delta(1) = 1 \otimes 1$, $\Delta(x) = x \otimes 1 + 1 \otimes x$, $\Delta(x^2) = x^2 \otimes 1 + 1 \otimes x^2$, $\Delta(x^3) = 1 \otimes x^3 + x \otimes x^2 + x^2 \otimes x + x^3 \otimes 1$ and counit $\varepsilon(1) = 1$, $\varepsilon(x) = \varepsilon(x^2) = \varepsilon(x^3) = 0$.

If one takes $\phi : B_4 \to k$ defined as $\phi(x^m) = \delta_{m,3}$, and $t = x^3$ in [4] the authors prove that (B_4, ϕ, t) is a biFrobenius algebra. As the antipode is antimultiplicative and $S(x) = \sum \phi(t_1 x)t_2 = x$, we conclude that $S = id$. Then $(S \star id)(x) = 2x$ and then if the base field has characteristic different from 2, B_4 is not a Hopf algebra.

In order to show that for a biFrobenius algebra we can define the Fourier transform, first we need to establish a few general results.

Theorem 5.1. *Let (H, t, ϕ) be a biFrobenius algebra, then:*

a. *For all $x \in H$, $\sum t_1 x \otimes t_2 = \sum t_1 \otimes t_2 S(x)$.*

b. *For all $f, g \in H^*$, $\sum (f_1 g)(t) f_2 = \sum (f g_1)(t)(g_2 \circ S)$.*

PROOF:

a. The equality $\sum t_1 x \otimes t_2 = \sum t_1 \otimes t_2 S(x)$ is equivalent to $\sum (\phi \leftharpoonup t_1 x) \otimes t_2 = \sum (\phi \leftharpoonup t_1) \otimes t_2 S(x) \in H^* \otimes H$. The corresponding endomorphisms of H evaluated at $z \in H$ are: $\sum (\phi \leftharpoonup t_1 x)(z) t_2 = \sum \phi(t_1 x z) t_2 = S(xz)$ and $\sum (\phi \leftharpoonup t_1)(z) t_2 S(x) = S(z)S(x)$.

b. If we evaluate a generic element $f \otimes g \in H^* \otimes H^*$ at both sides of the equality just proved we obtain that: $\sum f(t_1 x) g(t_2) = \sum f_1(t_1) f_2(x) g(t_2) = \sum (f_1 g)(t) f_2(x)$, and $\sum f(t_1) g(t_2 S(x)) = \sum f(t_1) g_1(t_2) g_2(S(x)) = \sum (f g_1)(t)(g_2 \circ S)(x)$.

□

Observation 5.1. By duality–see Example 5.2– we obtain from Theorem 5.1– that:

a. For all $f \in H^*$, $\sum \phi_1 f \otimes \phi_2 = \sum \phi_1 \otimes \phi_2 (f \circ S)$.

b. For all $x, y \in H$, $\sum \phi(x_1 y) x_2 = \sum \phi(x y_1) S(y_2)$.

Corollary 5.2. *Let H be a biFrobenius algebra then for all $x, y \in H$, $\sum \phi((Sx)y_1)y_2 = \sum x_1\phi((Sx_2)y)$ and for all $f, g \in H^*$, $\sum((f \circ S)g_1)(t)g_2 = \sum f_1((S \circ f_2)g)(t)$.*

The theorems that follow can be proved using the same arguments than for the case of Hopf algebras –see Observations 4.1 and 4.2–.

Theorem 5.3. *Let H be a biFrobenius algebra, then $x \cdot y = \sum \phi((Sx)y_1)y_2 = \sum x_1\phi((Sx_2)y)$ defines a Fourier product in H.*

Explicitly, $x \cdot y = \sum \phi(t_1x)\phi(t_2y_1)y_2 = \sum x_1\phi(t_1x_2)\phi(t_2y)$. In particular we have that $1 \cdot 1 = \phi(1)1$, $x \cdot 1 = \phi(Sx)1$ and $1 \cdot y = \phi(y)1$.

Theorem 5.4. *If H is a biFrobenius algebra, then $k \in \mathcal{M}^H$ is injective if and only if $k \in {}^H\mathcal{M}^H$ is injective and this happens if and only if $\phi(1) = 1$.*

In terms of the Fourier product one can say that k is injective if and only if $1 \cdot 1 = 1$

Theorem 5.5. *If H is a biFrobenius algebra, then $(H, \Delta^2) \in {}^H\mathcal{M}^H$ is injective if and only if for all $x \in H$, $x = \sum \phi((Sx_1)x_2)x_3$ or equivalently if for all $x \in H$, $\varepsilon(x) = \sum \phi((Sx_1)x_2)$.*

In the case of the biFrobenius algebra B_4, the bicomodule k is not injective because $\phi(1) = 0$. In this context (B_4, Δ^2) cannot be injective –as the injectivity of (B_4, Δ^2) would imply the semisimplicity of the category and hence the injectivity of k–. This is coincident with the fact that for $x = 1$, $\sum \phi((Sx_1)x_2)x_3 = \phi(1)1 = 0$.

We want to look at the injectivity of (H, Δ^2) as a bicomdule in more detail– assuming of course that $\phi(1) = 1$–.

Corollary 5.6. *Let (H, ϕ, t) be a biFrobenius algebra such that $\phi(1) = 1$, then $(H, \Delta^2) \in {}^H\mathcal{M}^H$ is injective if and only if for all $x \in H$, $\sum(Sx_1)x_2 - \varepsilon(x)1 \in \mathrm{Ker}(\phi)$.*

Observation 5.2. In the case where $\phi(1) = 1$ and the antipode S is the convolution inverse of the identity map on H, we conclude that (H, Δ^2) is injective as a bicomodule and hence, for this type of biFrobenius algebras the injectivity of k is equivalent to the semisimplicity of the category of bicomodules. In particular, this establishes Theorem 4.2–for a finite dimensional Hopf algebra–.

Next we exhibit an example of a biFrobenius algebra that has a total integral but (H, Δ^2) is not injective as a bi–comodule.

Example 5.4. Let $B_2 = k[X]/(X^2)$ with the obvious commutative product. Call $x = X + (X^2)$ and endow B_2 with the following coproduct and counit: $\Delta(1) = 1 \otimes 1$, $\Delta(x) = 1 \otimes x + x \otimes 1 - x \otimes x$, $\varepsilon(x) = 0$. Taking $t = x$ and $\phi : H \to k$ defined as $\phi(1) = 1$ and $\phi(x) = 1$, it is easy to show that $\phi \leftharpoonup t = \varepsilon$ and $t \leftharpoonup \phi = 1$. Then $S(x) = \sum \phi(t_1 x) t_2 = \phi(x)x + \phi(x^2)1 - \phi(x^2)x = x$ and $S = \mathrm{id}$. Clearly (B_2, ϕ, t) is a biFrobenius algebra. Computing $\sum S(x_1)x_2 - \varepsilon(x)1 = \sum x_1 x_2 = 2x \notin \mathrm{Ker}(\phi)$, we conclude that –unless the base field has characteristic 2–, (B_2, Δ^2) is not injective as a bicomodule. Observe also that in the situation that the characteristic is 2, B_2 is in fact a Hopf algebra.

References

1. E. Abe, *Hopf algebras*, Cambridge University Press, Cambridge, 1980, Translated from the Japanese by Hisae Kinoshita and Hiroko Tanaka.
2. M. Aguiar, *Infinitesimal Hopf algebras*, New trends in Hopf algebra theory (La Falda, 1999) (N. Andruskiewitsch, W. Ferrer Santos, and H-J. Schneider, eds.), Amer. Math. Soc., Providence, RI, 2000, pp. 1–29.
3. N. Andruskiewitsch and S. Natale, *Harmonic analysis on semisimple Hopf algebras*, Algebra i Analiz **12** (2000), no. 5, 3–27.
4. Y. Doi and M. Takeuchi, *Bi-Frobenius algebras*, New trends in Hopf algebra theory (La Falda, 1999) (N. Andruskiewitsch, W. Ferrer Santos, and H-J. Schneider, eds.), Amer. Math. Soc., Providence, RI, 2000, pp. 67–97.
5. M. Koppinen, *On algebras with two multiplications, including Hopf algebras and Bose-Mesner algebras*, J. Algebra **182** (1996), no. 1, 256–273.

6. ———, *On Nakayama automorphisms of double Frobenius algebras*, J. Algebra **214** (1999), no. 1, 22–40.

7. M. Takeuchi, *Survey of braided Hopf algebras*, New trends in Hopf algebra theory (La Falda, 1999) (N. Andruskiewitsch, W. Ferrer Santos, and H-J. Schneider, eds.), Amer. Math. Soc., Providence, RI, 2000, pp. 301–323.

NOTES ON THE CLASSIFICATION OF HOPF ALGEBRAS OF DIMENSION pq

AXEL SCHÜLER, INSTITUT FÜR MATHEMATIK, UNIVERSITÄT LEIPZIG, AUGUSTUSPLATZ 10, 04109 LEIPZIG, GERMANY

ABSTRACT. In this note we summarize results on the classification of Hopf algebras of dimension pq over an algebraically closed field of characteristic zero. Here p and q denote different odd prime numbers. We also include a short proof of the p^2 case given by Schneider and Radford which restates the result of Ng. It is known that semisimple or pointed or quasitriangular Hopf algebras of dimension pq are trivial, i.e. isomorphic to a group algebra or its dual. We conjecture that this is true in general and give some facts pointing in this direction. However we can not give the desired general proof.

Introduction. In the last years there has been some progress in the classification of finite dimensional Hopf algebras over an algebraically closed field of characteristic zero. Many results have been found especially in the semisimple and the pointed cases. The complete classification is known only when the dimension is a prime number p or the square of a prime. In the first case there is only one isomorphism type, the group algebra $\Bbbk[\mathbb{Z}_p]$ of the cyclic group of order p, see [21]. In the second case a semisimple Hopf algebra of dimension p^2 is isomorphic to the group algebra $\Bbbk[\mathbb{Z}_{p^2}]$ or to $\Bbbk[\mathbb{Z}_p \times \mathbb{Z}_p]$, see [7]. The only non-semisimple Hopf algebras of dimension p^2 are isomorphic to a Taft algebra. This is a recent result obtained by Ng [11] using arguments from [2].

Andruskiewitsch and Natale [1] obtained a series of general results which are listed below as Proposition 12, Corollary 13, Lemma 14, and Corollary 15. Their main idea was to consider the decomposition of H as an H_0-bicomodule, where H_0 is the coradical of H. This, in particular, led to the classification in dimensions 15, 21, 35 and 49.

Further, Beattie and Dăscălescu [3] showed that Hopf algebras of dimension 14, 55 and 77 are trivial, i.e. isomorphic to a group algebra or the dual of a group algebra.

Supported by the Schloeßmann foundation, email: axel.schueler@math.uni-leipzig.de.

Also, Ng [11, Section 5] obtained a series of general results which back up the conjecture that Hopf algebras of dimension pq are trivial. They are listed below as Lemma 7 and Proposition 10. He used the eigenspace decomposition of H with respect to S^2 and with respect to the right multiplication by the distinguished group-like element of H.

The semisimple case. The semisimple case with $\dim H = pq$ where p and q are distinct prime numbers was studied in the papers of Etingof and Gelaki [4] and Gelaki and Westreich [6]. It was shown therein that H is trivial. However, there was an error in [6] corrected in [5]. Meanwhile Sommerhäuser [19] had a correct proof in this case. The main part is to show that the Hopf algebra H is of Frobenius type, i. e. the dimension of any irreducible representation of H divides $\dim H$. A shorter proof of the same fact is given by Natale in [8, Corollary 2.2.2]. We add some notes of Natale [10, Lemma 3.1] and [9].

Lemma 1. *Let H be a Hopf algebra of dimension pq and let $p < q$.*
(i) *Let $p = 2$. If $|G(H)| = q$ then H is semisimple.*
(ii) *Suppose that $|G(H)| = |G(H^*)| = q$. Then H is semisimple.*
(iii) *Let H be non-semisimple and K a Hopf subalgebra of H. Then $K = H$ or $K = \mathbf{k}G(H)$ or $K = \mathbf{k}1$.*

The quasitriangular case. It was shown by Natale [10, Theorem 1.1] that a quasitriangular Hopf algebra of dimension pq with odd prime numbers p and q is semisimple and isomorphic to a group algebra.

Notations and Preliminaries. Throughout the ground field $\mathbf{k}$ is algebraically closed of characteristic zero and H is a finite dimensional Hopf algebra with antipode S. Its comultiplication and counit are denoted by Δ and ε, respectively. We write $\mathrm{tr}(T)$ for the trace of a linear map T. The set of group-like elements of the Hopf algebra H form a group denoted by $G(H)$. By the Nichols–Zoeller theorem it is immediate that $\dim H$ is divisible by $|G(H)|$.
Let $\lambda \in H^*$ be a non-zero right integral of H^* and $\Lambda \in H$ a non-zero left integral of H. Then there exists $\alpha \in G(H^*)$ (independently of the choice of Λ) such that $\Lambda a = \alpha(a)\Lambda$ for $a \in H$. Similarly, there is a group-like element $g \in G(H)$ such that $\beta\lambda = \beta(g)\lambda$ for $\beta \in H^*$. We call g the distinguished group-like element of H and α the distinguished group-like element of H^*. Then Radford's formula for S^4 in terms of α and g reads as follows [13, Proposition 6]

$$S^4(a) = g(\alpha \rightharpoonup a \leftharpoonup \alpha^{-1})g^{-1},$$

where $\rightharpoonup$ and $\leftharpoonup$ denote the right and left regular actions of H^* on H, respectively.

The use of the trace. Lots of arguments are mainly based on the following simple lemma from linear algebra which is given in [2, Lemma 2.6]

Lemma 2. *Let T be a linear endomorphism of the finite dimensional vector space V over $\mathbf{k}$, p an odd prime number and ω a primitive pth root of* 1.

(i) *If* $\operatorname{tr}(T) = 0$ *and* $T^p = \mathrm{id}$ *then all* V_i *have the same dimension where* V_i *is the eigenspace of* T *to the eigenvalue* ω^i. *In particular* $p \mid \dim V$.

(ii) *If* $\operatorname{tr}(T) = 0$ *and* $T^{2p} = \mathrm{id}$ *then* $p \mid \operatorname{tr}(T^p)$.

(iii) *If* $\operatorname{tr}(T) = 0$, $T^{2p} = \mathrm{id}$, *and* $\dim V = p$, *then* $T^p = \mathrm{id}$ *or* $T^p = -\mathrm{id}$.

We say that a set $\{e_i \mid i \in I\}$ of idempotents is *complete* if $\sum_{i \in I} e_i = 1$.

Lemma 3. *Let H be a finite dimensional Hopf algebra, $t\colon H \to H$ an algebra map, and $\{e_i \mid i \in I\}$ a complete orthogonal set of idempotents of H with $t(e_i) = e_i$ for all i. Suppose $S(\Lambda_{(2)})t(\Lambda_{(1)}) = 0$, where $\Lambda \in H$ is a left integral.*
Then He_i is a t-invariant subspace of H with $\operatorname{tr}(t|_{He_i}) = 0$ for all i.

Proof. The invariance of He_i is obvious. Let $\lambda \in H^*$ be a right integral. By Radford's trace formula $\operatorname{tr}(t) = \sum \lambda(S(\Lambda_{(2)})t(\Lambda_{(1)}))$ [15, Theorem 1] and by the assumption of the lemma we obtain

$$\operatorname{tr}(t \circ r(e_i)) = \lambda(S(\Lambda_{(2)})t(\Lambda_{(1)})e_i) = 0,$$

where $r(e_i)$ denotes the right multiplication by e_i. Note that the map $t \circ r(e_i)\colon H \to H$ is zero on He_j for all $j \neq i$ since $e_i e_j = 0$ and its image is in He_i. Moreover $t \circ r(e_i)$ coincides with t on He_i. Hence, $\operatorname{tr}(t \circ r(e_i)) = \operatorname{tr}(t|_{He_i})$ for all i. Consequently, $\operatorname{tr}(t|_{He_i}) = 0$. □

Corollary 4. *Let H be a non-semisimple Hopf algebra and $\{e_i\}$ a complete orthogonal set of idempotents of H with $S^2(e_i) = e_i$. Then $\operatorname{tr}(S^2|_{He_i}) = 0$ for all i.*

Proof. Since H is non-semisimple,

$$0 = \varepsilon(\Lambda) = S\left(S(\Lambda_{(1)})\Lambda_{(2)}\right) = S(\Lambda_{(2)})S^2(\Lambda_{(1)})$$

and the Lemma applies to $t = S^2$. □

Now we can complete the proof of Radford and Schneider [17] of Ng's result ($\dim H = p^2$).

Proposition 5. [11, Theorem 5.5] *Let H be a non-semisimple Hopf algebra of dimension p^2 where p is an odd prime number. Then H is isomorphic to one of the Taft Hopf algebras.*

Proof. Andruskiewitsch and Schneider showed in [2, Theorem A (ii)] that it is enough to prove that S^2 has order p. Let g and α be the distinguished (left-modular) group-like elements of H and H^*, respectively. If both are trivial, by Radford's formula $S^4 = \mathrm{id}$. This is impossible since $0 = \mathrm{tr}(S^2) = n_+ - n_-$ by non-semisimplicity of H where $n_\pm := \{x \in H \mid S^2(x) = \pm x\}$. It contradicts $\dim H = n_+ + n_-$ is odd. On the other hand neither g nor α can have order p^2 since H is a group algebra otherwise. Hence, we can assume that the order of g or of α is p. By Radford's formula $S^{4p} = \mathrm{id}$.
Let us assume that g has order p. The case that α has order p can be treated similarly since H^* is also non-semisimple. Consider a complete set of orthogonal idempotents $\{f_j \mid j = 0, \dots, p-1\}$ with $f_j \in \Bbbk\langle g\rangle$ where $\Bbbk\langle g\rangle$ is the Hopf subalgebra of H generated by g. Set $T_j := S^2|_{Hf_j}$. By Corollary 4, $\mathrm{tr}(T_j) = 0$. By Lemma 6, $\dim Hf_j = p$ for all j. Since $T_j^{2p} = \mathrm{id}$, Lemma 2 (iii) applies and we have $T_j^p = \pm\,\mathrm{id}$. Since $T_j(f_j) = f_j$ we have $T_j^p = \mathrm{id}$. Consequently, $S^{2p} = \mathrm{id}$ and we are done. □

From now on we suppose that H is non-semisimple of dimension pq with different odd primes p and q. If the order of both distinguished group-likes g and α is 1, then by Radford's formula $S^4 = \mathrm{id}$ which is impossible. If one of the orders is pq then H is a group algebra; a contradiction. Hence, at least one of the distinguished elements has prime order; say g has order p. By [11, Lemma 5.1] which is stated below as Lemma 7, the order of α is 1 or p. In particular, $S^{4p} = \mathrm{id}$.
Since S^2 is a semisimple map it can be diagonalized with eigenvalues $(-1)^a\omega^i$, $a = 0, 1$, $i = 0, \dots, p-1$ where ω is a primitive p-th root of 1. Similarly, since $g^p = 1$ the right multiplication $r(g)$ has eigenvalues ω^j, $j = 0, \dots, p-1$.

We use a similar notation as in [11].

$$H_{a,i,j} = \{u \in H \mid S^2(u) = (-1)^a\omega^i u,\ ug = \omega^j u\}, \tag{1}$$

$$n_{a,i,j} = \dim H_{a,i,j} \tag{2}$$

Further, we use the following notation

$$H_{\bullet,i,j} = H_{0,i,j} \oplus H_{1,i,j}, \qquad n_{\bullet,i,j} = \dim H_{\bullet,i,j}$$

$$H_{a,i,\bullet} = \bigoplus_{j=0}^{p-1} H_{a,i,j}, \qquad n_{a,i,\bullet} = \dim H_{a,i,\bullet}.$$

In a similar way one defines $H_{\bullet,\bullet,j}$ and the corresponding dimension $n_{\bullet,\bullet,j}$. But we write $H_0 = \bigoplus_{i,j\in\mathbb{Z}_p} H_{0,i,j}$, $H_1 = \bigoplus_{i,j\in\mathbb{Z}_p} H_{1,i,j}$ and n_0 and n_1 for their dimensions, respectively. Note that $n_0 + n_1 = pq$.

For a group-like element g of order p let $B = \mathbb{k}\langle g \rangle$ the p-dimensional Hopf subalgebra. Moreover set

$$(3) \quad e_i = \frac{1}{p}\sum_{k=0}^{p-1} \omega^{-ik} g^k, \quad i = 0, \ldots, p-1; \quad \text{and hence} \quad e_i g = \omega^i e_i.$$

The set $\{e_i\}$ is a complete orthogonal set of idempotents of B. We will see that the eigenspace to $r(g)$ to the eigenvalue ω^i is exactly He_i.

Lemma 6. *Let H be a Hopf algebra of dimension pm where p is a prime. Further let g be group-like of order p and let $\{e_i \mid i = 0, \ldots, p-1\}$ be the mentioned above complete set of idempotents of $B = \mathbb{k}\langle g \rangle$. Then for every i we have*

$$\dim He_i = \dim e_i H = m,$$
$$He_i = H_{\bullet,\bullet,i}.$$

Proof. Since B is a semisimple commutative algebra, the idempotent e_i generates a one-dimensional two-sided ideal in B. Since H is a free right B-module by Nichols–Zoeller it has a free basis $\{h_1, \ldots, h_m\}$. Multiplying $H = \bigoplus_{j=1}^m h_j B$ from the right by e_i gives $He_i = \bigoplus_{j=1}^m h_j e_i$. Hence $\dim He_i = m$. Since H is also a free left B-module the second statement also follows.
By (3) we have $He_i \subseteq H_{\bullet,\bullet,i}$. Since

$$H = \bigoplus_{i \in \mathbb{Z}_p} He_i \subseteq \bigoplus_{i \in \mathbb{Z}_p} H_{\bullet,\bullet,j} = H,$$

all inclusions must be equalities. □

Lemma 7 ([11, Lemma 5.1]). *Let p and q be two distinct prime numbers. Then there is no Hopf algebra of dimension pq such that $|G(H)| = p$ and $|G(H^*)| = q$.*

The proof is mainly based on the fact that in the above situation $\mathbb{k}G(H)$ is a semisimple normal Hopf subalgebra with a semisimple quotient. Then H itself must be semisimple. Hence it is trivial which contradicts the assumptions $|G(H)| = p$ and $|G(H^*)| = q$ of the lemma.
We adopt the notation of Ng and define $x \in \mathbb{Z}_p$ via $\alpha(g) = \omega^x$.

Lemma 8. (i)[13] *Under the above assumptions there exists a non-degenerate bilinear form $\beta\colon H_{a,i,\bullet} \times H_{a,-i+x,\bullet} \to \mathbb{k}$ for all a, i. In particular, $n_{a,i,\bullet} = n_{a,-i+x,\bullet}$.*
(ii) *If in addition H is unimodular, then there exists a non-degenerate bilinear form $\beta\colon H_{a,i,j} \times H_{a,-i,j+2i} \to \mathbb{k}$ for all a, i, j. In particular, $n_{a,i,j} = n_{a,-i,j+2i}$.*

Proof. (i) This is exactly the content of [13, Theorem 2 (c), p. 349] and [13, formula 1.2, p. 335]. The bilinear form is $\beta(a,b) = \lambda(ab)$ where λ denotes a left integral in H^*. By [13, formula 1.2] the bilinear form $\beta\colon H_{a,i,\bullet} \times H_{b,j,\bullet} \to \mathbb{k}$ is nonzero if and only if $a = b$ and $j = -i + x$. We also use the fact that $S^2(\lambda) = \alpha(g)\lambda = \omega^x\lambda$.
(ii) We use the idea of the proof of [11, Lemma 3.4]. Since $\alpha = \varepsilon$, $x = 0$ and $S^4(a) = gag^{-1}$ by Radford's formula. In particular, by (i), the pairing $\beta\colon H_{a,i,\bullet} \times H_{a,i',\bullet} \to \mathbb{k}$ is nonzero if and only if $i' = -i$. Using $S^4(a) = gag^{-1}$ we compute the eigenvalue of $\ell(g)$, the left multiplication by g, on $H_{a,i,j}$. For, let $u \in H_{a,i,j}$, then

$$gu = S^4(u)g = \omega^{2i}ug = \omega^{2i+j}u. \tag{4}$$

Suppose further that $v \in H_{a,-i,k}$. Since $u \in H_{\bullet,\bullet,j}$ we have

$$\beta(r(g)u, v) = \beta(\omega^j u, v) = \omega^j\lambda(uv).$$

By (4) and $v \in H_{\bullet,-i,k}$ we compute

$$\beta(r(g)u, v) = \lambda(u\,gv) = \omega^{-2i+k}\lambda(uv).$$

Hence, the pairing $\beta\colon H_{a,i,j} \times H_{a,-i,k} \to \mathbb{k}$ is identically zero except for $j+2i \equiv k \pmod p$. Now it is easy to see that $\beta\colon H_{a,i,j} \times H_{a,-i,j+2i} \to \mathbb{k}$ is non-degenerate. Suppose that $\beta(u,v) = 0$ for a fixed $u \in H_{a,i,j}$ and all $v \in H_{a,-i,j+2i}$. Then $\beta(u,w)$ is zero for all $w \in H$ since it is already zero on all other components $H_{a,i',j'}$ except $H_{a,-i,j+2i}$. Since $\beta\colon H \times H \to \mathbb{k}$ is non-degenerate $u = 0$ and the claim follows. □

We show that $\alpha(g) = 1$ implies $\alpha = \varepsilon$.

Lemma 9. *Let H be a non-semisimple Hopf algebra of dimension pq with distinct odd primes p and q and with distinguished group-likes g and α where $g^p = 1$. Then exactly one of the following two cases occurs.*
(i) $\alpha(g) = \omega$ is a primitive p-th root of 1. Then the order of α is also p. There exists a Hopf algebra projection $\pi\colon H \to B$ with $\pi\iota = \mathrm{id}_B$ where $B = \mathbb{k}G(H)$ and $\iota\colon B \to H$ is the natural embedding. Moreover $H = R\#B$ is a Radford biproduct of B and $R = H^{\mathrm{co}B} = \{r \in H \mid r_1 \otimes \pi(r_2) = r \otimes 1\}$. Further, $R \in^B_B \mathcal{YD}$ is a Yetter–Drinfeld Hopf algebra.
(ii) $\alpha(g) = 1$. Then $\alpha = \varepsilon$, and H is unimodular.

Proof. (i) is the content of the proof of Lemma 5.3 in [11]. The complete description of Radford's biproduct $R\#B$ is given in [14].
(ii) We apply [8, Lemma 1.1.8] with $B = \mathbb{k}\langle g\rangle$ and $\eta = \alpha$. Since $\alpha(g) = 1$ we have indeed $\alpha|_B = \varepsilon_B$. Hence $\mathrm{ord}(\alpha) \mid [H : B]$ i.e. $\mathrm{ord}(\alpha) \mid q$ Since $\mathrm{ord}(\alpha) = q$ is impossible by Lemma 7, we have $\mathrm{ord}(\alpha) = 1$ and therefore $\alpha = \varepsilon$. □

The following Proposition is our main result. It is essentially the same as the result of Ng [11, Theorem 5.4]. Our proof of the unimodular case uses slightly different methods.

Proposition 10. *Let H be a non-semisimple Hopf algebra of dimension pq and suppose that the distinguished group-like g has order p. Then there exists an integer $d \in \mathbb{Z}$ such that*

$$n_{0,i,j} - n_{1,i,j} = d \quad \textit{for all } i,\ j = 0, \ldots, p-1.$$

In particular $p^2 \mid \mathrm{tr}(S^{2p})$, and $p < q$.

Proof. We use the idempotents e_j, $j = 0, \ldots, p-1$ defined in (3) and fix j. Since $e_j \in \mathbf{k}G(H)$ we have $S^2(e_j) = e_j$ and the assumptions of Corollary 4 are fulfilled. We abbreviate $T_j := S^2|_{He_j}$. Hence $\mathrm{tr}(T_j) = 0$ and $T_j^{2p} = \mathrm{id}$. By the definition of $n_{a,i,j}$ and by Lemma 6 we have

$$T_j = \bigoplus_{a,i} (-1)^a \omega^i I_{n_{a,i,j}}, \tag{5}$$

where I_m is the $m \times m$ identity matrix. Taking the trace we get

$$\mathrm{tr}(T_j) = \sum_{a,i} (-1)^a \omega^i n_{a,i,j} = \sum_i (n_{0,i,j} - n_{1,i,j})\, \omega^i = 0.$$

Since ω is also a root of the irreducible cyclotomic polynomial $1 + x + \cdots + x^{p-1}$, we conclude that there is d_j such that

$$n_{0,i,j} - n_{1,i,j} = d_j \quad \text{for all } i = 0, \ldots, p-1.$$

Since all $n_{a,i,j}$ are integers so is d_j. We have to show that all d_j, $j = 0, \ldots, p-1$, coincide.

Suppose first that $\alpha(g) = 1$. By Lemma 9 (ii), $\alpha = \varepsilon$ and H is unimodular. By Lemma 8 (ii) and by the first part of this proof

$$n_{0,i,j} - n_{1,i,j} = d_j = n_{0,-i,j+2i} - n_{1,-i,j+2i} = d_{j+2i}.$$

Since this is true for all i and since $\{2i \mid i \in \mathbb{Z}_p\}$ is a complete residue set modulo p, we find $d_j = d = \text{const.}$ for all j.

Now suppose that $\alpha(g)$ is a primitive p-th root of 1. By Lemma 9 (i) $H = R\#B$ with $\dim B = p$, $\dim R = q$ and R is S^2-invariant by [2, (4.4)]; further $S^2 = T \otimes \mathrm{id}_B$ by [2, (4.5)] where $T = S^2|_R$. Since $\mathrm{tr}(S^2) = 0$ and $S^{4p} = \mathrm{id}$ one has $\mathrm{tr}(T) = 0$ and $T^{2p} = \mathrm{id}$. Using multiplication we can identify

$$H_{a,i,j} = R_{a,i} \# \mathbf{k}\, e_j,$$

where $R_{a,i} = \{r \in R \mid S^2(r) = (-1)^a \omega^i\}$. Setting $r_{a,i} = \dim R_{a,i}$ we have $n_{a,i,j} = r_{a,i}$ for all j and consequently d_j doesn't depend on j in the second case as well.

We show the last statement. By (5),

$$\text{(6)}\qquad \operatorname{tr}(S^{2p}) = \sum_{j\in\mathbb{Z}_p} \operatorname{tr}T_j^p = \sum_{i,j\in\mathbb{Z}_p} (n_{0,i,j} - n_{1,i,j}) = n_0 - n_1 = p^2 d.$$

Using $n_0 + n_1 = pq$ it follows $n_0 = p(q+pd)/2$ and $n_1 = p(q-pd)/2$. Since n_0 and n_1 are non-negative integers, $q \geq |pd| \geq p$. The last inequality holds since d is odd. Since p and q are different $q > p$. □

Remark 1. Using the properties of primitive roots of 1, by direct computation one obtains

$$\operatorname{tr}(S^{2k}) = 0, \quad \text{for all } k \text{ relatively prime to 2 and } p.$$

This also follows from [18, Proposition 2 (b)].

Lemma 11. *Let H be a non-semisimple unimodular Hopf algebra of dimension pq and the idempotents e_i and g as above. Then we have*
(a) $e_iH = \bigoplus_{k=0}^{p-1} H_{\bullet,k,i-2k}$,
(b) $e_iHe_j = e_iH \cap He_j = H_{\bullet,\frac{i-j}{2},j}$.

Proof. (a) As in the proof of Lemma 6 one can show that e_iH is contained in the eigenspace of $\ell(g)$ with respect to the eigenvalue ω^i which is of dimension q. On the other hand e_iH is of dimension q by Lemma 6, so that the eigenspace for ω^i coincides with e_iH. Now let $u \in H_{\bullet,k,i-2k}$; then we have by (4) $gu = \omega^i u$; hence $u \in e_iH$. Therefore, $\bigoplus_{k=0}^{p-1} H_{\bullet,k,i-2k} \subseteq e_iH$. But all the other spaces $H_{\bullet,k,i'-2k}$ are eigenspaces with respect to $\omega^{i'}$. Statement (a) follows.
(b) First note that by (a) and Lemma 6

$$e_iH \cap He_j = \bigoplus_k H_{\bullet,k,i-2k} \cap H_{\bullet,\bullet,j} = H_{\bullet,\frac{i-j}{2},j}.$$

Since obviously $e_iHe_j \subseteq e_iH \cap He_j$, we have

$$e_iH = \bigoplus_{j\in\mathbb{Z}_p} e_iHe_j \subseteq \bigoplus_{j\in\mathbb{Z}_p} e_iH \cap He_j \subseteq e_iH \cap \bigoplus_{j\in\mathbb{Z}_p} He_j = e_iH.$$

Since both sides coincide we always have equalities; in particular (b) follows.

□

The coradical filtration. The main statement of this subsection is [1, Proposition 1.8]. It can also be found in variations in [12, 2, 20, 16]. To state this result (Proposition 12 below) we need the following notation.

Let M and N be non-negative integers such that M divides N and let $\xi \in \Bbbk$ be a primitive M-th root of 1. Consider the algebra $K_\mu(N,\xi)$

generated by elements x and g with relations

$$x^M = \mu(1-g^M), \quad g^N = 1, \quad gx = \xi xg,$$

where $\mu = 0$ if $M = N$, and $\mu \in \{0,1\}$ if $M \neq N$. The fact that g is group-like, $\Delta(x) = 1 \otimes x + x \otimes g$, and $\varepsilon(x) = 0$ determine a Hopf algebra structure in $K_\mu(N,\xi)$. Its dimension is MN. If $M = N$ then $K_\mu(N,\xi)$ is isomorphic to the Taft algebra.

Proposition 12. *Let H be a non-semisimple finite dimensional Hopf algebra over $\mathbf{k}$. Suppose that there exists a non-trivial skew-primitive element. Then H contains a Hopf subalgebra K isomorphic to $K_\mu(N,\xi)$, for some root of unity $\xi \in \mathbf{k}$ and some $\mu \in \{0,1\}$.*
In particular, if $\dim H$ is square-free then H has only trivial skew-primitive elements.

Corollary 13. *A finite dimensional non-semisimple Hopf algebra H of dimension pq with distinct prime numbers p and q has only trivial skew-primitive elements. In particular, H is not pointed.*

There is another very useful approach of Andruskiewitsch and Natale [1, Lemma 1.7] which is based on the decomposition of H as an H_0-bicomodule. Here $H = \bigcup_{n\geq 0} H_n$ denotes the coradical filtration of H with coradical H_0. We restate the lemma and its consequences.

Lemma 14. *Let H be a non-cosemisimple finite dimensional Hopf algebra with a non-zero left integral $\lambda \in H^*$.*
(i) *Suppose that H has only trivial skew-primitive elements. Then $H_1 \subseteq (\lambda \leftharpoonup \mathbf{k}G(H))^\perp$. In particular $|G(H)| \leq \dim H - \dim H_1$.*
(ii) *If $H_1 = H$ then H has a non-trivial skew-primitive element.*

Corollary 15. *Let H be a non-semisimple Hopf algebra of dimension pq with different odd primes p and q and $p = |G(H)|$.*
Then we have $H_0 \subsetneq H_1 \subsetneq H$ where $\dim H - \dim H_1 \geq p$. Moreover $p \mid \dim H_0$ and $p \mid \dim H_1$.

Proof. Since H contains only trivial skew-primitive elements by Corollary 13, we conclude from (ii) that $H_1 \subsetneq H$. Then $H_0 \subsetneq H_1$ is a consequence. The last statement follows since H_n is a $(\mathbf{k}G(H), H)$-Hopf module and the Nichols–Zoeller theorem applies. □

We conclude our review with some result of Radford and Schneider [18, Lemma 4] which may be a bridge between the two approaches: the use of S^2 and traces and the coradical filtration.

Lemma 16. *Let C be a finite-dimensional coalgebra over $\mathbf{k}$ and suppose that t is a semisimple coalgebra automorphism of C. Then the order of t is the order of the coalgebra automorphism $t|_{C_1}$ of C_1.*

Our assumptions on $\Bbbk$ assure that $t = S^2$ is a semisimple map and the lemma applies in our situation.

Acknowledgment. I would like to thank Susan Montgomery for drawing my attention to Hopf algebras of dimension pq, for many suggestions, and warm hospitality during my visit at the University of Southern California Los Angeles. I also want to thank S.-H. Ng and H.-J. Schneider for valuable discussions.

References

[1] N. Andruskiewitsch and S. Natale, *Counting arguments for Hopf algebras of low dimension*, Tsukuba J. Math. **25** (2001), no. 1, 187–201.

[2] N. Andruskiewitsch and H.-J. Schneider, *Hopf algebras of order p^2 and braided Hopf algebras of order p*, J. Algebra **199** (1998), no. 2, 430–454.

[3] M. Beattie and S. Dăscălescu, *Hopf algebras of dimension* 14, math.QA/0205243 (2002).

[4] P. Etingof and S. Gelaki, *Semisimple Hopf algebras of dimension pq are trivial*, J. Algebra **210** (1998), no. 2, 664–669.

[5] S. Gelaki and S. Westreich, *Errata to "On semisimple Hopf algebras of dimension pq"*, Proc. Amer. Math. Soc. **128** (2000), no. 9, 2829–2831.

[6] ———, *On semisimple Hopf algebras of dimension pq*, Proc. Amer. Math. Soc. **128** (2000), no. 1, 39–47.

[7] A. Masuoka, *The p^n theorem for semisimple Hopf algebras*, Proc. Amer. Math. Soc. **124** (1996), no. 3, 735–737.

[8] S. Natale, *On semisimple Hopf algebras of dimension pq^2*, J. Algebra **221** (1999), no. 1, 242–278.

[9] ———, *On Hopf algebras of dimension pq*, private communication (2002).

[10] ———, *Quasitriangular Hopf algebras of dimension pq*, Bull. London Math. Soc. **34** (2002), no. 3, 301–307.

[11] S.-H. Ng, *Non-semisimple Hopf algebras of dimension p^2*, J. Algebra **255** (2002), no. 1, 182–197.

[12] W. D. Nichols, *Bialgebras of type one*, Comm. Algebra **6** (1978), no. 15, 1521–1552.

[13] D. E. Radford, *The order of the antipode of a finite dimensional Hopf algebra is finite*, Amer. J. Math. **98** (1976), no. 2, 333–355.

[14] ———, *The structure of Hopf algebras with a projection*, J. Algebra **92** (1985), no. 2, 322–347.

[15] ———, *The group of automorphisms of a semisimple Hopf algebra over a field of characteristic* 0 *is finite*, Amer. J. Math. **112** (1990), no. 2, 331–357.

[16] ———, *Finite-dimensional simple-pointed Hopf algebras*, J. Algebra **211** (1999), no. 2, 686–710.

[17] D. E. Radford and H.-J. Schneider, *Non-semisimple Hopf algebras of dimension p^2*, private communication (2002).

[18] ———, *On the even powers of the antipode of a finite-dimensional Hopf algebra*, J. Algebra **251** (2002), no. 1, 185–212.

[19] Y. Sommerhäuser, *Yetter–Drinfel'd Hopf algebras over groups of prime order*, Lecture Notes in Mathematics, no. 1789, Springer-Verlag, Berlin, 2002.

[20] D. Stefan, *Hopf subalgebras of pointed Hopf algebras and applications*, Proc. Amer. Math. Soc. **125** (1997), 3191–3193.
[21] Y. Zhu, *Hopf algebras of prime dimension*, Internat. Math. Res. Notices **1** (1994), 53–59.

A HOPF ALGEBRAIC MORITA INVARIANT

MITSUHIRO TAKEUCHI

INSTITUTE OF MATHEMATICS, UNIVERSITY OF TSUKUBA, TSUKUBA, IBARAKI, 305
JAPAN

ABSTRACT. We construct a hyperalgebra $hy_{out}(A)$ arising from outer actions on an algebra A. We show if two algebras A and B are Morita equivalent, then the hyperalgebras $hy_{out}(A)$ and $hy_{out}(B)$ are isomorphic.

INTRODUCTION

Let A be an algebra over a field k. Put

$$HH^1(A) = Der(A)/\text{inner derivations}.$$

If two algebras A and B are Morita equivalent (over k), it is known that $HH^1(A) \cong HH^1(B)$ ([FGM], Theorem 4.1). When k is perfect, put

$$\int HH^1(A) = \text{integrable derivations of } A/\text{inner derivations}.$$

Theorem 4.2 of [FGM] says that if A and B are finite dimensional and $A \sim B$ Morita equivalent, the isomorphism $HH^1(A) \cong HH^1(B)$ induces $\int HH^1(A) \cong \int HH^1(B)$. The purpose of this paper is to generalize this to hyperalgebras. We are going to construct a hyperalgebra $hy_{out}(A)$ with an arbitrary k algebra A and to prove the following facts.

(i) If $A \sim B$, then $hy_{out}(A) \cong hy_{out}(B)$,

(ii) We have $P(hy_{out}(A)) = HH^1(A)$,

where P means the Lie algebra of primitive elements. If k is perfect, a hyperalgebra H with $P(H)$ finite dimensional has its reduced part H_{red} ([T1], 1.9.5).

(iii) $P(hy_{out}(A)_{red}) = \int HH^1(A)$ if k is perfect and A is finite dimensional.

Since the isomorphism $hy_{out}(A) \cong hy_{out}(B)$ of (i) induces $hy_{out}(A)_{red} \cong hy_{out}(B)_{red}$ if k is perfect and A, B are finite dimensional, the above mentioned results of [FGM] follow.

This paper is organized as follows. In §1, we prepare some elementary facts on group functors on $\mathbf{W}_k^{cn}$, the category of cocommutative connected coalgebras, and actions of a coalgebra on an algebra. In §2, we discuss inner actions. In §3, we recall the Theorem of smoothness from [T1]. In §4, we define outer actions as well as the hyperalgebra $hy_{out}(A)$. We prove that the group functor of outer actions $Out(-, A)$ is representable by $hy_{out}(A)$ and that $P(hy_{out}(A)) = HH^1(A)$. Theorem of smoothness is used effectively to deduce these facts. In §5, we prove that $hy_{out}(A)$ is a Morita invariant.

Throughout the paper, we work over a fixed field k. Algebras and coalgebras are all over k. We use the standard notation on coalgebras and Hopf algebras as given in [Sw] or [Mo]. If C is a coalgebra, Δ and ε denote the comultiplication and the counit. The sigma notation will be used in the form

$$\Delta(c) = \sum c_1 \otimes c_2, \quad c \in C.$$

1. Notations and preliminaries

We say a coalgebra C is *connected* if it is pointed irreducible. Its unique group-like element will be denoted by 1_C or 1. Let $\mathbf{W}_k^{cn}$ be the category of all cocommutative connected coalgebras over k. By *hyperalgebra*, we mean a cocommutative irreducible bialgebra ([T1], 1.3.5). It is nothing but a monoid object in $\mathbf{W}_k^{cn}$. It is necessarily a Hopf algebra, hence a group object in $\mathbf{W}_k^{cn}$. If H is a hyperalgebra, we have a group functor on $\mathbf{W}_k^{cn}$

$$Sp^* H : C \mapsto Coalg(C, H), \quad C \in \mathbf{W}_k^{cn}$$

where $Coalg(C, H)$ denotes the group of coalgebra maps $C \to H$. Such a group functor on $\mathbf{W}_k^{cn}$ is called *representable* or *formal* (cf. [T2]).

Let C be a connected coalgebra and A an algebra over k. A bilinear map $\omega : C \times A \to A$ is called an *action* if we have

$$\sum \omega(c_1, a)\omega(c_2, b) = \omega(c, ab), \tag{1.1}$$

$$\omega(c, 1) = \varepsilon(c), \tag{1.2}$$

$$\omega(1, a) = a \tag{1.3}$$

for $a, b \in A$, $c \in C$. If ω and η are actions of C on A, we put

$$\omega \cdot \eta(c, a) = \sum \omega(c_1, \eta(c_2, a)), \quad c \in C, \quad a \in A. \tag{1.4}$$

If C is cocommutative, $\omega \cdot \eta$ is an action. Let $Act(C, A)$ be the set of all actions of C on A. If $C \in \mathbf{W}_k^{cn}$, $Act(C, A)$ becomes a monoid with product (1.4). Its unit is the trivial action $(c, a) \mapsto \varepsilon(c)a$. The monoid functor on $\mathbf{W}_k^{cn}$

$$C \mapsto Act(C, A), \quad C \in \mathbf{W}_k^{cn}$$

is representable by a hyperalgebra ([T1], 1.5.11). This means there is a hyperalgebra $hy_{aut}(A)$ and an action

$$hy_{aut}(A) \times A \to A, \quad (h, a) \mapsto ha \tag{1.5}$$

with the following universal property: For any $C \in \mathbf{W}_k^{cn}$ and $\omega \in Act(C, A)$, there is a unique coalgebra map $f : C \to hy_{aut}(A)$ such that $\omega(c, a) = f(c)a$, $c \in C$, $a \in A$. Further, the algebra structure of $hy_{aut}(A)$ arises from the product (1.4), hence A becomes a left $hy_{aut}(A)$ module algebra with action (1.5). It follows in particular that $Act(C, A)$ is a group for $C \in \mathbf{W}_k^{cn}$. ($hy_{aut}(A)$ is the irreducible component of $M_c(A, A)$ in the notation of [T1], 1.5.11 and the hyperalgebra of the automorphism group functor $\mathbf{Aut}(A)$ ([T1], 3.2.6)).

Recall that $Hom(C, A)$ has the structure of an algebra with the $*$ product

$$(f * g)(c) = \sum f(c_1)g(c_2), \quad c \in C, \quad f, g \in Hom(C, A)$$

([Sw], 4.0). When C is connected, we put

$$Hom^1(C, A) = \{f \in Hom(C, A) \mid f(1) = 1\}$$

which is a subgroup of units in $Hom(C, A)$ ([Sw], Lem. 9.2.3). The group functor on $\mathbf{W}_k^{cn}$

$$C \mapsto Hom^1(C, A), \quad C \in \mathbf{W}_k^{cn}$$

is representable ([T1], 1.5.6). The representing hyperalgebra is denoted by $B_m(A)$. There is an algebra map

$$\pi : B_m(A) \to A$$

such that for any $C \in \mathbf{W}_k^{cn}$, any linear map $f : C \to A$ with $f(1) = 1$ can be uniquely lifted to a coalgebra map $\tilde{f} : C \to B_m(A)$ such that $f = \pi\tilde{f}$.

The connected cocommutative coalgebra $B(U)$ was introduced in [Sw], 12.2. A connected cocommutative coalgebra is called *smooth* if it is isomorphic to $B(U)$ for some U ([T1], 1.9.5). The hyperalgebra $B_m(A)$ is isomorphic to $B(A)$ as coalgebra, hence smooth.

2. Inner actions

Let A be an algebra and $C \in \mathbf{W}_k^{cn}$. If $f \in Hom^1(C, A)$, we put

$$inn(f)(c, a) = \sum f(c_1)af^{-1}(c_2), \quad c \in C, \quad a \in A, \tag{2.1}$$

where f^{-1} is the inverse in the group $Hom^1(C, A)$. One checks easily that $inn(f)$ is an action of C on A and that the map

$$inn : Hom^1(C, A) \to Act(C, A) \tag{2.2}$$

is a group homomorphism. In fact, we have

$$\begin{aligned} inn(f) \cdot inn(g)(c, a) &= \sum inn(f)(c_1, inn(g)(c_2, a)) \\ &= \sum f(c_1)g(c_2)ag^{-1}(c_3)f^{-1}(c_4) = \sum (f * g)(c_1)a(g^{-1} * f^{-1})(c_2), \end{aligned}$$

for $f, g \in Hom^1(C, A)$. The actions of the form $inn(f)$ are called *inner*. Since the group homomorphism (2.2) is natural in C, it induces a homomorphism of group functors

$$Sp^* B_m(A) \cong Hom^1(-, A) \xrightarrow{inn} Act(-, A) \cong Sp^* hy_{aut}(A). \tag{2.3}$$

Hence it corresponds to a uniquely determined hyperalgebra map

$$inn : B_m(A) \to hy_{aut}(A) \tag{2.4}$$

which is characterized by the property:

$$inn(h)a = \sum \pi(h_1)a\pi(S(h_2)), \quad h \in B_m(A), \quad a \in A \tag{2.5}$$

where S denotes the antipode of $B_m(A)$.

For $\omega \in Act(C, A)$ and $f \in Hom^1(C, A)$, we put

$$(\omega \rightharpoonup f)(c) = \sum \omega(c_1, f(c_2)), \quad c \in C. \tag{2.6}$$

Obviously, $\omega \rightharpoonup f \in H^1(C, A)$.

Proposition 2.7. *We have*

(2.7.2.1) $$\omega \rightharpoonup (\eta \rightharpoonup f) = \omega \cdot \eta \rightharpoonup f,$$

(2.7.2.2) $$\omega \rightharpoonup (f * g) = (\omega \rightharpoonup f) * (\omega \rightharpoonup g)$$

for $\omega, \eta \in Act(C, A)$ *and* $f, g \in Hom^1(C, A)$. *Thus the group* $Act(C, A)$ *acts on the group* $Hom^1(C, A)$ *as group automorphisms.*

Proof. (2.7.2) is easy. (2.7.1) follows from:

$$\begin{aligned}(\omega \rightharpoonup (f * g))(c) &= \sum \omega(c_1, f(c_2)g(c_3)) \\ &= \sum \omega(c_1, f(c_2))\omega(c_3, g(c_4)) = \sum(\omega \rightharpoonup f)(c_1)(\omega \rightharpoonup g)(c_2), \quad c \in C.\end{aligned}$$

□

Proposition 2.8. *We have*

$$inn(\omega \rightharpoonup f) = \omega \cdot inn(f) \cdot \omega^{-1}, \quad \omega \in Act(C, A), \quad f \in Hom^1(C, A).$$

Proof. For $c \in C$ and $a \in A$, we have

$$\begin{aligned}(\omega \cdot inn(f) \cdot \omega^{-1})(c, a) &= \sum \omega(c_1, f(c_2)\omega^{-1}(c_3, a)f^{-1}(c_4)) \\ &= \sum \omega(c_1, f(c_2))\omega(c_3, \omega^{-1}(c_4, a))\omega(c_5, f^{-1}(c_6)) \\ &= \sum \omega(c_1, f(c_2))a\omega(c_3, f^{-1}(c_4)) = \sum(\omega \rightharpoonup f)(c_1)a(\omega \rightharpoonup f^{-1})(c_2).\end{aligned}$$

Hence it is enough to note that $\omega \rightharpoonup f^{-1} = (\omega \rightharpoonup f)^{-1}$. □

Corollary 2.9. *The image of inn (2.2) is a normal subgroup of* $Act(C, A)$.

By the functoriality, the action (2.6) corresponds to an action of hyperalgebra $hy_{aut}(A)$ on hyperalgebra $B_m(A)$ in the sense of [T1], 1.10:

(2.2.9.2.10) $$\begin{aligned} hy_{aut}(A) \otimes B_m(A) &\to B_m(A) \\ g \otimes h &\mapsto g \rightharpoonup h \end{aligned}$$

which is characterized as follows: Let $C \in \mathbf{W}_k^{cn}$ and $\phi : C \to hy_{aut}(A)$, $\psi : C \to B_m(A)$ be coalgebra maps. Let $\omega \in Act(C, A)$ and $f \in Hom^1(C, A)$ be the elements corresponding to ϕ and ψ respectively. Then the coalgebra map $C \to B_m(A)$, $c \mapsto \sum \phi(c_1) \rightharpoonup \psi(c_2)$ corresponds to $\omega \rightharpoonup f$.

Proposition 2.11. *The map (2.10) is the unique coalgebra map which makes the following diagram commute*

(2.2.11.2.12) $$\begin{array}{ccc} hy_{aut}(A) \otimes B_m(A) & \xrightarrow{(2.10)} & B_m(A) \\ \downarrow{\scriptstyle id \otimes \pi} & & \downarrow{\scriptstyle \pi} \\ hy_{aut}(A) \otimes A & \xrightarrow{(1.5)} & A \end{array}.$$

It follows in particular that $\pi : B_m(A) \to A$ *is a left* $hy_{aut}(A)$ *module algebra map.*

Proof. With the above notation, we have

$$\begin{aligned}\sum \pi(\phi(c_1) \rightharpoonup \psi(c_2)) &= (\omega \rightharpoonup f)(c) = \sum \omega(c_1, f(c_2)) \\ &= \sum \phi(c_1)f(c_2) = \sum \phi(c_1)\pi(\psi(c_2)), \quad c \in C.\end{aligned}$$

Since C, ϕ, ψ are arbitrary, it follows that (2.12) commutes. By the universality of $(B_m(A), \pi)$, the map (2.10) is unique as coalgebra map. □

Corollary 2.13 (to Proposition 2.8). *We have*

$$inn(g \rightharpoonup h) = \sum g_1 inn(h) S(g_2), \quad g \in hy_{aut}(A), \quad h \in B_m(A).$$

Corollary 2.14. *The image of the hyperalgebra map inn (2.4) is a normal sub-hyperalgebra of $hy_{aut}(A)$, i.e., a sub-hyperalgebra which is stable by the adjoint action ([T1], 1.10.5).*

3. Theorem of smoothness

If $f : C \to D$ is a morphism in $\mathbf{W}_k^{cn}$,

$$E = \{c \in C \mid \sum c_1 \otimes f(c_2) = c \otimes 1_D\} \tag{3.3.0.3.1}$$

is the largest subcoalgebra of C such that $f(E) = k1_D$. We put $E = Ker_c(f)$ and call it the *coalgebraic kernel* of f ([T1], 1.2.2). If $g : H \to L$ is a homomorphism of hyperalgebras, $Ker_c(g)$ is a sub-hyperalgebra of H and it represents the kernel of group functor map Sp^*g ([T1], 1.3.3). Hence we have

$$1 \to Sp^*Ker_c(g) \to Sp^*H \xrightarrow{Sp^*g} Sp^*L. \tag{3.3.0.3.2}$$

It is known that g is injective if and only if Sp^*g is a monomorphism if and only if $Ker_c(g) = k$ ([T1], 1.2.3).

Let K be a sub-hyperalgebra of a hyperalgebra H. We put $H//K = H/HK^+$, a quotient coalgebra of H, where $K^+ = Ker(\varepsilon : K \to k)$. If K is normal, it is a quotient hyperalgebra. It is known that H is a faithfully coflat $H//K$ comodule and that $K = Ker_c(\pi)$ with projection $\pi : H \to H//K$ ([T3], Theorem 4). We note that if C is a coalgebra, a C comodule is faithfully coflat if and only if it is an injective cogenerator ([T2], A.2.1). If C is cocommutative connected, a C comodule is coflat if and only if it is free, i.e., a direct sum of copies of C ([T2], A.2.2).

Proposition 3.3. *Let $g : H \to L$ be a homomorphism of hyperalgebras. Then $K = Ker_c(g)$ is a normal sub-hyperalgebra of H and g induces an injective hyperalgebra map $\bar{g} : H//K \hookrightarrow L$.*

Proof. The first part is obvious. It is enough to show that $\bar{g}$ is injective. Let $\bar{K} = Ker_c(\bar{g})$. Let us make a pullback diagram of formal schemes

$$\begin{array}{ccc} Sp^*\bar{K} & \subset & Sp^*H//K \\ {\scriptstyle Sp^*\gamma}\uparrow & & \uparrow{\scriptstyle Sp^*\pi} \\ Sp^*F & \subset & Sp^*H \end{array}$$

Since $Sp^*\pi$ is faithfully coflat, so is $Sp^*\gamma$ ([T2], §2). Since $F \subset K$ obviously, it follows that $Sp^*\bar{K} = \{e\}$. Hence $\bar{K} = k$. □

We recall the Theorem of smoothness ([T1], 1.8.1) which plays an important role in the sequel. Let K be a smooth sub-hyperalgebra of a hyperalgebra H. There is a coalgebra isomorphism $\theta : H \xrightarrow{\sim} K \otimes H//K$ such that the diagram

$$
\begin{array}{ccccc}
K & \subset & H & & \\
\| & & \wr\downarrow\theta & \searrow^{\pi} & \\
K & \xleftarrow{id\otimes\varepsilon} & K\otimes H//K & \xrightarrow{\varepsilon\otimes id} & H//K
\end{array}
$$

commutes. In particular, any coalgebra map $f : C \to H//K$ can be lifted to a coalgebra map $\tilde{f} : C \to H$ in such a way that $\pi\tilde{f} = f$.

Corollary 3.4. *Let K be a smooth normal sub-hyperalgebra of a hyperalgebra H. Then we have an exact sequence of group functors on $\mathbf{W}_k^{cn}$:*

$$1 \to Sp^*K \to Sp^*H \xrightarrow{Sp^*\pi} Sp^*H//K \to 1.$$

This means π induces a surjective homomorphism $Coalg(C, H) \to Coalg(C, H//K)$ for any $C \in \mathbf{W}_k^{cn}$.

Let $P(H)$ be the set of primitive elements in a hyperalgebra H. We have

$$P(H) \cong \mathbf{W}_k^{cn}(B_1, H)$$

where $B_1 = kb_0 + kb_1$ with b_0 group-like and b_1 primitive ([T1], 1.3b.1).

Corollary 3.5. *With the assumptions in 3.4, we have*

$$P(H//K) = P(H)/P(K).$$

4. Outer actions

Let A be an algebra with center $Z(A)$.

Proposition 4.1. *The hyperalgebra map $inn : B_m(A) \to hy_{aut}(A)$ (2.4) has coalgebraic kernel $B_m(Z(A))$. It induces an injective hyperalgebra map $B_m(A)//B_m(Z(A)) \hookrightarrow hy_{aut}(A)$.*

Proof. It is enough to show we have an exact sequence

$$1 \to Sp^*B_m(Z(A)) \to Sp^*B_m(A) \xrightarrow{inn} Sp^*hy_{aut}(A),$$

i.e.,

$$1 \to Hom^1(C, Z(A)) \to Hom^1(C, A) \xrightarrow{inn} Act(C, A)$$

for any $C \in \mathbf{W}_k^{cn}$. Let $f \in Hom^1(C, A)$. We have

$$
\begin{aligned}
& inn(f)(c, a) = \sum f(c_1)af^{-1}(c_2) = \varepsilon(c)a, \quad c \in C, \quad a \in A \\
\Leftrightarrow \quad & f(c)a = af(c), \quad c \in C, \quad a \in A \\
\Leftrightarrow \quad & f(c) \in Z(A), \quad c \in C.
\end{aligned}
$$

Since $f^{-1}(c) \in Z(A)$ in this case, the claim follows. □

We note that the hyperalgebra $B_m(A)//B_m(Z(A))$ is smooth by [T1], 1.9.4.

Definition 4.2. We put for $C \in \mathbf{W}_k^{cn}$

$$Out(C, A) = Act(C, A)/inn(Hom^1(C, A)).$$

Its elements are called *outer* actions of C on A.

It is a quotient group of $Act(C, A)$ by Corollary 2.9.

Definition 4.3. We put

$$hy_{out}(A) = hy_{aut}(A)//inn(B_m(A))$$

which is a quotient hyperalgebra by Corollary 2.14.

Proposition 4.4. *The group functor $C \in \mathbf{W}_k^n \mapsto Out(C, A)$ is representable by the hyperalgebra $hy_{out}(A)$, i.e., we have*

$$Sp^* hy_{out}(A) \cong Out(-, A).$$

Proof. Since $B_m(Z(A))$ and $B_m(A)//B_m(Z(A))$ are smooth hyperalgebras, it follows from Corollary 3.4 that we have exact sequences of group functors on $\mathbf{W}_k^{cn}$

$$1 \to Sp^* B_m(Z(A)) \to Sp^* B_m(A) \to Sp^* B_m(A)//B_m(Z(A)) \to 1,$$

$$1 \to Sp^* B_m(A)//B_m(Z(A)) \xrightarrow{inn} Sp^* hy_{aut}(A) \to Sp^* hy_{out}(A) \to 1.$$

Thus we have

$$Hom^1(C, A) \xrightarrow{inn} Act(C, A) \to Coalg(C, hy_{out}(A)) \to 1.$$

This yields $Out(C, A) \cong Coalg(C, hy_{out}(A))$. □

Corollary 4.5. $P(hy_{out}(A)) = Der(A)/$*inner derivations.*

Proof. Since A is a $hy_{aut}(A)$ module algebra, $P(hy_{aut}(A))$ acts on A as derivations. It follows easily from the universality of $hy_{aut}(A)$ that this induces an isomorphism of Lie algebras $P(hy_{aut}(A)) \xrightarrow{\sim} Der(A)$. On the other hand, we may identify $A = P(B_m(A))$ ([T1], 1.5.5). The map

$$A = P(B_m(A)) \xrightarrow{inn} P(hy_{aut}(A)) \cong Der(A)$$

associates the inner derivation $[a, -]$ to $a \in A$. Hence, the claim follows by applying Corollary 3.5. □

5. Morita invariance

Let A and B be algebras. Assume they are Morita equivalent over k. We show there is a hyperalgebra isomorphism $hy_{out}(A) \cong hy_{out}(B)$. By Proposition 4.4, this is equivalent to saying that the group functors $C \in \mathbf{W}_k^{cn} \mapsto Out(C, A)$ and $Out(C, B)$ are isomorphic, i.e., it is enough to show there is an isomorphism of groups $Out(C, A) \cong Out(C, B)$ which is natural in $C \in \mathbf{W}_k^{cn}$.

Since $A \sim B$ Morita equivalent, there is a Morita context $({}_AP_B, {}_BQ_A, [\ ,\ ], (\ ,\ ))$ where P (resp. Q) is an (A, B) (resp. (B, A)) bimodule over k, and

$$[\ ,\ ] : P \otimes_B Q \to A \quad \text{an } A\text{-bimodule map,}$$
$$(\ ,\ ) : Q \otimes_A P \to B \quad \text{a } B\text{-bimodule map}$$

which are surjective satisfying

$$[x, y]x' = x(y, x'), \quad (y, x)y' = y[x, y'] \quad \text{for } x, x' \in P \text{ and } y, y' \in Q.$$

Let $C \in \mathbf{W}_k^{cn}$. A linear map $\omega : C \otimes P \to P$ is called a *weak action* if $\omega(1, x) = x$ for all $x \in P$. Let $w.Act(C, P)$ be the set of all weak actions. We define the same product as (1.4) on it. Then it makes a group. In fact, we may identify

$$w.Act(C,P) \cong Hom^1(C, End(P)) \qquad (5.5.0.5.1)$$

$$\omega \leftrightarrow c \mapsto \omega(c, -).$$

Let $\alpha \in Act(C, A)$, $\beta \in Act(C, B)$ and $\omega \in w.Act(C, P)$. We say ω is *left α-twisted* if

$$\omega(c, ax) = \sum \alpha(c_1, a)\omega(c_2, x), \quad c \in C, \quad a \in A, \quad x \in P \qquad (5.5.0.5.2)$$

and ω is *right β-twisted* if

$$\omega(c, xb) = \sum \omega(c_1, x)\beta(c_2, b), \quad c \in C, \quad x \in P, \quad b \in B. \qquad (5.5.0.5.3)$$

If we have both (5.2) and (5.3), we say α and β are *ω-related* and write $\alpha \sim_\omega \beta$.

Lemma 5.4. *If $\alpha \in Act(C, A)$, there is $\omega \in w.Act(C, P)$ which is left α-twisted.*

Proof. Take $y_i \in Q$, $x_i \in P$ in such a way that $\sum_i (y_i, x_i) = 1$. It is enough to put

$$\omega(c, x) = \sum_i \alpha(c, [x, y_i])x_i, \quad c \in C, \quad x \in P.$$

□

Corollary 5.5. *If $\beta \in Act(C, B)$, there is $\omega \in w.Act(C, P)$ which is right β-twisted.*

This follows from left-right symmetry.

Lemma 5.6. *Let $\omega \in w.Act(C, P)$ and $\alpha \in Act(C, A)$. If ω is left α-twisted, there is a unique $\beta \in Act(C, B)$ such that ω is right β-twisted. Thus $\alpha \sim_\omega \beta$.*

Proof. Let ω^{-1} and α^{-1} be inverses of ω and α in the groups $w.Act(C, P)$ and $Act(C, A)$. It is easy to see that ω^{-1} is left α^{-1}-twisted. Hence, if $f \in End_A(P)$, then the map

$$x \in P \mapsto \sum \omega(c_1, f(\omega^{-1}(c_2, x))) \qquad (5.6.5.5.6.5.1)$$

is A-linear for any $c \in C$. We have $B^{op} \cong End_A(P)$ by Morita theory. When f is the right multiplication of $b \in B$, the map (5.6.1) is the right multiplication of a uniquely determined element $\beta(c, b)$ in B, i.e.,

$$\sum \omega(c_1, \omega^{-1}(c_2, x)b) = x\beta(c, b), \quad c \in C, \quad x \in P, \quad b \in B. \qquad (5.6.5.5.6.5.2)$$

This is equivalent to

$$\omega(c, xb) = \sum \omega(c_1, x)\beta(c_2, b), \quad c \in C, \quad x \in P, \quad b \in B, \qquad (5.6.5.5.6.5.3)$$

i.e., ω is right β-twisted. It is easy to check that β is an action of C on B. □

Corollary 5.7. *If $\omega \in w.Act(C, P)$ is right β-twisted with $\beta \in Act(C, B)$, then there is a unique $\alpha \in Act(C, A)$ such that ω is left α-twisted.*

Lemma 5.8. *Let $\alpha_i \in Act(C, A)$, $\beta_i \in Act(C, B)$ and $\omega_i \in w.Act(C, P)$, $i = 1, 2$. If $\alpha_i \sim_{\omega_i} \beta_i$, then $\alpha_1 \cdot \alpha_2 \sim_{\omega_1 \cdot \omega_2} \beta_1 \cdot \beta_2$ and $\alpha_i^{-1} \sim_{\omega_i^{-1}} \beta_i^{-1}$.*

This is easily checked.

Lemma 5.9. *Assume we have $\alpha \sim_\omega \beta$. Then α is inner if and only if β is inner.*

Proof. Assume $\alpha = inn(f)$ with $f \in Hom^1(C, A)$. Then

$$\omega(c, ax) = \sum f(c_1)af^{-1}(c_2)\omega(c_3, x), \quad c \in C, \quad a \in A, \quad x \in P. \tag{5.9.5.5.9.5.1}$$

Hence the map $x \mapsto \sum f^{-1}(c_1)\omega(c_2, x)$, $x \in P$ is A-linear for any $c \in C$. (If $c = 1$, this is the identity). Hence there is $g \in Hom^1(C, B)$ such that

$$\sum f^{-1}(c_1)\omega(c_2, x) = xg^{-1}(c), \quad c \in C, \quad x \in P. \tag{5.9.5.5.9.5.2}$$

We have

$$\begin{aligned} xbg^{-1}(c) &= \sum f^{-1}(c_1)\omega(c_2, xb) = \sum f^{-1}(c_1)\omega(c_2, x)\beta(c_3, b) \\ &= \sum xg^{-1}(c_1)\beta(c_2, b), \quad x \in P, \quad b \in B, \quad c \in C. \end{aligned}$$

It follows that $\beta(c, b) = \sum g(c_1)bg^{-1}(c_2)$, i.e., $\beta = inn(g)$. Similarly, if β is inner, so is α. □

If $\alpha \in Act(C, A)$ (resp. $\beta \in Act(C, B)$), let $\bar{\alpha}$ (resp. $\bar{\beta}$) denote the class of α (resp. β) in $Out(C, A)$ (resp. $Out(C, B)$).

Theorem 5.10. *The relation*

$$\alpha \sim_\omega \beta \quad \textit{for some} \quad \omega \in w.Act(C, P)$$

where $\alpha \in Act(C, A)$ and $\beta \in Act(C, B)$, determines a well-defined isomorphism of groups

$$\bar{\alpha} \leftrightarrow \bar{\beta}, \quad Out(C, A) \cong Out(C, B)$$

which is natural in $C \in \mathbf{W}_k^{cn}$.

Proof. With the notation of 5.8, we have

$$\alpha_1 \cdot \alpha_2^{-1} \sim_{\omega_1 \cdot \omega_2^{-1}} \beta_1 \cdot \beta_2^{-1}.$$

By Lemma 5.9, $\alpha_1 \cdot \alpha_2^{-1}$ is inner if and only if $\beta_1 \cdot \beta_2^{-1}$ is inner. Hence the correspondence $\bar{\alpha} \leftrightarrow \bar{\beta}$ is well-defined. This gives rise to a bijection $Out(C, A) \cong Out(C, B)$ by Lemmas 5.4–5.7. This is a group isomorphism by Lemma 5.8. Obviously, it is natural in $C \in \mathbf{W}_k^{cn}$. □

Corollary 5.11. *If A and B are algebras which are Morita equivalent over k, the group functors $Out(-, A)$ and $Out(-, B)$ on $\mathbf{W}_k^{cn}$ are isomorphic with each other. Hence there is a hyperalgebra isomorphism $hy_{out}(A) \cong hy_{out}(B)$.*

Corollary 5.12. *Let A and B be commutative algebras. If $A \sim B$, then $hy_{aut}(A) \cong hy_{aut}(B)$.*

We end the paper with some comments. If we take the primitive parts of the above hyperalgebra isomorphism, we get a Lie algebra isomorphism $P(hy_{out}(A)) \cong P(hy_{out}(B))$. Hence by Corollary 4.5,

$$Der(A)/\text{inner derivations} \cong Der(B)/\text{inner derivations}.$$

This is [FGM], Theorem 4.1.

Assume k is a perfect field. Then a hyperalgebra H is smooth if and only if it is reduced. If $P(H)$ is finite dimensional, there is the largest reduced sub-hyperalgebra

H_{red} of H ([T1], 1.9.5). Sweedler's theorem tells that $P(H_{red})$ consists of all elements in $P(H)$, over which there are ∞-sequences of divided powers ([T1], 1.9.6). In other words $P(H_{red})$ consists of all *integrable* primitive elements in the terminology of [FGM].

If A and B are finite dimensional, $hy_{out}(A)$ and $hy_{out}(B)$ have finite dimensional primitive parts, hence they have reduced parts $hy_{out}(A)_{red}$ and $hy_{out}(B)_{red}$. We have

$$\begin{aligned} A \sim B \quad &\Rightarrow \quad hy_{out}(A) \cong hy_{out}(B) \\ &\Rightarrow \quad hy_{out}(A)_{red} \cong hy_{out}(B)_{red}. \end{aligned}$$

Taking the primitive parts of the last isomorphism, we get an isomorphism of Lie algebras

$$Der_{int}(A)/\text{inner derivations} \cong Der_{int}(B)/\text{inner derivations},$$

where Der_{int} means integrable derivations. This is [FGM], Theorem 4.2.

REFERENCES

[FGM] D. R. Farkas, C. Geiss and E. N. Marcos, *Smooth automorphism group schemes*, Representations of algebras (São Paulo 1999), 71–89, L. N. in Pure and Appl. Math., 224, Dekker, New York, 2002.

[Mo] S. Montgomery, *Hopf algebras and their actions on rings*, CBMS Reg. Conf. Ser. in Math. 82, AMS, Providence, 1993.

[Sw] M. E. Sweedler, *Hopf algebras*, Benjamin, Inc., New York, 1969.

[T1] M. Takeuchi, *Tangent coalgebras and hyperalgebras I*, Jap. J. Math. 42 (1974), 1–143.

[T2] M. Takeuchi, *Formal schemes over fields*, Comm. in Algebra 5(14) (1977), 1483–1528.

[T3] M. Takeuchi, *Relative Hopf modules—Equivalences and freeness criteria—*, J. Algebra 60 (1979), 452–471.

Milton Keynes UK
Ingram Content Group UK Ltd.
UKHW052018071024
449327UK00027B/2324

9 780367 394578